ANALYTICAL FOOD MICROBIOLOGY

ANALYTICAL FOOD MICROBIOLOGY
A Laboratory Manual

Second Edition

AHMED E. YOUSEF

The Ohio State University
Columbus, Ohio, USA

JOY G. WAITE-CUSIC

Oregon State University
Corvallis, Oregon, USA

JENNIFER J. PERRY

University of Maine
Orono, Maine, USA

Registered Office
John Wiley & Sons, Inc., 111 River Street, Hoboken, NJ 07030, USA

Editorial Office
111 River Street, Hoboken, NJ 07030, USA

For details of our global editorial offices, customer services, and more information about Wiley products visit us at www.wiley.com.

Wiley also publishes its books in a variety of electronic formats and by print-on-demand. Some content that appears in standard print versions of this book may not be available in other formats.

Library of Congress Cataloging-in-Publication Data

Names: Yousef, Ahmed Elmeleigy, author. | Waite-Cusic, Joy G., author. | Perry, Jennifer J., author.
Title: Analytical food microbiology : a laboratory manual / Ahmed E. Yousef, The Ohio State University, Columbus, OH, USA, Joy G. Waite-Cusic, Oregon State University, Corvallis, OR, USA, Jennifer J. Perry, University of Maine, Orono, ME, USA.
Other titles: Food microbiology
Description: Second edition. | Hoboken, NJ : Wiley, [2022] | Revison of: Food microbiology / Ahmed E. Yousef, Carolyn Carlstrom. c2003. | Includes bibliographical references and index.
Identifiers: LCCN 2021033026 (print) | LCCN 2021033027 (ebook) | ISBN 9780470425114 (paperback) | ISBN 9781119428039 (adobe pdf) | ISBN 9781119428015 (epub)
Subjects: LCSH: Food–Microbiology–Laboratory manuals.
Classification: LCC QR115 .Y686 2021 (print) | LCC QR115 (ebook) | DDC 664.001/579–dc23
LC record available at https://lccn.loc.gov/2021033026
LC ebook record available at https://lccn.loc.gov/2021033027

Cover Design: Wiley
Cover Image: Courtesy of Ahmed Yousef

Set in 10.5/12pt TimesTen by Straive, Pondicherry, India

10 9 8 7 6 5 4 3 2 1

CONTENTS

PREFACE

Microbiological testing of food is an important component of the global food system. Results from these tests support food safety and quality programs by helping researchers and processors monitor fermentations, validate and verify the efficacy of processing treatments, and demonstrate the efficacy of sanitation programs. Well-trained food microbiologists are needed in the food industry, supporting industries, and regulatory agencies to conduct laboratory analyses that serve these functions.

There is no shortage of published methods for enumerating food microbiota or detecting foodborne pathogens. However, using these methods in teaching or training settings requires considerable adaptation and simplification. Professional analysts have an eight-hour work shift, which provides ample and flexible time to complete a variety of microbiological analyses. On the contrary, most teaching laboratory or training sessions are short, often less than three hours; therefore, the use of official and standard methods in teaching and training environments is impractical.

Many of the methods presented in this book were designed from the authors' combined experiences as microbiologists in academic, regulatory, and industry laboratories. Starting decades ago, with simple teaching exercises at the Ohio State University, we built full analytical protocols that echo recent advances in science yet are executable in diversely equipped laboratories, albeit to different degrees of completeness. The laboratory exercises have been structured so that each provides students or trainees an opportunity to learn a new technique or approach in microbiology. The learning objectives are listed explicitly below each exercise title and explained in detail within the body of the exercises. The book's initial exercises meet the needs for basic training in food microbiology; these activities also prepare students and trainees for more advanced training in subsequent exercises.

The book has four main parts. The first part covers safety considerations and reviews basic microbiological techniques that may have been covered in previous introductory biology or microbiology courses. Included in Part I are simple exercises to help students with limited background in applied microbiology or to refresh experienced students with essential microbiological techniques. This part also emphasizes terminology and sets the stage for the approaches used in the remainder of the book. Part II includes exercises to evaluate various microbiological groups of significance to food quality, starting with mesophilic aerobic bacteria and ending with foodborne fungi. Pathogen detection is covered in Part III of the book. Both culture- and molecular-based approaches are included in this section. Part IV covers the microbiological aspects of technologies used in the control of food microbiota. Included in this part are two exercises that familiarize students with microbiological control by thermal treatments and antimicrobial peptides.

This book evolved from simple exercises that have been offered as a food microbiology laboratory course at the Ohio State University (OSU), Columbus, Ohio, USA, since the early 1990s. As the course became popular, interest in documenting these exercises into a published book was expressed by OSU alumni as well as colleagues from other institutions, including those from developing countries. In response to this interest, the teaching methods were compiled in the *Food Microbiology Laboratory Manual*, which was authored by A. E. Yousef and C. Carlstrom and published in 2003 by John Wiley. The book became a popular textbook in several universities and its Spanish translation was published in 2006 in Spain. Considering the continuous advances in the food microbiology field, it took great effort to evolve course methods and test them repeatedly before offering them to the students as teaching exercises. Although many teaching assistants over the years helped with this effort, the contributions of Joy Waite-Cusic and Jennifer Perry were the most significant. They continued to develop and test these exercises post-graduation from OSU, and while they serve as faculty members at Oregon State University and University of Maine, respectively. Both J. Waite-Cusic and J. Perry are co-authors of the current book.

Methods included in this book were customized to fit two-hour laboratory periods and for a class that meets two or three times per week for at least 12 weeks per semester. Like any methodology publication, users may find errors or points of confusion despite our best efforts. These will be reported and corrected when future versions of the book are developed.

We recognize and appreciate the help we received from many teaching assistants, staff members, and visiting scholars while preparing and refining the exercises in this book. Many teaching assistants have contributed to the annual offerings of the Food Microbiology Laboratory course at OSU. They alerted us to exercise shortcomings, suggested solutions, and even tested corrective measures. We particularly appreciate the help we received from C. Carlstrom, A. Abdelhamid, E. Huang, Y-K. Chung, and B. Lado-Diono. Matthew Mezydlo, Department of Microbiology, OSU, shared his excellent microbiology expertise and provided valuable advice that was integral to the success of our effort. We also thank the hundreds of students at OSU who have practiced these methods over the past three decades and pushed us to provide a better compilation of exercises. A special thank you goes to Dr. Patrick Dunne, who was a great supporter of our effort in putting this book together.

Ahmed E. Yousef
Professor, The Ohio State University, Columbus, Ohio, USA

Joy G. Waite-Cusic
Associate Professor, Oregon State University, Corvallis, Oregon, USA

Jennifer J. Perry
Associate Professor, University of Maine, Orono, Maine, USA

PART I

BASICS OF FOOD MICROBIOLOGY LABORATORY

The first part of this book introduces the laboratory aspects of microbiology as applied to food. It includes four chapters, covering laboratory safety, food sampling, microbial enumeration, and an exercise in practicing the knowledge gained in the previous three chapters. This part also serves as a review of techniques and applications covered in introductory microbiology courses. Students will execute basic exercises that will help them recall and sharpen the skills gained in previous microbiology courses.

Analytical Food Microbiology: A Laboratory Manual, Second Edition. Ahmed E. Yousef, Joy G. Waite-Cusic, and Jennifer J. Perry.
© 2022 John Wiley & Sons, Inc. Published 2022 by John Wiley & Sons, Inc.

BASICS OF FOOD MICROBIOLOGY LABORATORY

CHAPTER 1

LABORATORY SAFETY

This chapter is intended to help students and other analysts maintain safety for themselves and coworkers while receiving microbiological training or executing laboratory exercises. Although the chapter does not cover every aspect of laboratory safety, it familiarizes laboratory workers with essential components of that safety. Successful and safe execution of a laboratory exercise by a student or a professional analyst requires the compliance with safety guidelines presented here, as well as those found in other resources. Additionally, it is important to emphasize that observance of common-sense safety precautions is important in many situations.

BACKGROUND

Live microorganisms that are handled in microbiology laboratories may cause laboratory-acquired infections (LAI). Safety of students in these laboratories requires observing and obeying a set of laboratory safety rules. It is essential that one becomes familiar with the food microbiology laboratory setup and safety guidelines before conducting any exercises. This knowledge not only minimizes the risk of LAI, but also leads to efficient use of time and laboratory resources. Once these rules are read and understood, students should sign a form indicating that they have read, understood, and are willing to comply with these rules. Completion of this step is needed before students can practice basic techniques and the instructor can ensure their compliance with the safety guidelines. Additionally, online or in-person training may be required before students use certain microbiology laboratories.

Analytical Food Microbiology: A Laboratory Manual, Second Edition. Ahmed E. Yousef, Joy G. Waite-Cusic, and Jennifer J. Perry.
© 2022 John Wiley & Sons, Inc. Published 2022 by John Wiley & Sons, Inc.

Laboratory Environment and Personal Safety

There are many facets to the laboratory environment, ranging from tangible items such as fixtures, equipment, supplies, and waste disposal containers to more conceptual aspects such as safety. Laboratory instructors and students should be familiar enough with the laboratory environment to respond appropriately to safety issues and emergencies.

Microbiology laboratory equipment includes basic items such as incubators, refrigerators, water baths, autoclaves, centrifuges, and microscopes, to more contemporary equipment such as gel electrophoresis systems, multiple-well (or microtiter) plate readers, and polymerase chain reaction (PCR) thermocyclers. Microbiology laboratories often contain biological safety cabinets and chemical fume hoods. A large array of small laboratory tools that are used routinely include pipettes, Bunsen burners, cell spreaders, streaking loops, and thermometers. Supplies and disposable items used in various exercises are also part of the laboratory environment.

Students will be able to complete various laboratory exercises successfully when all required equipment is available and in working order. To this end, a list of required equipment is included with each laboratory exercise in this manual. The students must understand the function of each piece of equipment and how to use it safely and correctly.

Food microbiology involves the study of numerous organisms, including those known to cause human diseases. Therefore, careful work habits are important to prevent the spread of disease to analysts or other workers who may use the laboratory space. Familiarity with the laboratory environment itself and with the procedures required to keep that environment safe and clean is a key component in good microbiology laboratory work.

Biosafety Levels

The health authorities in the US have created a publication regarding biosafety in microbiological and biomedical laboratories (see the reference listed at the end of this chapter). This manual sets forth guidelines for best practices for four biosafety levels. Each biosafety level has a particular set of protocols that laboratory users must follow to minimize risks to the laboratory workers and the public. There are protocols for primary containment (protection of individual workers) and secondary containment (protection of the public). The protocols provide guidance on laboratory security, laboratory practice and technique, required safety precautions, facility design and construction, and required training of supervisors and workers, as well as specific information regarding animal facilities, clinical facilities, and transportation of materials. The publication also contains listings of specific organisms and their assignment to risk groups (RG) with their recommended biosafety level. The following is a brief description of the four biosafety levels.

- Biosafety level 1 (BSL-1)
  - Appropriate for handling agents (RG-1) that are not known to cause disease and are well characterized; examples include *Bacillus subtilis* and non-pathogenic *Escherichia coli*.

- o Required protective devices include doors, sinks for handwashing, easily cleaned work surfaces, screened windows, and bench tops that are impervious to water.
- o No special construction or ventilation is required.
- o Primary barriers (lab coat and gloves) are required.
- Biosafety level 2 (BSL-2)
  - o Appropriate for handling moderate-risk agents (RG-2). These include agents associated with human disease, but for which immunization or antibiotic treatment is available; examples include *Salmonella* and Measles virus.
  - o All precautions for BSL-1 are observed, plus doors must be lockable and marked with biohazard signs, eyewash station is available, air does not recirculate to non-laboratory areas, and autoclave is available.
  - o Special work areas (e.g., biosafety cabinets) should be assigned for activities that might generate aerosols or splashing and for handling large volume or high concentration of organisms.
- Biosafety level 3 (BSL-3)
  - o Appropriate for handling agents that may cause serious and potentially lethal infections (RG-3). These include agents that are transmittable by aerosols. Examples of these agents are *Mycobacterium tuberculosis* and St. Louis encephalitis virus.
  - o All precautions for BSL-2 are observed plus separate facility or zone with double door entry; inward only airflow with 10–12 air changes/hour; water-resistant walls, floors, and ceilings; filtered vacuum lines; and respiratory protection may be required.
- Biosafety level 4 (BSL-4)
  - o Appropriate for handling exotic agents (RG-4) that (a) pose high risk of life-threatening diseases, (b) are transmittable by infectious aerosols, and (c) for which no treatment is available. An example of these agents is Ebola virus.
  - o All precautions for BSL-3 are observed plus single-pass dedicated air system; walls, ceilings, and floors create an internal seal; all liquid effluent and solid waste are decontaminated; entry doors cannot be opened simultaneously; communication system; emergency generator, positive pressure personnel suit, and showering upon exit are required.

Note: Food microbiology laboratories that accommodate pathogen work are classified as BSL-2.

Decontamination and Waste Disposal

Contaminated but reusable laboratory utensils and glassware should be decontaminated before cleaning. Chemical disinfection or autoclaving are often used in this case. Contaminated disposable items (e.g., disposable gloves, pipettes, and agar plates) should be placed in designated biohazard boxes with liners. Disposal of these biohazard boxes should be managed by a professional service, which may subject these items to incineration or other validated decontamination method.

PERSONAL SAFETY IN THE LABORATORY

It is important that students become familiar with the personal protective equipment required for working in a microbiology laboratory as well as safety-associated procedures and etiquette.

Personal Protective Equipment

Personal protective equipment (PPE) is needed to protect against physical, chemical, and biological laboratory hazards. Availability of this equipment is important but adequate training on how to use it is equally important. The following is a partial list of PPE, but others may be needed:

Safety glasses: These should be cleaned and sanitized before and after use. Alcohol wipes may be adequate for the sanitization.

Laboratory gowns or coats: These should be used and kept in the laboratory.

Face masks and face shields: Masks and shields protect against splashes and aerosolized droplets and particles. These are also essential to prevent the spread of infectious airborne agents.

Gloves: Use of disposable gloves is essential for most laboratory activities.

Observing Personal Safety

- **Appropriate attire.** Wear appropriate clothing for laboratory work. Closed-toe shoes must be worn during each laboratory period. No sandals, open-toe shoes, or bare feet are permitted. Shorts may be disallowed.

- **Winter coats and backpacks are stored away from the bench.** Keep these items at a safe distance from the laboratory bench and preferably outside of the laboratory. Coat hangers, cupboards, or preferably lockers should be used to store these items temporarily while students are working in the laboratory.

- **Items brought to the bench are subject to contamination; these should be kept to a minimum.** Keep away all books, notebooks, calculators, laptop computers, and similar items. Technically, two sheets of paper and a pencil are all that need to be brought to the bench. One of these papers contains the laboratory exercise summary, outline, or flow chart, and other is used for recording data. It is advisable that these two sheets are kept in a transparent plastic sleeve that can be sanitized at the end of the exercise.

- **Mobile phones should not be handled during the laboratory period.**

- **Laboratory coat.** It is mandatory that a laboratory coat or similar protective covering is worn during each laboratory period. The coat must have either buttons or a zipper. The laboratory coat should be labeled with the student's name and be kept completely buttoned or zipped for the duration of the laboratory period. After completing the work, coats should be kept in the laboratory. The instructor will point out appropriate coat storage, if available. Before coats are removed from the laboratory, they should be properly decontaminated; this can be accomplished by autoclaving.

- **Eye protection.** Safety goggles should be worn all the time in the laboratory.
- **Washing hands.** Washing hands minimizes or prevents the transfer of organisms between the analyst and the food to be analyzed and vice versa. Hands should be washed before starting any exercise (to avoid contaminating items being analyzed) and after the exercise (to prevent spreading contaminants outside the laboratory).
- **Use disposable gloves.** Even though disposable gloves may not be required for some experiments, it is advisable to wear these gloves for all activities in the laboratory. Analysts who have allergies to latex should wear gloves made of alternative materials (e.g., nitrile rubber). Once the work is completed, the gloves should be disposed of properly in the appropriate biohazard containers. Analysts should never leave the laboratory with gloves on. It is a sign of great carelessness when analysts are seen in hallways or elevators wearing disposable gloves. This can also be the cause of serious cross-contamination on non-laboratory surfaces, such as doorknobs.
- **Clean and sanitize the laboratory bench.** Use disinfectant and paper towels to wipe the laboratory bench both before and after any exercise. These paper towels should be disposed of in the regular trash, unless directed otherwise by the instructor.
- **Never begin laboratory work without the prior permission of the laboratory instructor or supervisor.** Generally, students are not allowed to work until *after* the instructor's presentation on the day's activities. If arriving early, the student may use this time to change into appropriate dress, review the exercises to be completed, inspect the laboratory for locations of needed equipment, and similar activities.
- **Eating or drinking in the laboratory is forbidden.** The laboratory environment is not an appropriate place for eating or drinking. In fact, any activity that might involve putting something into the mouth, (e.g., chewing gum, chewing tobacco, using a throat lozenge, smoking, habitually chewing on a pencil) may provide an opportunity for a pathogen to infect the analyst.
- **Applying cosmetics in the laboratory, including lip balm or lotion, is not allowed.** Anything that is applied may trap contaminants on the skin or introduce contaminants into the laboratory environment. Insertion of contact lenses is not permitted in the laboratory.
- **Avoid touching eyes, skin, or hair, particularly with worn gloves.** These activities can lead to body contamination with harmful microorganisms.
- **Miscellaneous.** Sitting on the laboratory bench is not permitted. Keep the laboratory as neat as possible at all times. At the end of each laboratory period, check and arrange all materials neatly. Return all materials to their proper places or dispose of them appropriately when your work is finished.
- **Never remove equipment, media, or microbial cultures from the laboratory.**
- **Label all materials properly so that they can be identified easily.** Tubes should be labeled using label tape and a marker. Petri dishes should be

labeled on the bottom (the side with the agar) with student name, the organism, type of medium, incubation temperature, and date.

- **Use pipettes carefully.** Pipettes can be hazardous if not used properly. Mouth pipetting is both a poor technique and a safety hazard; therefore, it is not permitted. Pipette bulbs, manual pipette aids, and semiautomatic pipetters (with pipette tips) are available for use. Forcing a pipette into either a bulb or a pipette aid may lead to breakage and should be avoided. It should be cautioned that improper use of pipettes can lead to dripping or generation of hazardous aerosols.

- **Be familiar with the available safety equipment and supplies.** Know the locations of the first-aid kit, safety showers, eyewash stations, fire extinguishers, fire blankets, and fire alarms.

- **Avoid fire hazards.** Hair that is shoulder length or longer must be tied back or pinned up to minimize the risk of it catching on fire. Similarly, hats with brims should be avoided as the brim might come near the flame. Hats such as baseball caps may be worn facing backward to keep the brim away from flames. For safety, constantly be aware of any burners near you. Always use your own burner. Do NOT reach across the bench to use someone else's burner. Some of the liquids present in the laboratory are flammable; keep these away from the Bunsen burner.

- **Handling fire emergencies.**

  a. Students should be aware of the location of available fire safety equipment (e.g., fire extinguishers, fire blankets) and the nearest exits in case of larger fires.

  b. Alcohol fires are among the most common laboratory fires. Should a jar of alcohol catch fire, placing the lid over the jar quickly may suffocate the fire. Alternatively, cover the burning jar with a slightly bigger glass jar, such as a beaker. Keep flame away from staining bottles as these often contain alcohol.

  c. If anyone's hair or clothing should catch on fire, obtain a fire blanket, wrap the person in the blanket, and have them roll on the floor to extinguish the flames.

  d. Any fire should be reported immediately to the laboratory supervisor.

  e. If a major fire occurs, proceed to the nearest exit. DO NOT USE ELEVATORS!

- **Handling first-aid emergencies.**

  a. Students should be aware of the location of the first-aid kit in the laboratory. The kit should contain gauze bandages, adhesive bandages, bandage tape, sterile swabs, burn cream, antiseptic wipes, and hydrogen peroxide.

  b. Get the instructor's assistance before using the first-aid kit.

- **Mercury spills.** While many laboratories have switched from mercury to alcohol thermometers, some laboratories may still be using mercury thermometers. Mercury is a hazardous material that requires special cleanup procedures. If a mercury thermometer is broken, the analyst should notify the instructor immediately. The students in the vicinity of the spill should not touch the mercury, should move away from the area, and should prevent others from entering the area.

- **No bicycles, skateboards, roller skates, or similar devices are permitted in the laboratory or the hallway.** These items can create a tripping hazard in the laboratory, to passersby in the hall, or during an emergency.
- **Report any personal injuries to the laboratory instructor/supervisor.** In case of accidents, mandatory accident reporting forms must be filled out as soon as possible.

MATERIALS IN MICROBIOLOGY LABORATORY

- In many teaching laboratories, each student (or a group of students) is assigned a storage drawer (or a similar compartment) containing materials commonly used in the laboratory. Students should be sure that the drawer contains all the materials indicated in the course instructions and that all materials are returned and stored at the end of each session. Typical tools contained in this storage space may include inoculating loop, inoculating needle, microscope slides, cover slips, microscope lens cleaner, lens cleaning paper, lens oil, wax marking pencils or permanent markers, pipette bulbs, bibulous paper, and matches or a striker for lighting the Bunsen burner. Some of the consumable materials may be used up during the course of the term and students should learn where replacement materials are kept. At times, the items from a storage location may be misplaced. If this occurs, the student should not take supplies from someone else's drawer.
- Some laboratory communal supplies may also be used up during the course of the school term. These items may include paper towels, disinfectant solutions, Gram stain reagents, other staining agents or reagents, adhesive tape, and other frequently used materials. Students should determine where these items are stocked so that they can replenish supplies.
- Students should know where to obtain distilled water. In most laboratories, special distilled water taps are used; these are often located near the regular hot- and cold-water taps. The distilled water taps are frequently spring-loaded to prevent anyone from leaving the tap open and wasting water. Often these taps have a tab on the handle labeled "DW."
- At the beginning of every laboratory session, the student should determine the location of all water baths, incubators, or other equipment that will be shared during that session. Students should collect all media and supplies required to perform the experiment. Many microbiological growth media look similar; therefore, caution should be taken to carefully and correctly label media. Students should not collect more media than will be used during the exercise. Careful reading of the laboratory exercise should allow students to determine the correct number of plates and tubes needed for each exercise.
- The microbiology laboratory contains many materials that are potentially dangerous if used outside the laboratory environment. Students should never remove slides, plates, or tubes from the laboratory. After use in the laboratory, materials are either prepared for reuse or discarded. Each laboratory has a system for material disposal, protocols for which items are reused and which are discarded, locations where reusable materials should be placed at the end of the laboratory, expectations for what to clean manually by students,

etc. Students must be familiar with proper disposal and proper clean up to ensure that materials are not wasted, biohazard containers do not contain excess materials, and everyone's safety is preserved.

- Used culture tubes should not be returned to the laboratory exercise set-up area, unless the instructor specifically tells students to do so. Only unused media should be returned to the set-up area.

- Reusable materials may include some glassware, such as test tubes, bottles, and flasks. This reusable glassware should then be placed in the designated location for each type of item. Depending on their contents, tubes, bottles, and flasks may need to be autoclaved before washing. These items should be separated from items that do not require autoclaving. Some other items, such as blender jars, may not require autoclaving and may be manually washed by students. These items should be washed according to the designated protocol and placed in the designated drying area.

- Non-reusable materials are disposed of in either hazardous or non-hazardous waste containers. Paper towels used with disinfectant to wipe off laboratory benches may be placed in the containers for non-hazardous waste (i.e., regular trash). Gauze or lens paper used to clean microscope lenses before or after use is also safe to be placed in the regular trash. Items that have not been exposed to microorganisms do not require special disposal.

- *Biohazard containers* (e.g., special marked bags in cardboard boxes or cans with plastic liners) are used to discard contaminated materials. These materials include all disposable gloves, culture-containing disposable Petri plates, and disposable test tubes. Contaminated materials (i.e., those exposed to laboratory microorganisms) are typically autoclaved or incinerated. Contaminated broken glassware should be disposed of in the sharps container. Broken, uncontaminated glassware should be placed in a receptacle designated for that purpose (e.g., the broken glass box). If syringes are used for a laboratory exercise, they should be disposed of in the sharps container designated by the instructor.

- Spilling or splashing of cultures can happen. In case of small spills, the student needs to encircle and flood the area with excess disinfectant, allow disinfectant to sit for the proper amount of time, and wipe the area with paper towels or other provided absorbent towels, wiping toward the center to prevent spread of the contaminant. These towels are considered contaminated and should be disposed of in the biohazard container. In case of larger spills, the instructor should be notified immediately. Any broken glassware should be disposed of in a sharps container. Appropriate disposable gloves should be worn during the cleaning process and should be discarded after the spill has been addressed.

- It is recommended that students tape their inoculated agar plates together at the end of each exercise or keep them in a designated group container (small plastic bin) to make retrieving the group's plates easier at the beginning of the subsequent session and make them easier to handle and inspect by instructors. Plates should be placed in the correct orientation and in the designated location for incubation.

PRACTICAL ASPECTS

In addition to keeping yourself and other laboratory members safe, the proper exercise of safety protocols and etiquette allows for the timely completion of laboratory sessions. Lack of preparation before arriving to the laboratory may prevent students from finishing the exercise within the allotted time.

Before the Laboratory Session

1. Carefully read the laboratory exercise and understand why and how it is executed.
2. Summarize the practical steps to be carried out during the session on a single sheet of paper. This "exercise summary" should be one of only two papers allowed on the bench during the execution of the exercise. The second is a blank paper for writing observations and recording results. As indicated earlier, the exercise summary (plus the recording sheet) are ideally kept in a plastic sleeve and placed on the bench while executing the exercise. The plastic sleeve should be sanitized properly before leaving the laboratory.

Immediately Before Entering the Laboratory

1. Finish or dispose of any food or drink items/containers.
2. Turn off electric devices (e.g., mobile phones, laptop computers, tablets, etc.); these should be stowed appropriately for the duration of the laboratory session.

Immediately After Entering the Laboratory

1. Enter the teaching laboratory when the instructor/supervisor is available; more than one instructor should be available to supervise the session.
2. Keep your belongings (e.g., backpack, winter coat, etc.) in the designated area, which is preferably outside the laboratory; bring only the exercise summary and a blank sheet of paper to the bench.
3. Put the exercise summary (and the blank sheet) in the provided plastic sleeve; this is a sanitizable pocket for protection against spills. Present the exercise summary on the bench to be reviewed by the instructor.
4. Hair that is longer than shoulder length must be tied up.
5. Wash hands in the laboratory sink using the soap and disposable towels provided.
6. Put on a lab coat; when not in use, these should be stored in the laboratory throughout the course.
7. Put on disposable gloves and sanitize the bench; a quaternary ammonium solution or alcohol is often used for bench sanitization.
8. Listen carefully to the instructor's short presentation; this presentation may include seating chart, assignment for the food to be analyzed, potential pitfalls, etc.
9. Start the exercise when instructed to do so.

While Executing the Exercise

1. Be aware of whether you are working individually or in groups of two or more. If working in groups, part of the work could be carried out individually and the other part is done cooperatively. If working in a group, make sure you communicate clearly with laboratory partner(s) before starting the exercise.
2. Start the laboratory exercise and observe the safety rules described earlier.
3. Do your best to complete the work efficiently and diligently.
4. Make sure you share the progress of the exercise or problems encountered with one of the instructors.
5. Record your observations or results. The exercise summary sheet or a separate sheet of paper may be used for recording. Alternatively, hand-held electronic notepads may be provided by instructors for note taking and data collection.

Immediately After Completing the Laboratory Exercise

1. Show your work (mounted microscope slide, reaction results, colony counts, etc.) to the instructor.
2. If asked, transfer the data collected to the class computer or class data sheet.
3. Dispose of work items correctly.
4. Sanitize the bench using the sanitizer provided (often a quaternary ammonium sanitizer or alcohol).
5. Remove disposable gloves and place them in the biohazard container.
6. Store lab coat appropriately.
7. Wash hands.
8. Take your belongings and exit the laboratory.

REFERENCE

Centers for Disease Control and Prevention. (2020). *Biosafety in microbiological and biomedical laboratories.* 6th ed. U.S. Department of Health and Human Services, Washington, DC, USA.

CHAPTER 2

SAMPLING FOR MICROBIOLOGICAL ANALYSIS OF FOOD AND PROCESSING ENVIRONMENT

It is a challenge to be able to assess the microbiological quality and safety of food accurately. The approach often used is a stepwise procedure that includes sampling, sample preparation, laboratory analysis, data collection, and result interpretation. Errors in each of these steps cumulatively determine the reliability of the overall procedure. Sampling can be an elaborate exercise (Figure 2.1), and analysts consider it the most error-prone step. Poorly planned and executed sampling operations compromise the analyst's ability to assess the quality or safety of food. This chapter includes two main sections: "Theoretical aspects," which provides the knowledge needed for proper sampling, and "Practicing sampling and sample preparation," which is a simplified practical exercise.

THEORETICAL ASPECTS

This section covers the theoretical principles of sampling and sample-size calculations. Additionally, techniques that may be followed during sampling and sample preparation of food or processing environment are covered.

Sampling Principles

Introduction

In the simplest sense, a "sample" may be defined as a small and manageable quantity intended to represent the whole. The whole is commonly referred to as the "population," and in relevance to the subject matter of this book, the

Analytical Food Microbiology: A Laboratory Manual, Second Edition. Ahmed E. Yousef, Joy G. Waite-Cusic, and Jennifer J. Perry.
© 2022 John Wiley & Sons, Inc. Published 2022 by John Wiley & Sons, Inc.

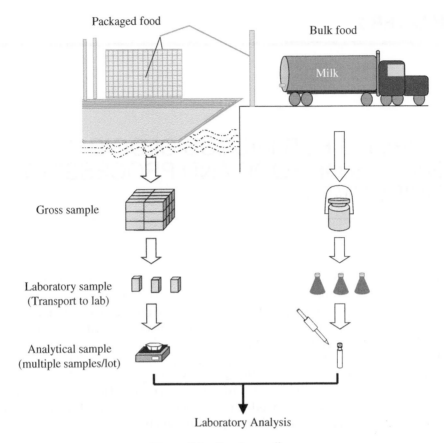

Figure 2.1   Food sampling.

population is the food lot. Foods vary considerably in physical, chemical, and microbiological characteristics. These variations dictate the way a food is sampled and analyzed. Physically, food could be in a solid, gel, or liquid state, at different degrees of hydrophobicity, with ingredients in homogeneous or heterogeneous distribution. Compositionally, foods vary in water content, pH, presence of antimicrobial ingredients, and many other attributes. Foods differ in microbial burden and profile; this depends on whether the food is raw or processed, and the type of processing it received. The goal of the analysis also varies. Some foods are analyzed for the enumeration of indicator microorganisms (e.g., coliforms), whereas others are tested for the presence of pathogens (e.g., *Listeria monocytogenes*). These factors must be taken into account to determine appropriate sampling, sample preparation, and microbial recovery methodology.

Foods subject to analysis are usually found in sizable quantities located in a storage facility, ship container, tanker, retail display case, vendor stand in open market, etc. With consideration of the size of the food lot and expected variations among multiple samples from the same lot, a number of samples are collected for analysis on the hope that they accurately represent the entire lot. If samples taken are not a good representation of the whole, whether it is the fault of the sample collector or due to an error inherent in the sampling plan, the laboratory results will be misleading. Therefore, sampling operations should be planned well. Additionally, every effort should be made to avoid mishandling or contamination of the collected sample. Correctly withdrawn, handled, and analyzed samples may serve as evidence of the quality or the safety of the whole.

Although food is emphasized in this book, sampling and analysis of water and processing environment will also be addressed. A sample of water from a stream is described as a "specimen." Similarly, samples from circulating cleaning or rinsing solution or swabs from a moving conveyer belt are also considered specimens. In these situations, the population sampled is not static and thus getting a representative portion is a challenging task.

Preparing A Sampling Plan

Sampling is an essential step in any procedure for assessing the microbiological quality or safety of food. Sampling is an integral part of food inspection, which is practiced for commercial or legal reasons. Researchers experimenting with food need sampling schemes that lead to statistically meaningful results. Regardless of the ultimate goal of the analysis, sampling should be planned and executed properly. The following are steps used in preparing a sound sampling plan.

1. **Identify the Reasons for Sampling**

   Sampling is a key and critical step in microbiological analyses that are done for many reasons including: (i) assessing the general microbiological quality of a raw product or an ingredient; (ii) validating a food processing operation; (iii) assuring the safety of the processed food; and (iv) evaluating the sanitary condition of a food processing environment. Each of these cases require a carefully considered sampling plan.

2. **Assess the Size, Nature, and Uniformity of the Lot to be Sampled**

   The population (i.e., the food lot) from which the samples are to be taken could be made of discrete units or bulk in a container. For example, a food lot of half-and-half coffee cream could be a stack of wholesale boxes, each containing multiple smaller retail boxes, and the latter containing multiple single-serve (0.4 oz.) units. Alternatively, the lot could be bulk flour in a store bin or sack, milk in a tanker, or loose grains in a silo.

   Sampling starts by taking a number of units from the stacks of the lot or a portion of the bulk; these are described as gross (or primary) samples (Figure 2.1). Subsets of gross samples constitute the laboratory samples. The analyst who receives laboratory samples should further reduce them to test samples. For example, a laboratory sample could be a 10 lb cheese block, from which a 25 g test sample is withdrawn. In this particular example, it is desirable to collect several test samples to overcome the lack of uniformity from the edge to the center of the cheese block. Once the test sample is subjected to laboratory analysis, it is no longer described as a "sample." Instead, the analyst should use descriptive words such as food homogenate, test solution, cell suspension, cell pellet, culture supernatant, isolate, etc. Note that the number of samples to be withdrawn from the lot is determined as described later.

3. Determine the Acceptable Quality Level or the Tolerable Safety Risk

The acceptable quality or safety level should be identified before sample size is determined. A supermarket chain importing strawberries may accept only a truckload that produces less than 1% moldy samples among all the samples analyzed. Determination of "moldiness" may be done subjectively (e.g., visual inspection) or analytically (e.g., fungi count on microbiological media). Note that the latter approach is time consuming, and the product could suffer significant quality damage while waiting for results to be obtained.

Sampling for assessing food safety risks should be planned carefully. In addition to the factors discussed earlier, this sampling plan also should consider food status within the supply chain (e.g., raw or ready-to-eat), degree of processing (e.g., minimally processed or retorted), intended consumer population (e.g., infants or adults), and other factors. Analysts, for example, may be asked to determine the prevalence of *Salmonella enterica* on the surface of raw shell eggs produced in a cluster of farms that switched from a caged to a cage-free or free-range hen system. In a different scenario, analysts may be sampling for *S. enterica* in pasteurized shell egg from a company that introduced pasteurization as a new technology in egg processing and compare that with normal incidence of *S. enterica* in raw shell eggs. Although the food product and the targeted pathogen is the same in both examples, some pathogen-positive samples are allowed in the first example (on-farm), but none is allowed in the second example. Zero tolerance is common for infectious pathogens in ready-to-eat foods, whereas some degree of contamination is allowed in raw foods that are supposed to be cooked or processed before consumption.

4. Determine the Number of Samples to be Withdrawn

Once acceptable quality level or tolerable safety risk is considered, sample size should be determined. Sample size is determined by a statistical approach called "power analysis." For explanation, let us consider the first of the two examples presented earlier: the contamination of raw shell eggs surfaces with *S. enterica*. The analysis can be viewed as a comparison between two groups: caged versus cage-free products. To estimate the appropriate sample size for this experiment, a systematic approach needs to be followed.

a. Define the hypothesis to be tested

  i. Null hypothesis: The rate of incidence of *Salmonella* in the two groups do not differ.

  ii. Alternative hypothesis: The rate of incidence of *Salmonella* in the two groups is different.

b. Determine the statistical parameters needed to calculate sample size

  i. Significance level (α). This parameter is commonly set at 0.05 (i.e., 5%). It is the possibility of falsely rejecting the null hypothesis (Type-I error), i.e., the "false positive" rate. In other words, with this α value, there is a 5% chance to conclude that there is a difference, while in fact there is no difference. The smaller the value of α, the larger the sample size needed to produce statistically meaningful experiments.

ii. False-negative rate (β). The chance of failing to reject the null hypothesis when the alternative hypothesis is true (Type-II error). The commonly used values of β are 0.1 and 0.2 (i.e., 10% and 20%). The smaller the value of β, the larger the sample size needed to produce statistically meaningful experiment.

iii. Power of the experiment (1-β): Once β is selected, the power is determined. Power is the probability that the analysis can detect a difference between the two groups, when this difference is truly present. Although it is always desirable to have a high power, 1-β is often set at 0.8 (80%) to make the sample size manageable.

iv. Effect size: A meaningful effect size should be considered. In the example above, one may assume that if the percentage of *Salmonella*-positive eggs from the less-infected hen group is 10% (p_1), the more infected group should produce *Salmonella*-positive eggs at the rate of 20% (p_2) or more, for the difference to be meaningful. Additionally, if 1-β is chosen to be 0.8, it means that we have 80% chance of detecting the stated difference between these two groups. Note that the smaller the p_1 and p_2 values, the larger the size of sample needed to produce statistically meaningful results.

v. Population standard deviation: This is needed for continuous data, such as changes in pathogen populations. However, in the example described above, the variation between the two groups is not continuous, it is dichotomous (i.e., results reported as *Salmonella*-positive or *Salmonella*-negative) and expressed as incidence rate or proportion. Therefore, determination of standard deviation is not relevant to the example presented here, but it will be needed for other types of analyses.

c. **Calculate sample size**

The information collected so far is sufficient to determine sample size per group, i.e., number of eggs to be tested for *S. enterica* for each of the two hen groups. Based on this information, $\alpha = 0.05$, 1-$\beta = 0.8$, $p_1 = 0.1$, and $p_2 = 0.2$. These values can be entered in an appropriate statistical equation (e.g., Dell et al., 2002) for calculating sample size. Alternatively, commercial statistical programs (e.g., Minitab or SAS) contain modules that automate the sample size calculation once the above parameters are determined. In the example at hand, 218 eggs for each of the two groups need to be analyzed to determine if incidence of this pathogen is different in the two groups of hens. The sample size in this example is quite large because of the small rate of contamination of *S. enterica* on eggs. The rate of incidence of the pathogen inside the egg (i.e., in the yolk) is even smaller and thus requires much larger sample size to detect any differences between the two groups.

It should be obvious from the previous discussion that sample planning is not limited to size determination. The plan should be tailored to address food characteristics, reason for sampling, lesson to be learned or action to be taken, and other factors. Food inspection, mandated by regulatory agencies, should be based on sound sampling plans considering the health and economic impact of the outcome of the analysis. The frequency of inspection and sampling may depend on the quality or safety history of the food produced or processed by a given establishment. Inspection and sample collection are minimized for establishments with a recent history of producing good quality food or food that posed no or low safety risks. Therefore, a vendor's recent history should be considered when developing a sampling plan.

SAMPLING TECHNIQUES AND SAMPLE PREPARATION

Sampling Tips

After developing the sampling plan, actual sampling should be carefully executed. The following are some tips to be observed during withdrawing, handling, and transporting of samples.

- **Collect Laboratory Sample in Original Container and Repackage Only if Necessary.** Samples are ideally submitted to the laboratory in the original unopened containers. In case of bulk food, or when the original container is too large for submission to the laboratory, a subsample is aseptically transferred to a clean, sterile container.

- **Use Sterile Sampling Utensils.** In the event of repackaging, suitable sterile plastic or metal containers are preferable over glass containers. These containers must be clean, dry, leak-proof, and of a size suitable for the sample. Sampling tools such as forceps, spatulas, and scissors should be appropriately wrapped and autoclaved prior to use.

- **Label Samples Appropriately and Create a Sample Record.** A proper label should be developed to identify sample contents, date of sampling, sample collector's name, and other pertinent information (e.g., sample temperature or storage conditions, and type of package from which subsample was taken). The simplest form of labeling is using masking or labeling tape, on which information is written with a permanent marker; this is preferred over writing directly on sample container. In addition to the information on the label, a record should be created to document additional pertinent information, such as the times of collection and of arrival at the laboratory, condition of sample at the time of arrival, etc. In some food inspection agencies, the label on the sample package is replaced with a barcode that is linked to a record in an electronic database.

- **Deliver Samples Promptly and Control Temperature During Transportation.** Samples of refrigerated food should be kept refrigerated and those of frozen food should be handled and transported in the frozen state. However, samples of refrigerated food should not be frozen at any time; freezing can alter sample microbiota. Holding these samples for considerable time before analysis may alter the microbial burden or profile.

Sample Preparation

The food inspector or sample collector delivers laboratory samples to the analytical facility. The delivered sample could be a retail package, a consumer-size container, or a portion of a food bulk. Sample preparation refers to the reduction of the laboratory sample into a test sample (or analytical sample) and preparation of the latter for analysis. Therefore, sample preparation includes: (i) withdrawal and measurement of a representative test sample from the laboratory sample; (ii) homogenization to distribute microorganism uniformly in the test sample; and (iii) dilution of the sample homogenate to decrease food microbiota to a countable or detectable level.

Withdrawing the test sample
Microbiological results are often reported quantitatively, therefore, sample mass (or volume) should be carefully measured and reported. The test sample, which is used directly in microbiological analysis, could weigh 10, 25, or 50 g, but a 25 g test sample is commonly used in the detection of pathogens. Larger sample size means more accurate representation of the food lot and greater ability to recover scarce contaminants. Many analysts, however, prefer smaller samples, since these are easier to handle and less costly to analyze.

If the recommended sample mass cannot be easily obtained (e.g., food difficult to mix before weighing), analysts should be able to modify the analytical procedure to accommodate this deviation. Analysts occasionally opt to combine several test samples into a single "composite sample." For example, if 15 portions (25 g each) are taken from 15 one-pound meat packages, and these packages are expected to be similar in microbiological quality, the analyst may combine these into a 375 g composite sample. The composite sample is then diluted (10^{-1}) and analyzed. Composite sampling is a cost-saving practice, but it could conceal an abnormally contaminated sample.

The physical characteristics of food dictate the technique suitable for withdrawing the test sample:

1. Pourable liquid, powder, and some shredded foods are easy to mix in original packages. Withdrawing a test sample form these types of food involves thoroughly mixing the contents of the package, aseptically measuring a predetermined portion, and transferring this portion to sterile container.
2. For pasty and thick products (e.g., packaged ground meat or multilayered cake), the package contents may be transferred to a bigger sterile container, and the contents are mixed using an appropriate sterile implement.
3. Solid foods that cannot be mixed manually include blocks or wheels of hard cheese and similar products. Pieces or wedges of the cheese may be cut with a sterile knife and aseptically shredded, using a sterile shredder. The shredded cheese should be mixed thoroughly before a test sample is withdrawn. Alternatively, a sterile cheese trier may be used to extract several core samples from the cheese wheel or block. The cheese pieces extracted by the trier can be easily cut into small pieces with a knife or a spatula, and the test sample can be taken from them.

4. Frozen bulky food (e.g., frozen meat) may be thawed in a refrigerator (2–4°C) overnight (~18 hrs) before a test sample is withdrawn. However, it is sometimes preferable if a test sample can be aseptically withdrawn directly from the frozen food package. In the latter case, sampling equipment may include a drill with sanitized pits, or a band saw with sanitized stage and blade.

5. Bulky food, with contaminants residing predominantly on the surface, is tricky to sample. Such foods include whole chicken carcasses and cantaloupes. Grinding a piece of chicken should distribute contaminants evenly and makes it easy to withdraw a 25-g test sample; however, results of analysis expressed as CFU/g could be misleading, since the microbiota were not evenly distributed in the initial sample. It may be preferable to take a surface sample from this type of food, by swabbing a defined area and expressing the results as CFU/cm². Limited surface area, however, may not produce a test sample sufficiently large enough to detect low levels of pathogen on these products. In this case, the whole food unit (e.g., the whole chicken carcass) may be thoroughly mixed with a diluent in a large sterile bag, and the pathogens washed off the surface are detected in a portion of the diluent. Note that in this case, results are reported per unit, not per area or mass.

6. Bulky but easy-to-cut foods such as vegetables and fruits require careful handling during sample preparation. These could be aseptically cut in a biological cabinet using a sterile cutting board and knife, and the cut pieces mixed in a sterile container or bag before a test sample is withdrawn.

The mass of the transferred portion is measured on an analytical or a top-loading balance. If the test sample is measured in volume, a sterile pipette or graduated cylinder is used. It should cautioned that fast but careful sample withdrawal and measurement is critical for minimizing contamination by the analyst or laboratory environment. For example, weighing the test sample to the nearest 0.1 gram takes less time and offers less chance for contamination than a process that produces a test sample weight with two-decimal digit accuracy.

Homogenizing the test sample

Most test samples require homogenization before analysis. The goal of homogenization is to release microorganisms from food into suspension. Blenders, stomachers, sonicators, shakers, and hand massaging of bagged food are different approaches for homogenizing a food sample. Liquid food samples are manually mixed before analysis, whereas solid foods commonly require mechanical stirring (homogenization) in a suitable diluent to break food clumps and release microorganisms from the food matrix. Methods of homogenization may vary in ability to recover entrapped microorganisms. Blenders and stomachers (Figure 2.2) are commonly used homogenizers to prepare the food sample for analysis. The revolving blades of the blender divide the food into small particles and mix them with the diluent. The same goal may be achieved using a stomacher, depending on the characteristics of the food. The stomacher is a mechanical device that agitates a food sample placed in sterile plastic bag. The back and

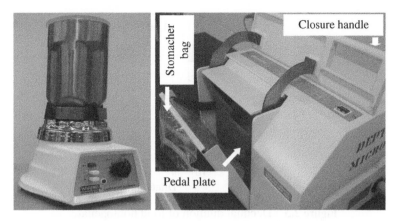

Figure 2.2   Food blender (left) and stomacher (right).

forth mechanical action of the stomacher paddles mimics the stomach action (hence the name) and helps incorporate the sample particles into the diluent.

Dilution of food homogenate

When pathogens are present in food, they are often found in very small numbers. Foods analyzed for detection of such pathogens are homogenized in an enrichment broth and the homogenate is not diluted further. The mixture is incubated to enrich the pathogen population so that it becomes easy to detect.

Other types of food are analyzed to determine the size of a certain microbial population (e.g., coliform count). In this case, the food is homogenized in a suitable diluent (e.g., sterile saline solution) and the homogenized food is appropriately diluted, usually decimally. A diluent is a liquid used to release the microorganisms from the food matrix, resulting in a suspension that represents the food. The suspension can be further analyzed by enumeration techniques. Ideally, the diluent should be compatible with the food system (i.e., allows for ease of homogenization) and should allow for maximum recovery of the microorganisms by not inducing biological stress (i.e., pH should be near neutral and osmolarity should be close to the microbial physiological level). Diluents include peptone water, citrate buffer, saline solution, and neutralizing buffer. Peptone water is the most commonly used diluent for food analysis; however, it may not be appropriate for some foods. Foods that are naturally acidic or contain antimicrobial compounds should be diluted with neutralizing buffer to prevent the inactivation of microorganisms during the recovery step. A buffer, such as warm (40°C) citrate solution, is the preferred diluent for microbial analysis of hard cheeses, which will not mix evenly in cold aqueous solutions.

A subset of dilutions is selected for plating on a microbiological medium suitable for sustaining the targeted population. The extent of dilution and selection of the dilutions to be plated depends greatly on the analyst's expectations of the size of microbial population in the food. The larger the population, the greater the degree of dilution required. An example of a dilution scheme is shown in Figure 2.3.

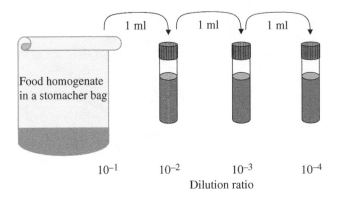

Figure 2.3 Decimal dilution of food homogenate.

ENVIRONMENTAL SAMPLING

Food microbiota include pre-harvest contaminants as well as microorganisms introduced from the processing environment. Assessing the microbial load and variety of microorganisms in the factory environment provides insight into the quality and safety of the finished product produced in the facility. Consequently, the processing environment is often sampled, and results of analysis are carefully considered by the facility management. Samples are often taken from floors, drains, and equipment surfaces, particularly those that are food contact surfaces. Refrigerators and other storage sites also should be sampled frequently. The sampling procedure depends on the nature of the site, degree of contamination, and the microbiological information sought. There are two broad categories of samples that are commonly taken from food processing environment: surface and air samples.

Surface Sampling

Surfaces in processing facilities are commonly designated by their likelihood to cause food contamination. Food contact surfaces (Zone 1) are important sites that are often considered during environmental sampling. A food contact surface could be part of a piece of equipment, packing material, storage tank, ripening room, conveyer belt, or any other item that is expected to touch the food item. Zone 2 is an area where surfaces do not directly contact food but are usually located in the same room as those of zone 1. If zone 2 is contaminated with pathogens, it is likely that zone 1 also becomes contaminated. Contaminant transfer is often caused by human or machine action. Surfaces in zone 2 include walls or floors located near processing equipment and overhanging pipes or equipment. Zone 3 is an area that may cause contamination of zone 2. Zone 3 includes warehouses, employee locker rooms, and loading docks. *Listeria monocytogenes* is an example of pathogens that are commonly transmitted to food from the processing environment.

Sampled surfaces could be rough or smooth, flat or with curves and corners, continuous or with cracks and crevices, and accessible or difficult to reach. Therefore, the choice of a method to sample a surface depends not only on the surface zone but also on its characteristics.

Swab method

A sterile cotton swab may be used for limited surfaces or on hard-to-reach crevices. The swab is typically made of a wound cotton head (~0.5 cm diameter and 2 cm long) and a 12–15 cm long wooden stick. The swab may be prepared in the laboratory, sterilized, and kept in a sterile container until the time of use. Alternatively, cotton swabs may be purchased as individually wrapped sterile units. In addition to the swab, a sterile rinse solution in a test tube is needed for surface sampling. Many commercially available products include a sterile swab packaged with appropriate diluent in a shatterproof (plastic) tube. A predetermined area (e.g., 100 cm^2) of the surface to be sampled is swabbed with the moistened cotton swab, which is returned to the rinse solution tube. The rinse solution may be diluted serially, and selected dilutions are spread on the surface of a suitable agar growth medium. The inoculated plates are incubated, and populations of the targeted microorganisms are counted. In this scenario, when quantitative results are desired, the rinse solution is considered the undiluted analytical sample.

Sponge method

If the surface to be swabbed is large, or if a microorganism of a low incidence rate in processing environment is sought, a sponge may be used instead of a cotton swab. A natural or synthetic sponge with ~5 × 5 cm contact surface that is free from antimicrobial agents is suitable for this purpose. The sponge can be packed in a heat-resistant bag or wrapped in aluminum foil and sterilized by the analyst, or purchased prepacked and presterilized from a commercial source. During sampling, the sponge is held aseptically, moistened with 10 ml rinse solution, rubbed against the surface to be sampled and returned to a sterile plastic bag. The sample should be transferred to the laboratory under refrigeration and analyzed without delay. If the purpose of sampling is to detect pathogens, the sponge is transferred to a suitable enrichment broth and the mixture is incubated. When sampling is carried out to quantify environment microbiota, the sponge is mixed with 50- or 100-ml diluent and further dilutions are made. Selected dilutions are then spread on the surface of a suitable agar growth medium, plates are incubated, and the population of the targeted microorganism is counted.

Replicate organism direct agar contact (RODAC) method

The replicate organism direct agar contact (RODAC) method may be used on easily accessible flat surfaces. In this method, Petri plates are filled with an agar medium suitable for the microbiota to be analyzed. These plates may be prepared in the laboratory or purchased from a commercial source. The RODAC plates should contain enough agar medium so that the surface of the medium is convex and rises above the rim of the plate. At the sampling site, the agar medium in the RODAC plate is exposed to the surface being sampled. This exposure is accomplished by pressing the plate against the sampled surface and rolling the plate while applying some pressure. The cover is replaced, and the plate is incubated at a temperature and for a time appropriate to the targeted microorganism or microbiota. After incubation, the colonies on each plate are counted and colony subculturing may follow. Since no sample dilution takes place, the RODAC method is suitable for sampling pre-cleaned or sanitized surfaces. If the surface is contaminated heavily, incubated RODAC plates will be crowded with colonies and results will be difficult to interpret.

Air Sampling

Microorganisms may become airborne due to activities such as water spraying, dry ingredient handling, and vigorous air movements. Air-suspended dust particles can carry microorganisms. Mold and bacterial spores are common contaminants of air since they survive dryness and other detrimental environmental factors. The microbiological quality of air in a processing facility impacts the quality and safety of perishable food processed in this facility. Improper filtration of air entering a facility or recycling air from the raw product area into the finished product area can result in food contamination. Air quality in the packaging area is particularly important for the control of post-processing contamination. Therefore, determining the microbial load in air is an important task.

Sedimentation is a simple method to measure air quality. It involves exposing agar media plates to air by leaving these plates uncovered in the location to be sampled. Air contaminants will sediment by the force of gravity during the exposure time (e.g., 15 min). The plates are incubated, and the colony count may be considered proportional to air contamination level.

Air in a particular environment may also be forcibly impacted onto the surface of agar media plates using mechanical means. Jets of air are directed over the media plates so that air load collides and sticks to agar surface. After receiving a measured air sample, the agar plates are incubated, and colonies are counted. Air streams also may be filtered through a microfilter. Microorganisms are released from the filter using a suitable diluent and the microbial load is counted.

SELECTED REFERENCES

Dell, R.B., Holleran, S., and Ramakrishnan, R. (2002). Sample size determination. *ILAR Journal* 43: 207–213.

Food and Drug Administration (FDA). (2017). *Draft guidance for industry: Control of Listeria monocytogenes in ready-to-eat foods*. Docket Number: FDA-2008-D-0096. Washington, DC: FDA.

International Commission on Microbiological Specifications for Foods (ICMSF). (1986). *Microorganisms in Foods 2. Sampling for Microbiological Analysis: Principles and Specific Applications*. 2e. Oxford, UK: Blackwell Scientific Publications.

Moberg, L. and Kornacki, J.L. (2015). Microbiological monitoring of the food processing environment. In: *Compendium of Methods for the Microbiological Examination of Foods*, 5e. (ed. Y. Salfinger and M.L. Tortorella), 27–43. Washington, DC: APHA Press.

PRACTICING SAMPLING AND SAMPLE PREPARATION

Accurate sampling, careful sample handling, and diligent preparation of sample for analysis are critical measures for successful recovery of microorganisms from food. Factors to consider during sampling and sample preparation have been described in the first section of this chapter. These factors should be reviewed carefully before starting this exercise. Students also should prepare, in advance, an exercise summary that includes a sample preparation approach specific to the food assigned to them. The single-sheet summary will be used on the bench as a guide while executing the exercise.

OBJECTIVES

Practice sampling of food and preparing the sample for microbiological analysis.

PROCEDURE OVERVIEW

Students will work in groups (two students per group). Each group will be assigned a food sample to prepare for a typical microbiological analysis. Ahead of the laboratory session, each group will be informed of the food sample assigned and members of the group are asked to prepare an exercise summary that includes a sample preparation approach specific to the food assigned to their group. Examples of foods to be provided are heads of iceberg or Romaine lettuce, and whole fish (e.g., tilapia). Members of each group will prepare their laboratory sample for analysis, withdraw two 10-g analytical samples, homogenize the samples, and prepare three decimal dilutions of each homogenate. Students will be evaluated on the basis of work efficiency, aseptic technique, accuracy of measurements, and ability to work cooperatively within the group.

MATERIALS

Per Group of Two

- Sterile peptone water (diluent):
  - Two bottles, 90 ml each
  - Six test tubes, 9 ml each
- Food sample (laboratory sample)
  - One of the following foods: iceberg lettuce, Romaine lettuce, whole fish, cantaloupe, etc.
- Pipettes
  - One 1000-μl automatic pipetter and a box of compatible sterile tips
  - Two 1-ml individually wrapped sterile serological pipettes, and two pipetting aids

Group-Shared

- Top-loading balance
- Stomacher and stomacher bags
- Sample preparation equipment (equipment is to be cleaned and sanitized with ethanol, from an alcohol squeeze bottle, between uses)
  - Cutting board
  - Knife and a pair of scissors
  - Spatula and pair of tongs
  - Two aluminum-covered sterile beakers: 250-ml or 400-ml capacity, depending on sample assigned

METHODS

Important Notes:
- *Aseptic technique is expected in this and all other exercises. This includes, but is not limited to working in proximity of a lit Bunsen burner, flaming the mouth of the medium container after removing the cap, protecting the sample during processing against contamination from the work environment or the analysts, and using common sense in handling sterile and contaminated tools.*
- *Wearing gloves and goggles is needed for personal safety.*

1. **Safety and Laboratory Etiquette Compliance**
   a. Read and follow laboratory safety guidelines described in Chapter 1.
   b. Give special attention to the "practical aspects" described in that chapter.
2. **Sample Information**
   a. Pick the food sample assigned to the group.
   b. Record any pertinent sample information.
3. **Sample Cutting** (group task)
   a. Using the plan prepared by the group in advance, cut the food sample (laboratory sample) into small pieces suitable for withdrawing 10-g analytical samples. Use the appropriate sample preparation equipment provided.
   Note: The analytical sample should represent the laboratory sample accurately.
4. **Withdrawing the Analytical Sample** (individual task: one sample per group member)
   a. Using the sterile beaker, weigh out one 10-g analytical sample.
   b. Aseptically transfer the analytical sample from the beaker into a sterile stomacher bag.
5. **Homogenization** (individual task)
   a. Add the 90-ml diluent to the contents of the stomacher bag. Carefully push most of the air out of the bag and close the bag with the provided fastener before stomaching.

   b. Homogenize the sample in the stomacher for two min. The contents of the stomacher bag (i.e., the homogenized sample) is the 1/10 (10^{-1}) dilution of the original analytical sample.

6. **Preparing Decimal Dilutions** (individual task, see Figure 2.3)

   a. Label three dilution tubes that will become the 10^{-2}, 10^{-3}, and 10^{-4} dilutions.

   b. Using the serological pipette, transfer 1 ml of food homogenate into the diluent tube labeled 10^{-2}. Vortex tube contents briefly.

   c. Using the automatic pipetter, transfer 1 ml of the 10^{-2} tube into the tube labeled 10^{-3}. Vortex tube contents briefly.

   d. Using the automatic pipetter, transfer 1 ml of the 10^{-3} tube into the tube labeled 10^{-4}. Vortex tube contents briefly.

7. **Recording Observations** (group task)

   a. Report work progress to instructor.

QUESTIONS

1. How similar or different is the sample preparation scheme you wrote from the one you executed? If you encountered irregularities, how do you plan to overcome these in future laboratory exercises?

2. A batch of alfalfa sprouts was produced by company A and temporarily stored in a walk-in refrigerator before shipping to a retail grocery store (company B). The lot is made of 1500 cardboard packages, each containing 250 g of sprouts. The criterion for acceptance or rejection of the sprouts by company B is based on the percentage of packages positive for coliforms. Company A claims their shipments will have no more than 5% of the packages positive for coliforms, and company B decided to reject any shipment when 10% (or more) of its packages are positive for coliforms. You were asked to develop a sampling plan, analyze the lot, and determine statistically if it will be accepted or rejected. To assist in developing this plan, answer the following:

   a. Determine the number of samples that should be analyzed to produce 80% statistical power.

   b. Determine the number of samples that should be analyzed to produce only 50% statistical power.

   c. Considering the large number of samples that needs to be analyzed, what could you do to decrease the number of samples analyzed yet still produce meaningful results that help company B accept or reject the lot?

CHAPTER 3

ENUMERATION OF MICROORGANISMS IN FOOD

DILUTION SERIES. COLONY COUNTING. MOST PROBABLE NUMBER.

INTRODUCTION

How can one evaluate the microbiological quality of a food? Enumeration of microorganisms in the food is the answer that often comes to mind. The important follow up question then is: Can this enumeration be done accurately so that the results are used reliably to measure the microbiological quality? Reliability of the results, obviously, depends on how the enumeration is executed. This chapter addresses this topic with the goal of familiarizing the analysts with the methods and techniques used in enumeration and helping them apply these to produce repeatable and reliably results.

Foods vary in microbial load depending on how they are produced, processed, transported, stored, and handled. Microbial load, which is sometimes referred to as microbial burden, bioburden, or microbiota, is made of a predominant microorganism, a group of microorganisms having common characteristics, or a number of unrelated microorganisms. Determining the microbial load involves counting or enumerating the population of microorganisms in the food. "Counting" or "enumeration" in food microbiology refers to process of determining the *concentration* of a population of a microorganism (or microorganisms) of interest in a food. The counting exercise, therefore, determines the number of microbial cells, the number of colony forming units (CFU), or the most probable number of CFU present in a unit volume, weight,

Analytical Food Microbiology: A Laboratory Manual, Second Edition. Ahmed E. Yousef,
Joy G. Waite-Cusic, and Jennifer J. Perry.
© 2022 John Wiley & Sons, Inc. Published 2022 by John Wiley & Sons, Inc.

or surface area of a given sample. There are several methods used in counting or enumerating microbial populations in food. Counting using the "plate count" and the "most probable number" methods will be discussed, as these are the primary means of enumeration presented in the remainder of this book. Other counting techniques are also used in quality control laboratories (e.g., direct microscopic count and spiral plating); however, these techniques will not be discussed in this chapter.

PLATE COUNT METHOD

The procedure for determining the count of a microbial population using the plate count method often involves homogenizing a sample, preparing dilutions of the homogenized product, plating appropriate dilutions on a suitable medium, incubating the inoculated medium, counting resulting colonies, and calculating the concentration of the targeted population. For determining the concentration of microbial population accurately, it is imperative that the analytical sample is appropriately obtained and prepared. Information about sampling and sample preparation (including homogenization) has been discussed in Chapter 2.

Dilution

Differences in food's microbial populations span several orders of magnitude, hence dilutions should be made before these populations can be measured with reasonable accuracy. To accomplish this task, an analytical sample is typically weighed, dilutions are made, and the count of microorganisms in the diluted sample is determined. The degree of dilution should be tracked carefully so that concentration of microorganisms in the undiluted food can be calculated. The degree of dilution (i.e., dilution factor) can be represented, generically, by the following equation:

$$\text{Dilution factor} = \frac{\text{Weight or volume to be diluted}}{\text{Final weight or volume of diluted product}} \quad (3.1)$$

Although weights can be measured with great accuracy, microbiologists prefer volumetric over gravimetric measurements because in the former, the analysis can be completed more quickly and aseptic techniques can be applied more easily. Furthermore, dilution is completed in multiple steps, typically in a decimal dilution series. To simplify the volumetric dilution procedure, the following approximations will be applied:

a. Food density is equal to 1 g/ml (at ambient temperature), therefore, food volume and mass will be considered numerically equal.
b. Final volume of diluted sample equals the sum of the volumes of the sample to be diluted and the diluent to be added.

With these practical considerations in mind, the equation above can be approximated as follows:

$$\text{Dilution factor} = \frac{\text{Volume (or weight) to be diluted}}{\text{Volume (or weight) to be diluted} + \text{Volume of added diluent}}$$

(3.2)

For example, a ten-fold dilution (i.e., decimal dilution) of a food sample is prepared by mixing one part of the food with nine parts of a diluent, which commonly is a physiological saline solution or peptone water. Applying equation 3.2, the "dilution factor" for this diluted sample is 1/10 (i.e., one tenth) or 10^{-1}.

The subsequent dilution, in a decimal dilution series, is made by mixing 1 ml of the first diluted sample with 9 ml diluent. The new mixture will have a total dilution of 1/100 (one hundredth) or 10^{-2}. Note that exponents in a ten-fold dilution scheme are additive, i.e., a 10^{-1} dilution followed by a subsequent 10^{-1} dilution yields a total dilution factor of 10^{-2}. Additional dilutions are prepared as needed and the dilution factor, at any step of the series, can be calculated using equation 3.3.

Dilution factor of the new mixture (or total dilution) =

$$\frac{\text{Volume to be diluted} \times \text{Its dilution factor relative to the original food sample}}{\text{Volume to be diluted} + \text{Volume of diluent added}}$$

(3.3)

Decimal dilution series are recommended for ease of calculation, but other ratios of weight (or volume) of a sample and diluent can be used. If a dilution in the series is not decimal (e.g., two-fold dilution), the dilution factor of the new mixture can also be calculated using equation 3.3.

Dilutions suitable for plating

Portions of the prepared dilutions are used to inoculate agar media in Petri plates and the process is called "plating" (Figure 3.1). The questions that should be considered by the analyst are: How many dilutions should be prepared, and which dilutions should be selected for plating? In other words, what dilution scheme should the analyst prepare and follow? Preparing all possible dilutions and plating these dilutions is a waste of resources and effort. On the contrary, preparing a limited number of dilutions may lead to the failure of the analyst to accurately determine the population of the organism in the food.

To answer the questions just presented, the dilution scheme should be based on the projected concentration of the targeted microorganism in the product. A product expected to contain a high load of the microorganism should be diluted further than that expected to contain a small load. The microbial load is sometimes easy to predict if the analyst has prior experience in analyzing the same product. In most cases, however, the analyst should check published literature for microbial populations expected in the product. In other situations, the analyst may seek information about the product from the producer, manufacturer,

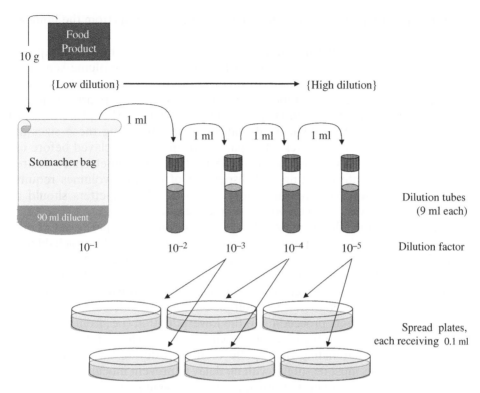

Figure 3.1 Example of a dilution scheme, showing the dilutions (prepared from a homogenized food) and the dilutions selected for plating, presuming the targeted population in the food is 1.0×10^7 CFU/g.

processor, or retailer. Considering that determining a dilution scheme is often based on the analyst's best guess, the scheme should accommodate a reasonable margin for error. For example, if the food is presumed to contain 10^7 CFU/g of the microorganism subject to analysis, the dilution scheme should allow determination of populations in the range of 10^6 to 10^8 CFU/g. This can be achieved by the scheme shown in Figure 3.1, assuming that an ideal countable plate contains 100 colonies. This scheme can be verified using the population count equations described later in this chapter.

Pipetting
Pipetting is an important activity for completing dilutions successfully. Quantitative transfer of broth culture, diluent, homogenized food, liquid food, or similar materials requires accurate pipetting. This can be accomplished by using a pipette in combination with a pipetting device (pipette dispenser, pipetaid, etc.). The pipetting device is an essential tool in analytical laboratories and may vary from simple rubber bulbs to automated pipetters. Simple mechanical pipetting aids are often used with glass or plastic pipettes. These pipettes, often referred to as serological pipettes, are maintained sterile in canisters or individual wrappers. In this book, it is suggested that individually wrapped sterile serological pipettes, along with a

hand-held pipette pump, are used to make the initial dilution (from the homogenized food) in most exercises.

For convenience and consistency, quantitative and aseptic transfer of liquid in microbiological laboratories is accomplished using variable-volume semiautomatic micropipetters (e.g., Eppendorf pipettes or Gilson Pipetman), in combination with matching sterile pipette tips. These micropipetters are capable of handling specific ranges of liquid volume and the 1000 μl and100 μl are the most popular sizes in microbiological laboratories. Tips matching these sizes are packed in autoclavable box-racks. The packaged tips are autoclaved before use and disposed of appropriately after use. Adjusting micropipetters to desired volumes and accurate pipetting and dispensing of pipetted volumes requires some practice before starting laboratory exercises. Micropipetters should be calibrated regularly to avoid errors in volume measurements.

When transferring a homogenized food (or a culture) to prepare a set of dilutions, a new clean and sterile pipette or pipette tip should be used for each dilution made. Transferring these dilutions to agar plates can be done using one of two approaches. Starting with the lowest dilution (i.e., most concentrated) requires the use of a new pipette or tip for each dilution transferred. However, a single pipette or pipette tip may be used to transfer multiple dilutions provided the analyst starts with the highest dilution, proceeding to the lowest dilution (i.e., from the least to the most concentrated). If the latter approach is followed, caution should be exercised to avoid contaminating the pipette or the tip during this multistep use. Additionally, plates must be spread with no delay to prevent inoculum from being absorbed into agar before proper distribution across agar surface.

Plating

"Plating" refers to the process of transferring and incorporating the sample to be analyzed, or its dilutions, into a suitable agar medium in a Petri plate. When the agar medium is poured and solidified in the Petri plate in advance, incorporation of a small volume of the sample dilution is done by spreading and the process is described as "spread-plating." Alternatively, a larger amount of the sample, or its dilution, may be dispensed first in an empty Petri plate into which warm molten agar is poured, and plate contents are mixed. This process is known as "pour-plating." Analyzing a food for a given microorganism may necessitate using pour-plating or spread-plating, but in other circumstances the two methods can be used interchangeably. Note that these two plating methods require different dilution schemes.

Spread-plating
After a set of dilutions is prepared from a homogenized food sample or a culture, portions of these dilutions are deposited and spread over the surface of agar plates. Spreading inocula (ideally 0.1 ml) on an agar plate requires the use of a cell spreader. This device can be as simple as a bent-end glass rod, made in the laboratory by a skilled technician. Alternatively, metal cell-spreaders are used for their durability. Glass or metal spreaders are decontaminated (sanitized) immediately before use as follows. Dip the spreader into a jar of alcohol, remove the spreader, and pass it quickly through the flame of a Bunsen burner to allow

remaining alcohol to catch fire. Notice that alcohol decontaminates the spreader and flaming does not heat the spreader enough to sterilize it; flaming is done only to remove excess alcohol (the spreader should never be held in the flame). Disposable sterile plastic spreaders are also available; these are preferred when the transferred inoculum is expected to contain bacterial spores, as the ethanol dipping (just described) inactivates cells but not spores. Using an alcohol jar, along with glass or metal spreader used to spread spores, is likely to result in contamination of subsequently spread plates.

Calibrated sterile inoculation loops (usually disposable) may also be used to spread a specimen or its dilution on an agar plate. This requires scanning the agar surface with the loop repeatedly in a systematic fashion. This spreading technique is used when a limited number of spread-plates are needed and the microbial load in the analyzed sample is relatively small. This technique may be used in conjunction with sterility testing.

Pour-plating

Pour-plating involves dispensing a portion of the sample or its dilution (commonly 1 ml) into a standard Petri plate, adding molten agar medium (10–15 ml, at 48–50°C), mixing plate contents carefully, and letting the mixture solidify. Using this technique requires that molten agar media be prepared ahead of the sample preparation and held in a water bath set at ~50°C until poured. The molten medium could be prepared in bulk in Erlenmeyer flasks or partitioned in test tubes. In the former case, a skilled analyst can pour the agar into multiple plates at quantities suitable for the analysis. In the latter case, the agar quantity in each tube should be sufficient to prepare one plate.

Incubation

Inoculated plates are incubated at a time-temperature combination appropriate for the growth and colony formation by the microorganism being counted. Microorganisms vary in their ability to grow at different temperatures. While psychrophiles prefer refrigeration temperatures (1 to 10°C), mesophiles grow optimally at temperatures close to that of the human body (37°C), and thermophiles grow best at higher temperatures (e.g., 55°C). Psychrotrophic bacteria grow optimally in the mesophilic range, but they are also capable of growing under refrigeration. Choice of incubation temperature, therefore, depends on the microorganisms of interest and their natural habitat, as well as presence of competing microbes.

In food microbiology, several incubation temperatures are typically used. For potentially pathogenic organisms, such as *Salmonella enterica*, incubation occurs at 32–37°C, a temperature range suitable for mesophiles. A somewhat cooler temperature (e.g., 25°C) is more preferred by spoilage organisms such as yeasts, molds, and psychrotrophic bacteria. "Room temperature" is typically taken to mean 22°C, but this temperature may vary depending on the room used for incubation and even the season and area of the world. Refrigeration at 4°C is typically used to maintain cultures without allowing further growth. A refrigerated incubation can also be used for cold enrichments of psychrotrophic microorganisms such as *Listeria monocytogenes*.

Plates containing inoculated agar media are typically inverted before incubation. If plates are incubated with lids upward, water condensate falls on the agar surface causing the spreading of colonies. When plates are inverted, condensed water (from moist agar) accumulates on the plate lid. Excessive water condensation on the lid, however, is undesirable and should always be minimized. Pouring hot agar ($> 50°C$) aggravates this problem. In the case of spread plates, it is preferable to pour the agar in these plates 24–48 hours before use. Some microbiologists choose to "dry" the spread plates soon after preparation for several hours in a warm clean incubator. For the safety of the analyst, plate lids with excessive water condensation should be replaced with dry sterile lids.

Colony Counting

"Counting" in food microbiology refers to the determination of the size of a microbial population within a specific quantity of food (i.e., population concentration). Enumerating the number of colonies on agar plates may also be referred to as "counting," therefore careful distinction between these two usages is urged. Throughout this manual, the former will be referred to as "population count" and the later as "colony count."

Some enumeration techniques, such as the direct microscopic counting method, allow determination of the number of cells per unit volume or weight of the sample. The plate count technique, however, determines the number of cells or cell clumps capable of forming colonies on agar plates. Since it is impossible to distinguish colonies arising from individual cells and those from cell clumps, the final population count determined by this method is expressed as colony forming units per unit volume or weight, i.e., CFU/ml or CFU/g.

To begin the counting process, the analyst should lay out the incubated plates in order of dilution to evaluate the executed dilution scheme and technique. The lowest dilution plated should have yielded the plate with the most colonies, and the number of colonies are expected to decrease by approximately a factor of ten as dilutions increase, provided that a decimal dilution scheme is used. If this is not the case, analysts should be cautious when interpreting results. Counting colonies on plates can be done visually, preferably with the help of a colony counter. Colony counters, such as the darkfield Quebec colony counter (Figure 3.2), provide background lighting and magnification so that small colonies are not overlooked. To carry out this process accurately and rapidly, the analyst should mark the counted colonies with a marker pen on the *bottom* of the plate, to make sure that the same colonies are not counted repeatedly.

Counting rules

In order to obtain counts that can be compared among different laboratories, it is necessary to establish consistent guidelines for counting colonies. In some circumstances, however, different counting or calculation methods may be used in place of, or in conjunction with, the standard counting rules. The following are the rules that will be applied throughout this book for counting bacterial colonies.

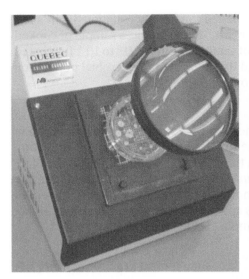

Figure 3.2   Darkfield Quebec colony counter with a Petri plate mounted for colony counting.

Plates with colonies in the range of 20–200 (the best possible scenario)

Although other references may use 30 to 300 or 25 to 250 colonies as <u>suitable countable bacterial colonies on a plate</u>, the range 20–200 will be used throughout this book. If plating yields plates with colony counts in the range of 20–200/ plate (as judged by a preliminary estimation), discard all remaining plates and count colonies only on the plates in this range. Calculate the CFU/g using equation 3.7. Familiarity with equation 3.7 is important. Examples of dilution factors are 1/10 (i.e., 10^{-1}) and 1/100 (i.e., 10^{-2}), and the volume plated is commonly 1 or 0.1 ml.

Duplicate plates with colony counts in the 20–200 range are ideally obtained from one dilution only. When this condition is not met, the following rules are applied.

- **One plate in the 20–200 range.** If the exercise yields only one plate with a colony count in the range of 20–200, calculate the CFU/g in the original sample using the number of colonies on that plate instead of an average.
- **Consecutive dilutions producing plates in the 20–200 range.** If plates from two consecutive dilutions yield 20–200 colonies, compute the CFU/g resulting for each of the two dilutions. If the two population counts are not appreciably different (e.g., 1.5×10^4 and 1.2×10^4 CFU/g), average the numbers and report the average as sample population count. If the numbers are substantially different (e.g., the higher CFU/g is more than twice the lower one), report only the lower computed CFU/g.

Plates with < 20 colonies

If the plating procedure results in only plates with fewer than 20 colonies, record the actual number of colonies on the plates receiving the lowest dilution (i.e., the highest concentration of sample plated) and apply equation 3.7. In addition, report the number as "estimated" or "est." For example, a sample of cooked meat was analyzed by pour-plating 1 ml of sample dilutions 10^{-2} to 10^{-4}. Incubated plates produced less than 20 colonies and thus the count of microbial population in meat is calculated and reported as shown in Table 3.1.

Plates with no colonies

When all plates produce no colonies, the count is estimated to be smaller than the minimum detection limit of the procedure followed. The minimum detection limit is the count that would result from the presence of one colony on the plate receiving the *lowest dilution of the sample*. Apply equation 3.4 to estimate the count using the lowest dilution plated and substitute the numerator of the equation with < 1. For example, if no colonies appeared on any of the plates receiving the dilutions shown in the previous example (Table 3.1), then

$$\text{Population count} = \frac{<1}{1 \times 10^{-2}} = <1 \times 10^2 \, \text{CFU} / \text{g} \qquad (3.4)$$

It is not necessary in this case to report the count as "est." because the presence of the less than (<) sign indicates the uncertainty of the count.

Plates with greater than 200 colonies

If the plating procedure yields only plates with greater than 200 colonies, obtain an estimated count as follows. Count colonies in representative portions (determined subjectively) of the plate receiving the highest dilution. Using a lighted colony counter with gridlines that are 1-cm apart (Figure 3.2) assists in choosing a representative area of the plate to count.

- If there are fewer than 10 colonies per cm^2, then count 13 squares chosen as follows: 7 consecutive horizontally and 6 consecutive vertically; see Figure 3.3 for a visual example of plate counting field. The sum of the colonies in the 13 squares multiplied by 5 equals the estimated count per 65 cm^2 plate, which is the area of a typical Petri dish, assuming its inner diameter is ~90 mm.

TABLE 3.1 Microbial population count in cooked meat.

Dilution Factor	Number of Colonies	
	Plate 1	Plate 2
10^{-2} {least dilute}	12	16
10^{-3}	3	1
10^{-4} {most dilute}	0	0

$$\text{Population count} = \frac{(12+16)/2}{10^{-2} \times 1} = 1.4 \times 10^3 \, \text{CFU} / \text{g} \, (\text{est.})$$

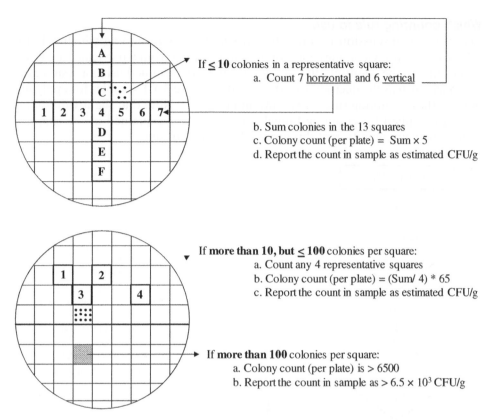

Figure 3.3   Procedure to estimate microbial population count in samples using crowded plates (> 200 colonies/plate).

- If the number of colonies per cm^2 are from 10 up to 100 colonies, count colonies in 4 representative squares and multiply the average by 65 to give estimated count per plate (Figure 3.3).
- If there are greater than 100 colonies per cm^2, then record the count as > 6500 CFU/plate (Figure 3.3).

In all cases, equation 3.7 is used to estimate population counts (CFU/g, est.). Never report the final count in the food sample as too numerous to count (TNTC). Plates with different surface areas, such as Petrifilms, are counted using the same principle. Count a minimum of 4 squares and average those counts, then multiply that count by the area of the plate being used.

Plates with spreaders
Count a chain of colonies that are not too distinctly separated as a single colony. If colonies can be distinguished, then it is not considered a spreader for counting purposes. If chains of colonies appear to originate from separate sources, count each chain as one colony. If the spreader is greater than 25% of the plate, report the results as spreaders (Spr.) rather than as a number.

Which counting rule to use

The previous discussion is generally sufficient for determining which rule to follow for counting colonies and population resulting from a particular analysis. Alternatively, the counting rule to follow may be determined using a systematic approach, such as the decision tree described in Figure 3.4. Examples for applying most of these counting rules are shown in Figure 3.5. Note that slightly different rules are applied for counting fungal colonies; this is explained in the chapter dealing with enumeration of this group of microorganisms.

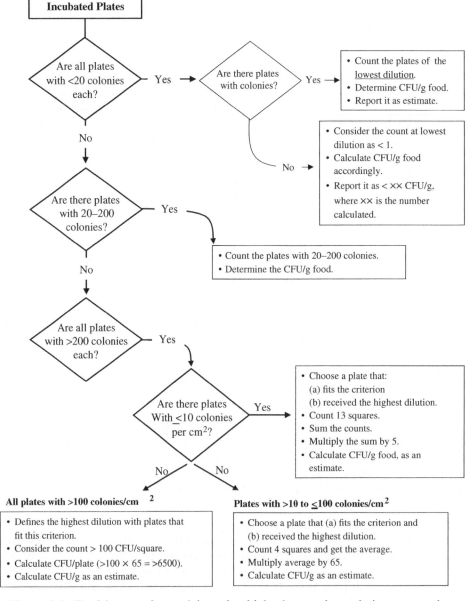

Figure 3.4 Decision tree for applying microbial colony and population count rules.

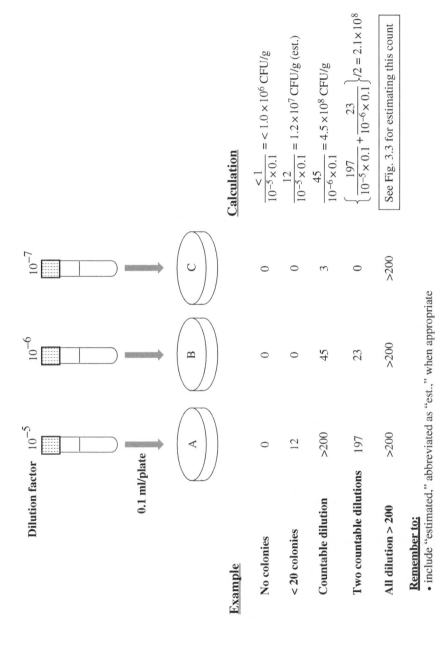

Figure 3.5 Applying colony and population counting rules.

Population Counting

After the colony count is determined, the analysist should be able to determine the concentration of a microorganism (or a microbial group) in the food. The concentration is often expressed as CFU per gram of that food. As inidicated before, this concentration will be referred to as "population count," and it is governed by the following general equation:

$$\text{Population count}\left(\text{CFU}/\text{g}\right) = \frac{\text{Average colony count from the duplicate plates}}{\text{Weight}\left(\text{g}\right)\text{of food dispensed in each plate}}$$

(3.5)

To calculate the denominator of equation 3.5, one should know the volume of diluted homogenate that has been plated and the concentration of food in that dilution.

$$\text{Population count}\left(\text{CFU}/\text{g}\right)$$
$$= \frac{\text{Average colony count from the duplicate countable plates}}{\text{Volume}\left(\text{ml}\right)\text{dispensed per plate}\times\text{Concentration of food in the plate dilution}}$$

(3.6)

The volume dispensed in a Petri plate is commonly 0.1 ml for spread-plating and 1.0 ml for pour-plating, but volumes other than these can be used. The concentration of food in this diluted homogenate can be considered to equal the dilution factor. For example, if a food has been homogenized and diluted 1:100 (i.e., 10^{-2} dilution factor), this results in a concentration of 0.01 g of food per g (or mL) of diluted homogenate. Therefore, equation 3.6 can be rewritten as follows:

$$\text{Population count}\left(\text{CFU}/\text{g}\right)$$
$$= \frac{\text{Average colony count from the duplicate countable plates}}{\text{Volume}\left(\text{ml}\right)\text{dispended in each plate}\times\text{Dilution factor for the plated dilution}}$$

(3.7)

As previously stated, the density of food, diluted homogenate and diluent is approximated as 1 g/ml. Note that volume of sample dispensed is quantified in ml. If analysts have dispensed a volume measured in µl, this value will need to be adjusted to yield a correct population count.

Important Considerations

The following reminders should help in avoiding the common pitfalls encountered during enumeration of microorganisms in food:

- Notice the difference in usage of "colony count" and "population count."
- When a specific microbial population is counted, the generic word "population" is replaced with a word or a phrase that describes the population. For example, in subsequent chapters, the "population count" is reported as aerobic mesophilic count, spore count, *Enterobacteriaceae* count, etc.

- Population count should be presented in *scientific notation*, with two significant digits only reported (e.g., 3.7×10^8).
- In all examples presented, population count was based on unit food weight (g); however, samples of many liquid foods are measured volumetrically. In the latter case, population count is more accurately represented as CFU/ml.

MOST PROBABLE NUMBER METHOD

Principles

The most probable number (MPN) technique is useful for enumerating low levels of microorganisms ($<10^3$ CFU/g) in water, milk, and foods whose structure interferes with accurate plate count methods. These small populations, generally, cannot be enumerated reliably with other techniques. The essence of the MPN technique is that the dilution of the sample into replicate tubes causes the viable population to decrease to a point where it becomes detectable in some of the diluted tubes (positive tubes) but not in the others (negative tubes). When positive and negative tubes in each dilution are counted, results can be used to estimate the population in the original sample. An estimate of population density in the original sample is based on probability formulas that take into account the following: number of dilutions included, the number of replicate tubes receiving each dilution, the number of positive tubes in each dilution, and the amount of original sample present in each dilution. Although equations are available for calculating MPN, tables that simplify the calculations are used (Table 3.2).

To enumerate microbial population in a solid food using the MPN technique, the sample is homogenized and decimally diluted in an appropriate diluent as previously discussed. The number of dilutions prepared are often limited, with 2–4 dilutions typically prepared. Measured portions (e.g., 1 ml) of the food homogenate and its subsequent dilutions are transferred into replicate tubes (commonly 3) of a selective, differential, or selective-differential broth medium depending on the target population. A larger number of replicate tubes (e.g., 5 tubes) may be used for increasing the proportion of the analytical sample subjected to analysis, thus improving the detection limit of the technique. The tubes are incubated at the appropriate temperature for the appropriate time, and the tubes are inspected for turbidity or for reactions specific to the microorganisms of interest. The results are used in calculating the MPN of the microorganism in g of food, as described later. An example of an MPN dilution and inoculation scheme is shown in Figure 3.6. To enumerate microbial population in a liquid food, the sample may be decimally diluted directly, an appropriate set of MPN tubes are prepared, and MPN/ml is calculated.

Scoring the Tubes and Calculating MPN

Following incubation, each tube is scored as positive or negative, depending on the target microorganism's characteristic of interest to the analyst. The positive tubes per each dilution are tallied and the number of positives from consecutive dilutions are reported. The outcome is reported as a series of numbers separated

TABLE 3.2 Most Probable Number (MPN) estimates[a] based on the number of positive tubes resulting from three consecutive dilutions (three tubes each) representing 0.1 g, 0.01 g, and 0.001 g of food.

Number of positive tubes			MPN/g	Number of positive tubes			MPN/g	Number of positive tubes			MPN/g	Number of positive tubes			MPN/g
0.1 g	0.01 g	0.001 g		0.1 g	0.01 g	0.001 g		0.1 g	0.01 g	0.001 g		0.1 g	0.01 g	0.001 g	
0	0	0	<3	1	0	0	3.6	2	0	0	9.2	3	0	0	23
0	0	1	3	1	0	1	7.2	2	0	1	14	3	0	1	38
0	0	2	6	1	0	2	11	2	0	2	20	3	0	2	64
0	0	3	9	1	0	3	15	2	0	3	26	3	0	3	95
0	1	0	3	1	1	0	7.4	2	1	0	15	3	1	0	43
0	1	1	6.1	1	1	1	11	2	1	1	20	3	1	1	75
0	1	2	9.2	1	1	2	15	2	1	2	27	3	1	2	120
0	1	3	12	1	1	3	19	2	1	3	34	3	1	3	160
0	2	0	6.2	1	2	0	11	2	2	0	21	3	2	0	93
0	2	1	9.3	1	2	1	15	2	2	1	28	3	2	1	150
0	2	2	12	1	2	2	20	2	2	2	35	3	2	2	210
0	2	3	16	1	2	3	24	2	2	3	42	3	2	3	290
0	3	0	9.4	1	3	0	16	2	3	0	29	3	3	0	240
0	3	1	13	1	3	1	20	2	3	1	36	3	3	1	460
0	3	2	16	1	3	2	24	2	3	2	44	3	3	2	1100
0	3	3	19	1	3	3	29	2	3	3	53	3	3	3	>1100

[a] Confidence intervals are not included to simplify the table (FDA Bacteriological Analytical Manual and other sources).

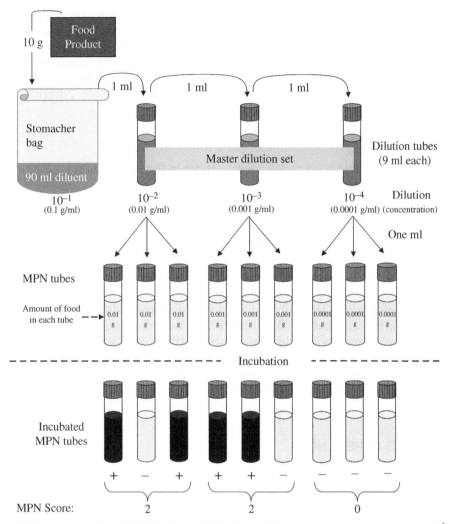

MPN/g food sample = 21 MPN/g (from MPN table) × 10 (sample mass adjustment) = 2.1×10^2

Figure 3.6 Dilution and inoculation scheme for most probable number technique.

with commas from least dilute to most dilute. Using the results presented in Figure 3.6, the MPN score would be 2, 2, 0.

After scoring MPN tubes, the score is converted into an MPN/g food sample using standard MPN tables (Table 3.2). There are equations and computer programs available for calculating MPN; online MPN calculators may also be useful. The MPN table is based on the amount of sample (i.e., food) that is present in the MPN tubes; therefore, column headings of MPN tables are usually expressed in grams. MPN tables are usually constructed for sample sizes of 0.1, 0.01, and 0.001 g, for three consecutive dilutions. In practice, these values would correspond to 1 ml transferred to each MPN tube from 10^{-1}, 10^{-2}, and 10^{-3} dilutions of the master dilution set. In the example shown in Figure 3.6,

1 ml of the 10^{-2}, 10^{-3}, and 10^{-4} dilutions were used to inoculate the MPN tubes. This inoculation would correspond to 0.01, 0.001, and 0.0001 g of sample in successive MPN tube sets. Therefore, the value from the MPN table, corresponding to the number of positive tubes, must be adjusted to accommodate the difference in amounts of the sample analyzed. Because the sample is ten times more dilute, the table values should be multiplied by ten. Alternatively, the following equation can be used:

$$\text{Most probable number per g food (MPN/g)} =$$
$$\frac{\text{Table reading for positive tubes in the 3 dilution set}}{1} \times \frac{\text{Table's food weight in first dilution tube}}{\text{Actual food weight in first dilution tube}} \qquad (3.8)$$

This pattern continues if higher dilutions are used to inoculate MPN tubes. MPN results should be reported in scientific notation with the units of MPN/g. For the example in Figure 3.6, the result after correction would be reported as 2.1×10^2 MPN/g.

In a situation where all incubated tubes were negative (MPN score: 0, 0, 0), the result would be reported as $< 3.0 \times 10^1$ MPN/g (assuming no adjustment for dilution scheme). Analysts should not report 0 MPN/g. Similarly, if all incubated tubes were positive (MPN score: 3, 3, 3) the result would be $> 1.1 \times 10^4$ MPN/g. These values do not require the "est." designation due to the presence of the greater than ($>$) or the less than ($<$) signs.

SELECTED REFERENCES

Blodgett, R. (2010). *Most Probable Number from Serial Dilutions: Bacteriological Analytical Manual Appendix 2.* Washington, DC: U.S. Food and Drug Administration.

Harrigan, W.F. (1988). *Laboratory Methods in Food Microbiology*, 2e. San Diego: Academic Press.

Sutton, S. (2006). Counting colonies. *Pharma Microbiology Forum Newsletter*, 12(9): 2–5.

QUESTIONS

1. An analytical facility received a laboratory sample containing ground meat. The product weighed 1000 g and the requested microbiological analysis included determining the product's aerobic mesophilic count. The vendor did not provide information about the microbiological quality or safety of product but according to laboratory's records, a similar product from the same vendor was analyzed in the previous month and its aerobic mesophilic count was 5.4×10^4 CFU/g.

  a. Develop a sample preparation plan that includes withdrawing analytical sample(s). Hint: Review information provided in Chapter 2.

  b. Prepare a dilution scheme that shows the dilutions to be prepared and dilutions to be plated. Consider that the spread-plating method is to be used.

  c. How different would the dilution scheme be if pour-plating was used instead of spread-plating?

  d. How different would the dilution scheme be if no information or record about this sample was available?

2. Develop a mathematical formula (or formulae) that the analyst can use to design dilution-plating scheme if the approximate concentration of the targeted population (CFU/g) is known.

CHAPTER 4

PRACTICING BASIC TECHNIQUES

POUR-PLATING. SPREAD-PLATING. COLONY COUNTING. GRAM STAINING. MICROSCOPY

INTRODUCTION

This chapter covers techniques that should help in navigating subsequent exercises in this book. The exercise described in this chapter allows students to practice preparing dilutions, plating, colony and population counting, isolation, and staining. Additionally, applying aseptic techniques is emphasized throughout the exercise. Students should read Chapters 2 and 3 thoroughly to become acquainted with some of the techniques applied this exercise.

Microbial Growth

Methods in food microbiology usually include growing microorganisms using culture media. Examples of microbial growth include formation of colonies on agar-based culture media and multiplication of a pathogen in an enrichment broth. Growth of a microorganism in a medium is expected when water and nutrients are available, the environment (temperature, oxygen level, pH, etc.) is fit and supportive, and enough time is permitted under these conditions. In laboratory settings, water and nutrients are provided by specially formulated culture media. The medium also is a conduit for the right environment such as temperature, oxygen, and pH.

Analytical Food Microbiology: A Laboratory Manual, Second Edition. Ahmed E. Yousef, Joy G. Waite-Cusic, and Jennifer J. Perry.

Culture media

Microbiological culture media contain nutrients to support growth of targeted microorganisms and may contain agents that aid in selection and differentiation of these microorganisms. Using the right media enables microbiologists to make significant inferences about the biochemical and physiological nature of the organisms investigated. The medium used for microbiological analysis should be sterile to ensure correct interpretation of results. Autoclaving, boiling, and occasionally microfiltration are used to achieve sterility. Care should be exercised during handling and use of sterile media to avoid introduction of contaminants; therefore, aseptic techniques should be followed. Using aseptic technique ensures that microorganisms of interest only, whether they are a part of the food or in pure cultures, are introduced (inoculated) to the medium.

Media used for culturing microorganisms are commonly described as non-selective, selective, differential, or selective-differential (Figure 4.1).

- Non-selective media are nutritionally rich and commonly used to enumerate food microbiota or to transfer and maintain isolated microorganisms.
- A selective medium allows the microorganism of interest (target microorganism) to grow and suppresses the growth of competing microorganisms. Appropriate selective agents (e.g., crystal violet, which selects against Gram-positive bacteria) are essential components of this class of media.
- Differential media are occasionally used to distinguish between sub-populations in food based on a biochemical characteristic. Proteolytic and non-proteolytic, lipolytic and non-lipolytic, and acid and non-acid producing bacteria are detected and/or enumerated using simple differential media containing appropriate differential agents.
- A medium containing both selective and differential agents provides an efficient means of selection and differentiation of the microorganism of interest when present in a complex microbiota.

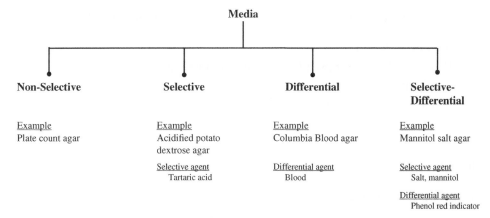

Figure 4.1   Types of culture media that are commonly used in microbiological analysis of food.

Inoculation of agar media includes streaking, stabbing, pour-plating, and spread-plating methods. Inoculation methods may affect interpretation of results; only qualitative information is obtained using streaking or stabbing, but pour- and spread-plating methods allow enumeration of organisms (quantitative information).

Sampling, Homogenization, Dilution, and Plating

Depending on the food analyzed, the analyst should determine the appropriate sampling plan, homogenization procedure, and dilution and plating scheme. Properly sampled food is diluted to the extent necessary to yield countable plates. Prepared dilutions are then introduced into a suitable agar medium by pour-plating, or spread onto the medium by spread-plating. These aspects of analysis have been presented in Chapters 2 and 3.

Incubation, Colony Counting, and Population Determination

Analysis of a food results in a series of inoculated agar plates that is incubated at a temperature between 4°C and 55°C, depending on the targeted microorganism or microbial group. Incubated plates, resulting from the analysis of a single sample, may have a wide range of colony counts. Despite the obviousness of the counting rules, it is sometimes beneficial to follow a systematic approach for determining which plates one should use for determining population count. Under these conditions, it is proposed to follow the decision tree shown in Chapter 3.

The tally (or estimation) of the number of colonies on a plate is termed "colony count" and is given the unit colony forming unit (CFU) per plate. The colony count is used in conjunction with the dilution factor to calculate the "population count," with the unit being CFU/g, CFU/mL, or CFU/cm² of the initial food or surface. Rules for colony counting and determining population count can be found in Chapter 3.

OBJECTIVES

This introductory laboratory exercise involves practicing the following:

1. Aseptic technique
2. Plating techniques: Spread-plating and pour-plating
3. Applying colony counting rules
4. Isolation by streaking
5. Staining and use of microscope

MEDIA

Plate Count Agar (PCA)

This is a general-purpose culture medium typically used for enumeration of organisms in food. Tryptone, glucose (also called dextrose), and yeast extract make this rich and complex medium suitable for growing a wide variety of microbes.

Peptone Water

This simple dilution medium is used frequently as a diluent. It contains 1% peptone and 0.5% NaCl, and provides a nearly isotonic environment for analyzed microorganisms. Some microbiologists prefer using peptone water over physiological saline (0.85% NaCl) as a diluent. Accidental contamination of the former is easier to detect because the solution turns turbid.

PROCEDURE OVERVIEW

This laboratory exercise will be executed during three laboratory sessions (Figure 4.2). The first session covers preparing dilutions of bacterial cultures, plating selected dilutions on an agar medium, and incubating the inoculated

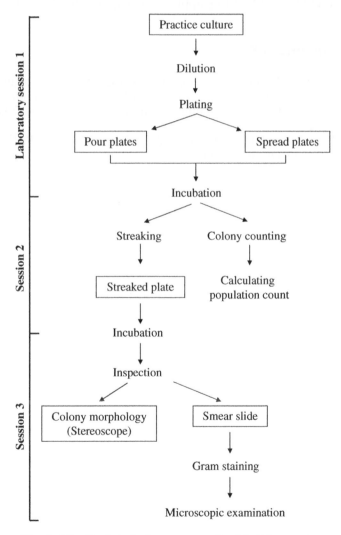

Figure 4.2 Basic techniques as practiced in this exercise.

agar plates. In the second session, incubated plates are used to count colonies and to determine population count of the tested bacterium. Additionally, isolation will be practiced through streaking of selected colonies. The third session will be dedicated to examining streaked cultures, Gram straining smears of the isolated bacteria, and examining cell and colony morphologies under the microscope. Aseptic technique will be observed throughout the exercise.

Students will work individually to complete this exercise. If needed, students may work in groups of two, and prior experience in basic techniques is shared between the partners. In preparation for this exercise, students should read the laboratory procedure carefully and should prepare a one-page exercise summary to be used on the bench during the analysis. A plastic cover sleeve will be provided to protect the summary page from contamination; the sleeve should be sanitizable using alcohol or other sanitizing agents provided in the laboratory. Annotated flow charts may serve as an exercise summary. For example, Figure 4.2 may be annotated by hand-writing appropriate experimental details.

SELECTED REFERENCES

Pertan, R.L., Grieme, L.E., and Foong-Cunningham, S. (2015). Culture method for enumeration of microorganisms. In: *Compendium of Methods for the Microbiological Examination of* Foods, 5e (ed. Y. Salfinger and M.L. Tortorella), 75–87. Washington, DC: APHA Press.

Zimbro, M.J., Power, D.A., Miller, S.M., et al (eds). (2009). *Difco™ & BBL™ Manual*, 2e. Sparks, MD: Becton, Dickinson and Company.

SESSION 1: DILUTION AND PLATING

Before completing this exercise, observe the instructor's demonstration of spread-plating, pour-plating, and aseptic transfer of media and cultures. To maintain aseptic technique, work in proximity to, but keep a safe distance from, the flame of a Bunsen burner. Tubes and plates should be kept covered except when actively in use.

MATERIALS AND EQUIPMENT

Per Student

- Test culture: Mixture (~3 ml) of overnight cultures of *Escherichia coli* and *Enterococcus faecium*.
  Notes:
  - *Cultures of the two bacteria are prepared separately and mixed by instructors ahead of the laboratory session.*
  - *Other bacterial combinations may be used (e.g.,* Enterobacter *sp. and* Enterococcus *sp., or* E. coli *and* Staphylococcus *sp.).*
  - *Individual cultures may be tested, instead of culture mixtures.*
- Eight 9.0-ml tubes of peptone water (dilution tubes)
- Four Petri plates containing PCA medium, for spread-plating
- Four 15-ml tubes of molten PCA medium, held in a water bath at ~50°C; these should be left in the water bath until the analyst is about to prepare the pour plates
- Four empty, sterile Petri plates, for pour-plating
- Sterile serological pipettes, 1-ml capacity (alternatively, use an automatic pipetter, 1000-µl capacity, with a box of sterile tips)
- Vortex mixer
- Alcohol jar

Class-Shared

- Water bath maintained at ~50°C, incubator maintained at 35°C, disposable gloves, and other common laboratory equipment and supplies; some of these will be needed throughout the exercise.

PROCEDURE

An overview of the procedure applied in this exercise is shown in Figure 4.2. Additionally, the dilution and plating scheme is illustrated in Figure 4.3.

Note: duplicate plates (per dilution) are ideally used; however, in this basic exercise, one plate/dilution is used to conserve supplies.

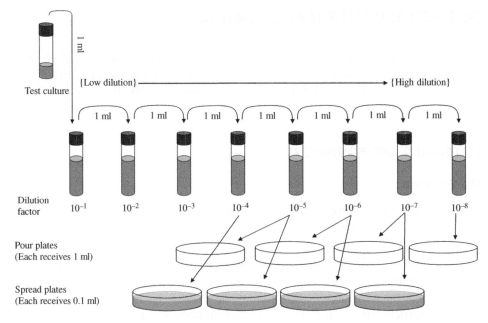

Figure 4.3 Dilution and plating scheme used in basic technique exercise.

Dilution

1. Prepare a rack of eight dilution tubes. Label the tubes with the dilution factors 10^{-1} to 10^{-8}, using labeling tape. Since the original tube (the test culture) is non-diluted, it would be given the dilution factor of 10^0.

2. Vortex the tube containing the test culture for 10 seconds and transfer 1 ml into the first 9-ml test tube using a sterile pipette (or automatic pipetter fitted with sterile pipette tip). The resulting tube has a dilution factor of 10^{-1}.

3. Vortex the 10^{-1} test tube for 10 seconds and transfer 1 ml into the second 9-ml test tube, using a new sterile pipette or tip. This tube now contains a culture dilution at the factor of 10^{-2}.

4. Repeat these steps to dilute the sample until 10^{-8} total dilution is reached (a total of eight test tubes).

Spread-Plating

1. Label the bottom, outer edge of four PCA plates with analyst's initials, date, medium name, the organism or sample, and the incubation temperature. Also include the dilution factor and volume plated (Figure 4.4). Notice these plates will be receiving the 10^{-4} to 10^{-7} dilutions.

2. Dispense and spread the culture dilutions on the four labeled PCA plates (Figure 4.3) as follows:

   a. Mix (using the vortex) the inoculated 10^{-4} dilution tube for 10 seconds and dispense 0.1 ml onto the surface of the agar using a sterile pipette (or automatic pipette with a sterile tip).

b. Dip the cell spreader into the alcohol jar, remove the spreader and let excess alcohol drip momentarily in the jar. Pass the spreader quickly through the flame to allow remaining alcohol to catch fire and burn off. This decontamination process will be referred to as "flaming the spreader." Note that the spreader does not become hot by the burning alcohol; therefore, no need to cool the spreader before use. To avoid fire hazard, never heat the spreader directly on the flame or dip it hot in the alcohol jar.

c. Open the plate cover partially and spread the culture onto the agar surface using the sterilized spreader. Rotate the plate during the spreading to ensure an even distribution of the culture. Spread evenly by making sure the spreader reaches the edges as well as the center of the plate.

d. Alcohol-dip the used spreader and then flame the excess alcohol before setting the spreader down.

e. Repeat steps (a) to (d) for the 10^{-5}–10^{-7} dilutions. Remember to replace the pipette or pipette tip and to flame the spreader for every dilution transferred.

3. Stack the plates, secure the stack with masking tape, invert the stack of plates, and incubate at 35°C for 24–48 hours.

Pour-Plating

Label the bottom, outer edge of four sterile empty Petri plates with analyst's initials, date, the medium name, the organism or sample, and the incubation temperature. Also include the dilution factor and volume plated (Figure 4.4). Notice these plates will be receiving the 10^{-5} to 10^{-8} dilutions, and will contain 1000 µl, as opposed to 100 µl, of culture dilution.

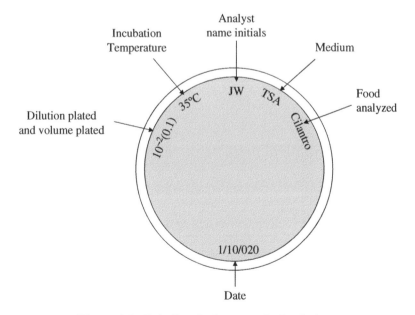

Figure 4.4 Labeling the bottom of a Petri plate.

4. Prepare the four labeled plates as follows:

   a. Mix (using the vortex) the inoculated 10^{-5} dilution tube for 10 seconds, pipette 1 ml and dispense it into the center of the base of appropriately labeled Petri plate. Use a sterile pipette (or pipette tip if you are using automatic pipette) to transfer the dilution to the plate.

   b. Pour the molten agar from one PCA tube into the plate. Gently slide the plate on the bench in a "figure-8" motion, about three times, to mix the culture into the medium and to spread the agar.

   c. Repeat using the 10^{-6} to 10^{-8} dilutions. Remember to replace the pipette or pipette tip and to flame the spreader for every dilution transferred.

   d. Allow agar to solidify before moving or inverting the plates.

5. Stack the plates, secure the stack with masking tape, invert the stack of plates, and incubate at 35°C for 24–48 hours.

SESSION 2: POPULATION COUNTING AND COLONY STREAKING

During the second session of this exercise, incubated plates are inspected, colonies developed on incubated plates are counted, and rules are applied for calculating population count. Counting results are recorded in the tables provided. Selected colonies will be streaked on fresh agar plates. Before completing this exercise, observe the instructor's demonstration of three-phase streaking.

MATERIALS AND EQUIPMENT

Per Student
- Incubated plates (from the previous session)
- Colony counter
- For streaking:
  - Two PCA plates
  - Inoculation loop

PROCEDURE

Determining Population Count

Spread plates
1. **Visual examination:** Judge the evenness of spreading by observing the distribution of colonies on the plates. Plates should contain colonies that are evenly distributed across the medium surface (no large clumps or empty spaces). Compare the appearance of the plates of different dilutions. Notice the relatively uniform colony morphology, compared to that present in the pour plates.
2. **Colony count:** Count colonies at each dilution. Ability to visualize colonies is enhanced with the aid of a lighted colony counter. To minimize errors, mark the colonies on the back of the plate, as they are counted, using a marker. Record colony counting results on the back of the plate and in Table 4.1.

TABLE 4.1 Counts of test culture population in the sample tube, based on serial dilution and plating on PCA[a] medium after incubation at 35°C for 48 hrs.

Method	Number of colonies (colony count)					Population count (CFU[b]/ml)
	10^{-4}	10^{-5}	10^{-6}	10^{-7}	10^{-8}	
					X[c]	
	X[c]					
Calculation example:						

[a] Plate count agar
[b] Colony forming units
[c] Dilution not plated

3. **Population count:** Determine population counts (CFU/ml in original culture) based only on the dilution producing 20–200 colonies, as described in the colony counting rules (Chapter 3). If no dilution produces colonies within this range, follow the counting decision tree described in Chapter 3 to select the dilution most suitable for determining or estimating population count. Highlight the dilution used to calculate population count. Record results in Table 4.1 and write down a sample of these calculations.

4. Save a plate containing well-isolated colonies for streaking practice, and dispose of all remaining plates in the biohazard container. Note, this plate does not have to be the one you counted to determine population count.

Pour plates

1. **Visual examination:** Judge the evenness of mixing by observing the distribution of colonies in and on the agar. Compare the appearance of the plates of various dilutions. Notice that colonies may have different morphologies depending on their location in the agar. Colonies at the air-agar and plastic-agar interfaces may look different, and these are different from the ones embedded within the agar layer.

2. **Colony count:** Count colonies at each dilution. Use the colony counter to enhance your ability to visualize and count colonies. Mark the colonies on the back of the plate, as they are counted, using a marker. Record colony count results in Table 4.1.

3. **Population count:** Determine population counts (CFU/ml in original culture) based only on the dilution producing 20–200 colonies, as described in the colony counting rules. If no dilution produces colonies within this range, follow the counting decision tree described in Chapter 3 to select the dilution most suitable for determining or estimating population count. Highlight the dilution used eventually to calculate population count. Record results in Table 4.1 and write down a sample of these calculations.

4. Save a plate containing well-isolated colonies for streaking practice and dispose of all remaining plates in the biohazard container.

Streaking

1. Label the bottom outer edge of two PCA plates with analyst's initials, date, the medium name, the organism or sample, and the incubation temperature (Figure 4.4).

2. Select an incubated plate (spread-plate or pour-plate) containing well-separated colonies.

3. Assuming two different colony morphologies are observed, select and label a well-isolated colony representing each morphology. Trace the selected colonies by marking the bottom of the plate using a marker.

4. Observe the morphological differences between these colonies and report these observations in Table 4.2.

TABLE 4.2 Microscope examination of isolated bacteria from streaked PCA plates, after incubation at 35°C for 48 hrs.

Isolate #	Colony morphology	Cell morphology

5. Complete three-phase streaks (Figure 4.5) of each marked colony, on separate PCA plates, as follows:

    a. Flame the inoculating loop until it is glowing red and allow it to air cool.

    Note: To speed or ensure loop cooling, some analysts use it to stab an area of the agar that is free from any colonies.

    b. Use the cooled, sterile loop to touch the surface of the marked colony and pick some of its mass.

    Note: Avoid digging out or transferring the whole colony, but the transferred amount should be visible by the naked eye on the loop.

    c. Prepare primary streak: Beginning at the outer edge of the agar, move the loop while touching agar surface, in a zigzag pattern. Three strokes are sufficient for primary streak. Flame the loop as described before.

    d. Complete the secondary streak by taking the flamed and cooled loop and crossing back into the primary streak 2-3 times to pick up the organism and then zigzagging toward plate center to spread the organism. Other than the first 2-3 times, do not go back into the primary streak area; avoid making the new streaks overlap. This portion of the streak uses half of the remaining portion of the plate.

    e. Repeat for the tertiary streak. This uses up the remaining portion of the plate. Tertiary streak should only cross into the secondary streak 1-2 times.

6. Replace the plate lids, invert the plates, and incubate at 35°C for 24–48 hr.

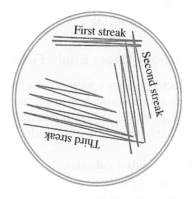

Figure 4.5 Three-phase streaking.

SESSION 3: MORPHOLOGICAL EXAMINATION

During this session of the exercise, incubated streak plates are inspected and different colony morphologies are observed. Selected colonies are Gram stained, examined microscopically, and their cell morphologies are described.

MATERIALS

- Incubated streak plates: Use the plates prepared in the previous session, which have been incubated at 35°C for 24–48 hours
- Dissecting microscope
- Bright-field light microscope with oil immersion lens
- Microscope slides
- Microscope oil and material to clean oiled lens (lens paper)

PROCEDURE

Examining Streak Plates

Visually examine the streaked isolates on agar surface. Look for the presence of well-isolated colonies. Notice the streaking phase that produced most of the isolates. Determine if there is a need to improve the procedure. Consult the instructor if you are unsure or have questions.

Examining Colony Morphology

1. Examine the colony morphology of the two isolates (on the incubated streak plates) using a dissecting microscope (stereomicroscope). This will allow observation of the colony in detail.
2. Observe colonies for differences in color, size, consistency, elevation, margin, or other characteristics.
3. Record these observations in Table 4.2.

Examining Cell Morphology Under Bright-Field Microscope

Morphology of the selected isolates will be examined under microscope after cell staining (Figure 4.6). Gram staining is most suitable for staining Eubacteria that often appear Gram-positive or Gram-negative. Use colonies on the incubated agar plates to prepare smears on glass slides and stain the dried smears.

1. **Preparing smears of isolated colonies**
   Prepare a smear from one colony of each of the two isolates (two colonies per a microscope slide) as follows:
   a. Identify the colonies to be transferred from the streaked plates and label their locations on the base of the agar plate.

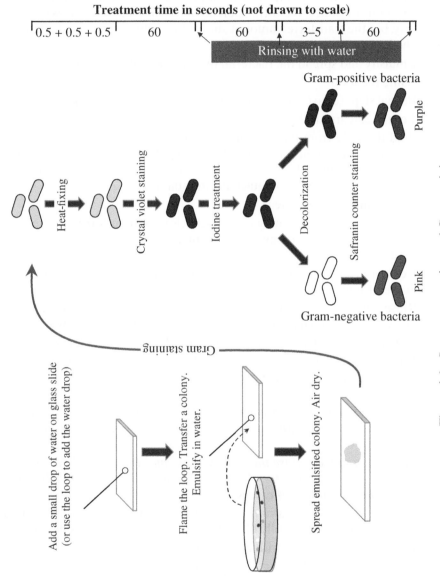

Treatment time in seconds (not drawn to scale)

0.5 + 0.5 + 0.5 | 60 | 60 | 3–5 | 60

Rinsing with water

Gram-positive bacteria

Purple

Heat-fixing

Crystal violet staining

Iodine treatment

Decolorization

Safranin counter staining

Gram-negative bacteria

Pink

Gram staining

Add a small drop of water on glass slide (or use the loop to add the water drop)

Flame the loop. Transfer a colony. Emulsify in water.

Spread emulsified colony. Air dry.

Figure 4.6 Smear preparation and Gram staining.

b. Label each end of a glass microscope slide with a colony number, using a wax pencil.

c. Using the loop, transfer a small drop of water onto one end of the slide; large water drops should be avoided.

d. Lightly touch the center of the appropriate colony with the tip of a sterile inoculation loop and transfer a portion of the colony to the water drop on the slide. Avoid transferring a sizeable mass or the whole colony, but the transferred amount should be visible with the naked eye on the loop.

e. Emulsify and spread the bacterial mass in the water drop using the loop.

f. Flame the loop.

g. Repeat steps (c) to (e) for the second colony using the other end of the slide.

h. Let the smears air dry.

i. Hold the slide from the edge using a clothespin or forceps and heat fix the smears. This is accomplished by passing the slide over a burner flame two or three times. Heat fixing before smears have been allowed to air dry will cause smears to detach during staining.

2. **Gram staining of fixed smear**

a. Place the slide on a staining rack. Use a clothespin or forceps to hold the slide during heat fixing and rinsing; this will help keep gloves stain-free.

b. Cover the smears with crystal violet (primary stain) and let sit undisturbed for 60 seconds.

c. Rinse with water. Do not squirt water directly on the smear; rather let the water run down gently over the smear. Drain off excess water.

d. Cover the smear with Gram's iodine (mordant) and hold for 60 seconds.

e. Rinse the slide with water as described earlier. Lightly blot the slide with bibulous paper to remove excess water; do not let the slide dry during blotting.

f. Tilt the slide and rinse with acetone-alcohol decolorizer for 3-5 seconds; immediately rinse with water to prevent further decolorization. Drain off excess water.

   Note: Alternatively, rinse the slide with ethanol (95%) until the draining alcohol appears colorless; this takes 15 seconds, approximately. Rinse with water and drain off excess water.

g. Cover the smear with safranin (secondary stain) for 60 seconds, and then rinse with water. Carefully blot the slide dry using bibulous paper.

3. **Examining stained smear with the microscope**

a. Place a low magnification objective lens (e.g., 4× or 10×) close to the slide surface using the coarse adjustment knob.

b. Locate the smear by observing though the eyepiece and raising the lens slowly to focus on the smear.

c. Switch to a higher magnification lens (e.g., 40×) and use the fine adjustment knob (if necessary) to focus on cells in the smear. It may be necessary to carefully move the microscope stage (left to right, NOT up or down) to find a suitable field to observe. It is advisable not to change condenser setting unless instructed to do so.

d. Rotate objective lens turret to clear the slide and apply a drop of immersion oil on the smear.

e. Examine using oil immersion objective lens (100×). Use the fine adjustment knob only (if needed) to bring stained cells to focus.

f. Determine the Gram reaction of the isolate. Gram-positive cells will appear blue to purple and Gram-negative cells will appear pink to red. It is also possible, but less likely, that stained cells appear Gram-variable. Record observed Gram reaction in Table 4.2.

g. Observe any distinct cellular arrangements (e.g., single cells, pairs, tetrads, chains, clusters). Record observed cell arrangements in Table 4.2.

h. In addition to the written record, include a drawing of the cells observed under the microscope. Alternatively, include a microscope picture of a field that represents the smear inspected, if the microscope is camera-equipped. Cameras of hand-held devices (e.g., mobile phones) with proper eyepiece adapter may be used to take the picture.

i. Repeat the previous steps with the second smear on the slide.

j. Show the microscope-mounted slide to the laboratory instructor for verification or advice on interpretation of the observation.

k. Clean the microscope lenses with lens paper or microfiber cloth lightly moistened in lens cleaner. Start the cleaning with the smaller magnification, ending with the oil-immersion lens. Store the microscope in its box or cabinet.

QUESTIONS

1. What is the main objective of streaking on agar culture media?
2. What are the advantages and disadvantages of pour-plating and spread-plating?
3. A microbiological culture was decimally diluted, 1 ml aliquots of selected dilutions were plated on duplicate plates of PCA medium, plates were incubated at 35°C for 48 hours, colonies on the plates were counted, and results were recorded in the following table.

	Colonies	
Dilution	Plate 1	Plate 2
10^{-1}	Not counted	
10^{-2}	489	520
10^{-3}	47	54
10^{-4}	2	8
10^{-5}	1	0

    a. Calculate the CFU/ml in the original culture.

    b. Why is the count reported as CFU/ml rather than cells/ml?

4. A sample of raw milk was decimally diluted, 0.1 ml aliquots of selected dilutions were plated on duplicate plates of PCA medium, plates were incubated at 25°C for 48 hours, colonies on the plates were counted, and results were recorded in the following table.

Dilution	Colonies	
	Plate 1	Plate 2
10^{-1}	Not counted	
10^{-2}	630	591
10^{-3}	Not counted	
10^{-4}	5	3

    a. Calculate the CFU/ml milk.

    b. How should this procedure have been modified to produce a more accurate count?

5. You are given a plate that has a large number of colonies (see the schematic below). Calculate the estimated original microbial count in the food sample, knowing that the plate was prepared by pour plating at a dilution of 10^{-2}. Show your calculations.

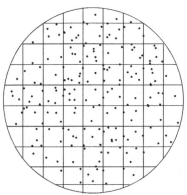

6. Bacterial population that may have survived juice pasteurization was determined as follows: Freshly pasteurized apple cider was sampled and 11-ml subsample was mixed with 99 ml diluent. Aliquots (0.1 ml each) of the mixture were plated on four Petri plates containing suitable agar culture medium. The plates were incubated at 35°C for 48 hours. When the plates were inspected in preparation for colony counting, no colonies were found on any of the four plates. Use this information to estimate the bacterial population (CFU/ml) in the cider.

PART II

FOOD MICROBIOTA

This part of the book covers the diverse groups of microorganisms that can be found in a food and that contribute to its microbiota. The term "microbiota" refers to the microbial community (bacteria, fungi, and other microscopic organisms) of a particular source. The source could be a food, sample of a body of water soil location, intestinal tract content, or material from another environment. Each ecological community is expected to include diverse species, and member species are expected to vary in relative abundance. A closely related term is "microbiome," which is the collection of genomes (the cumulative genetic material) from all microorganisms in a particular source.

Transient and Resident Food Microbiota

Production, harvesting, warehousing, processing, packaging, storage, distribution, retailing, and home preparation are important links in the typically extensive food supply chain (Figure II.1). Various microorganisms may intermingle with food at any point throughout this supply chain. These microorganisms may

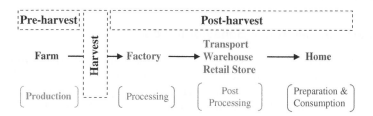

Figure II.1   Typical links in the food supply chain.

Analytical Food Microbiology: A Laboratory Manual, Second Edition. Ahmed E. Yousef,
Joy G. Waite-Cusic, and Jennifer J. Perry.

remain transient or become food inhabitants. Transient microorganisms do not interact with food; the food serves merely as a vehicle for their transmission. Contaminants that find the food a hospitable environment become natural inhabitants. For example, lactic acid bacteria are expected in artisan cheeses made from raw milk, even if these bacteria were not added as a starter culture during cheese making. Fresh produce seems to serve as a habitat for *Pseudomonas* spp., which is understandable considering that some members of this genus are plant pathogens. Similarly, cold-tolerant microorganisms (psychrotrophs) are often found in raw seafood, considering the cold environment from which these foods are harvested.

The transient members of food microbiota are often harmless commensals; they do not spoil the food, but sometimes include harmful infectious pathogens. On the contrary, resident microbiota are fit to grow in food (at the right conditions) and cause its spoilage. Growth of pathogenic microorganisms may be accompanied by toxin production or an increase in population to infectious dose levels. It should be cautioned that considering some members of food microbiota transient and others resident is a classification of convenience. Microorganisms may shift between these passive and active states upon changes in the environment under which the food is held, e.g., rise in temperatures due to refrigeration failure.

Food Characteristics as Determinants of Its Microbiota

Raw food may contain diverse and complex microbiota, but the species found to eventually prevail in the finished product depends principally on how the food was processed and on the food's innate characteristics. Dry or low-moisture foods, for example, are not commonly populated with large and diverse microbiota. The microbiota in these foods is limited by the ability to tolerate low osmotic pressure and to survive under dry conditions. Examples of these foods are breakfast cereals, dry nuts, and raisins. On the contrary, foods that are high in moisture, are low in acidity, undergo elaborate manipulation, and are minimally processed may contain large and diverse microbiota. Examples of these foods are bean sprouts and mold-ripened cheeses. Therefore, food characteristics and processing history determine, to a great extent, the prevalence and diversity of microorganisms in that food.

Temperature and Microbial Behavior in Food

Many factors affect the behavior (i.e., growth, survival, or death) of microorganisms residing in food; these factors include temperature, oxygen level, and food's pH and water activity. Considering the complexity of the influence of temperature on microbial behavior, this factor will be addressed in some detail. The discussion includes defining the terms associated with temperature as applied in this book.

Different microorganisms have been adapted to habitats having different temperatures. Examples of such habitats are arctic environments, mammals' intestinal tracts, and hot springs. Temperatures of habitats to which microorganisms have adapted range from sub-freezing to above water boiling point. Considering the diversity of Earth habitats, it is plausible to assume that for each temperature within this range, there are microorganisms that have been acclimatized to it. For simplicity, selected points along this continuum have been emphasized by researchers of different disciplines. Many microbiologists, including those specializing in food, emphasize temperatures representing refrigeration (4–7°C), room and laboratory ambient (20–22°C), and bodies of warm-bloodied animals (36–42°C). Additionally, scientists became interested in microbial behavior at other temperatures, such as water's freezing point (0°C), Earth's average temperature (~15°C), and water's boiling point (100°C).

The temperature of the habitat to which a given microorganism has adapted is considered the optimum for its growth, i.e., "optimum growth temperature." This term is defined here as the incubation temperature resulting in a growth curve exhibiting microorganism's shortest generation time, i.e., largest specific growth rate. One temperature lower and another higher than the optimum are designated as the minimum and the maximum growth temperatures, respectively. The minimum and maximum growth temperatures span the microorganism's "growth temperature range." A temperature greater than the maximum marks the lethality threshold, and the higher a temperature above that of the growth maximum, the more lethal it is.

Taking into account the temperature ranges and optimum growth temperatures reported by many microbiologists, foodborne microorganisms are grouped into psychrophiles (cold-loving), psychrotrophs (cold-tolerant), mesophiles (moderate-loving), and thermophiles (heat-loving).

- **Psychrophiles:** Can grow at or below 0°C with growth optimum at 15°C and a maximum growth temperature at 20°C.
- **Psychrotrophs:** Psychrotrophs can grow at refrigeration temperatures (0 to 7°C), but their optimum growth temperatures are greater than 15°C (in the range of 20 to 30°C) with growth maxima above this range. The growth temperature range for this group overlaps with the psychrophilic and mesophilic ranges; hence, these microorganisms are capable of growing in environments where the temperature fluctuates.
- **Mesophiles:** Mesophiles grow optimally at 30 to 40°C, with a minimum at ~10°C and a maximum growth temperature at 45 to 50°C.
- **Thermophiles:** Thermophiles have a growth temperature range of 45 to 70°C, with optima between 50 and 60°C.

Taxonomy of Food Microbiota

Considering the large number of microorganisms encountered in food, it is difficult to compile these in a comprehensive list. Revealing interrelationships among these microorganisms can serve as a basis for their categorization. Members of food microbiota may be grouped based on their taxonomic standing, role in food, or traits of importance, particularly to processors, consumers, or regulators.

- A grouping based on current taxonomic standing is illustrated in Figure II.2. This grouping covers many common foodborne bacteria, but foodborne fungi are described in a later chapter dedicated to this important group of

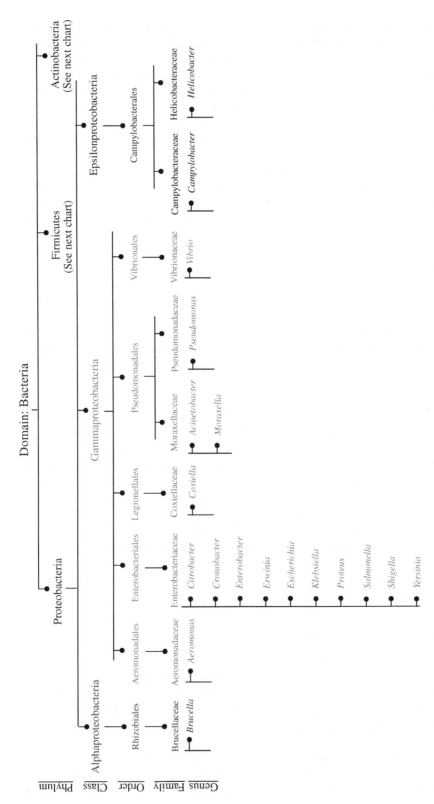

Figure II.2 Taxonomy of foodborne bacteria, drawn from Bergey's manual of systematic bacteriology, Volume 2. The classification takes into consideration the morphologic, biochemical, physiological, and genetic characteristics of the microorganisms. Based on this classification, all prokaryotic organisms belong to two domains: archaea and bacteria. The taxonomic ranks under bacteria are phyla (singular, phylum), classes, orders, families, genera (singular, genus), species (singular, species), and subspecies. The taxonomy of foodborne fungi is presented in Chapter 9.

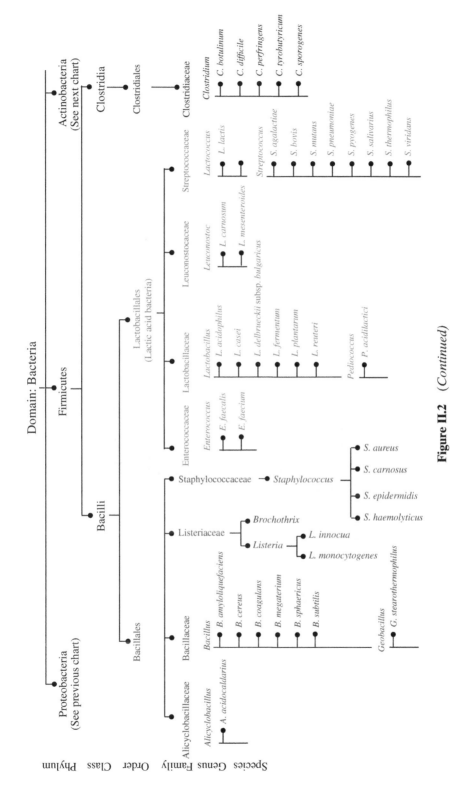

Figure II.2 (*Continued*)

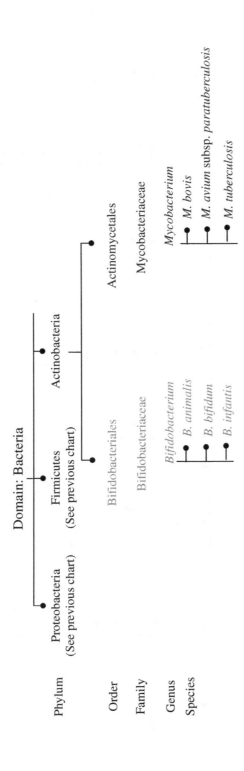

Figure II.2 (*Continued*)

food microbiota (Chapter 9). *Enterobacteriaceae* is an example of one of the important taxonomic groups of food microbiota; this family includes *Escherichia coli*, *Salmonella enterica*, and many other species.

- Based on their role, food microbiota include beneficial, spoilage, or pathogenic microorganisms. Commensal microorganisms are presumed to have a neutral relationship with the host or other members of food microbiota.
- Grouping on the basis of important microbial traits includes resistance to processing (e.g., heat-resistant microbiota or thermodurics), ability to grow in acid or low-acid foods, and utility as indicator organisms (e.g., the coliform group as indicators of fecal contamination).

This book covers several groups and individual species within food microbiota. In Part II, selected groups will be enumerated and characterized. While enumerating or detecting microorganisms in food, each exercise gives the student an opportunity to practice a new analytical technique. Each chapter title includes the newly introduced technique as a subtitle. Description of relevant microbiological media is included in each chapter and detailed compositions of these media are listed in Appendix III. When examining food microbiota, it is necessary to be familiar with the product sampled. Combining the analyst's observations about the characteristics of the microbiota with food's characteristics and processing history may help assess the product's quality and safety.

SELECTED REFERENCES

Brenner, D.J., Krieg, N.R., and Staley, J.T. (eds). (2005). *Bergey's Manual of Systematic Bacteriology*, 2e. Volume Two, The *Proteobacteria*: Part A introductory essays. New York, NY: Springer.

Moyer, C.L., Collins, R.E., and Morita, R.Y. (2017). Psychrophiles and Psychrotrophs. Reference module in life sciences. Elsevier Inc. doi.org/10.1016/B978-0-12-809633-8.02282-2.

Vasavada, P.C. and Critzer, F.J. (2015). Psychrotrophic microorganisms. In: *Compendium of Methods for the Microbiological Examination of Foods*, 5e. (ed. Y. Salfinger and M.L. Tortorello), 175–189. Washington DC: American Public Health Association Press.

Yousef, A.E. and Abdelhamid, A.G. (2019). Behavior of microorganisms in food: Growth, survival and death. p. 3-21. In: *Food Microbiology: Fundamentals and Frontiers*, 5e. (ed. M. Doyle, F. Diez-Gonzalez, and C. Hill), 3–21. Washington, DC: American Society for Microbiology Press.

CHAPTER 5

AEROBIC MESOPHILIC PLATE COUNT

APPLICATION OF BASIC TECHNIQUES TO FOOD

INTRODUCTION

Measuring the Microbiological Quality of Food

The quality of a food is contingent upon its physical, chemical, microbiological, nutritional, and sensory characteristics. The microbial load of a food (i.e., size of its microbiota) is often taken as a measure of its microbiological quality. Methods to quantify food microbiota were developed based on the following premises: (i) the larger the microbial population in food, the greater the propensity of food to spoil; and (ii) a population that represents the majority of the food's microbiota can be measured analytically.

Many exceptions to the first premise are known despite the abundance of supporting results in published literature. For example, a food may contain a large microbial load (e.g., 10^6 CFU/g), but if the population is relatively dormant, it is not expected to lead to imminent quality deterioration. On the contrary, if this food is held under conditions that allow active metabolism and multiplication of its microbial load, this leads to rapid product spoilage. A different situation is encountered in fermented food. Many of these foods (e.g., yogurt and cheese) naturally contain high microbial loads, mainly lactic acid bacteria, but this large population cannot be used to judge the microbiological quality of these products.

Similarly, there are many exceptions to the second premise. A food produced, processed, distributed, and handled under well-controlled conditions may contain a uniform microbial population that can be measured reasonably accurately using a method designed to enumerate this population. On the contrary, many

Analytical Food Microbiology: A Laboratory Manual, Second Edition. Ahmed E. Yousef, Joy G. Waite-Cusic, and Jennifer J. Perry.
© 2022 John Wiley & Sons, Inc. Published 2022 by John Wiley & Sons, Inc.

foods carry biologically diverse species within their microbial load; thus, a single analysis cannot provide accurate information about the size and diversity of the microbiota of that food.

Aerobic Mesophilic Count as a Measure of Food's Microbiological Quality

Despite the challenges just discussed, food analysts often use simple methods that approximate the total microbial load in food. Such methods involve homogenizing the food, plating its homogenate (or homogenate dilutions) on a nonselective culture medium, and incubating the inoculated plates *aerobically at 35°C*, the most common incubation conditions in a microbiological laboratory. Results of such analysis are often reported as "total plate count," "standard plate count," or "aerobic plate count," and are used as an indicator of food's microbiological quality. Considering the microbiological media and incubation conditions often used, results of such analysis can be described more accurately as aerobic mesophilic plate count (AMPC). Mesophilic microorganisms have been defined earlier (Part II: Food Microbiota) as those that grow at 10 to 45°C, but their optimum growth occurs at 30 to 40°C. Many foods are exposed to environments within these ranges during production, processing, handling, or storage, thus prevalence of mesophiles in these foods is expected. Additionally, aerobes and facultative anaerobes are favored considering that foods are often exposed to air and these microorganisms generally grow rapidly when compared to anaerobes. Growth of these contaminants, while food continues to be held under mesophilic-favoring conditions, could lead to spoilage. Therefore, it is plausible to assume that AMPC can be used as an indicator of microbiological quality of food.

Although AMPC has the advantage of being a quick and efficient method to evaluate the microbiological quality of food, this test has inherent disadvantages. For example, microbially spoiled food is expected to display high AMPC (often $\geq 10^7$ CFU/g), however, analysis of such spoiled food may reveal low AMPC. If this spoilage is truly microbial in origin, it may have been caused by a microbial group not well represented in AMPC, as discussed later. Alternatively, the food may have been spoiled by aerobic mesophilic population, but these microorganisms have entered the death phase at the time of food analysis.

The following are examples that show how AMPC underestimates certain microbial groups in food:

- Media used in AMPC determination often do not support rapid growth of fastidious microorganisms such as lactic acid bacteria.
- Incubation conditions do not favor strict anaerobic and psychrophilic bacteria.
- Molds and yeasts grow better at lower temperatures than those used in this test and require a longer incubation period.

TABLE 5.1 Approximate representation of different categories of foodborne microorganisms in the aerobic mesophilic plate count (AMPC).

Criteria	Groups	AMPC representation	Remarks
Microorganism	Bacteria	+++	
	Yeast	+	Fast-growing yeast represented
	Mold	–	
	Parasites	–	
Temperature	Thermophiles	–	Extremophiles also are not represented
	Mesophiles	+++	
	Psychrotrophs	+	
	Psychrophiles	–	
Oxygen	Aerobes	+++	Microaerophiles may not be represented
	Facultative anaerobes	+++	
	Anaerobes	–	
Nutrition	Fastidious	–	This grouping includes the lactic acid bacteria.
	Non-fastidious	+++	
Other factors	Xerophiles	–	Media with reduced water activity are used to enumerate this category
	Acidophiles	Unmeasurable	Loosely defined term implying acid tolerance rather than acid obligatory requirement

Based on this discussion, analysis of food for AMPC has benefits and limitations. Table 5.1 summarizes the approximate representation of different categories of food-related microorganisms in AMPC. Additional information can be reviewed in published literature including the references listed in this chapter. Testing a food for AMPC often involves the following steps: (i) sampling, (ii) sample homogenization, (iii) diluting the homogenized sample serially, (iv) plating selected dilutions on a relatively rich, non-selective agar medium, (v) incubating inoculated plates under conditions that favor aerobic mesophilic microbiota, (vi) enumerating colonies on plates, and (vii) calculating the aerobic mesophilic population in the food and reporting the results as CFU/g or CFU/ml. These steps are listed in Figure 5.1 as executed in this laboratory exercise.

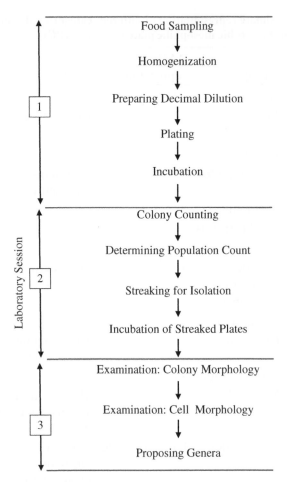

Figure 5.1 An overview of the aerobic mesophilic plate count method.

OBJECTIVES

1. Determine and compare the AMPC of different foods
2. Practice application of colony counting rules
3. Become familiar with the morphology of microorganisms commonly isolated from food

MEDIA

Plating method and medium used may affect the outcome of the analysis. The homogenized sample, with or without further dilutions, is mixed with molten agar medium (if using a pour-plating technique) or spread onto the surface of solidified agar (spread-plating technique). Tryptic soy agar (TSA) medium using the spread-plating technique will be employed in this exercise. It should be noted that laboratories may use alternative media or plating techniques to obtain AMPC. For example, pour-plating in plate count agar (PCA) is another common approach for enumerating AMPC in foods. The same general procedures (serial dilutions and spread-plating techniques) may be used with selective media or alternative incubation conditions to enumerate other populations of interest and the method will typically be named by the group of organism targeted (e.g., *Enterobacteriaceae*, yeast and mold, pseudomonads). These alternatives will be the focus of subsequent chapters in this book.

Tryptic Soy Agar (TSA)

TSA powder contains only four ingredients: pancreatic digest of casein; enzymatic digest of soy meal; sodium chloride; and agar. Despite its simplicity, the medium is often used for cultivation of a wide range of microorganisms. It is used in plate count methods, as is the case in this chapter, and for isolation of microorganisms and maintenance of cultures.

PROCEDURE OVERVIEW

The current laboratory exercise is designed to train students to analyze raw solid foods for AMPC. Foods of different types that are inherently different in microbial population size and profile will be analyzed. Food from animal sources (e.g., raw meat, poultry, and seafood) and plant sources (e.g., leafy greens and bagged, fresh-cut salad vegetables) are suggested. For example, raw meat and raw vegetables would provide different results and require different interpretations.

Samples of food will be prepared for analysis (i.e., homogenized) as described in Chapter 2. The blender or stomacher will be used to homogenize the food. The resulting homogenized samples will be diluted appropriately, depending on the microbial load expected in the food. Selected serial decimal dilutions will be spread-plated on tryptic soy agar or another appropriate medium. Plates will be incubated at 35°C for 48 h. Colonies appearing on incubated plates will be enumerated and population count (CFU/g) of food will be calculated (Chapter 3). Selected colonies representing the microbiota of the food sampled will be examined and colony and cell morphologies will be determined (Chapter 4). All safety protocols presented previously (Chapter 1) should be observed.

This exercise will be completed in three laboratory sessions as shown in the overview schematic (Figure 5.1). During the first session, students will homogenize food samples, prepare dilutions, plate on an agar medium, and incubate the inoculated plates. Incubated plates will be inspected during the second session and counts determined. Selected colonies will be isolated by transferring to fresh plates for further analysis. Colony and cell morphologies will be inspected during the third session.

Students will work in groups of two, each group sampling one food. Members of the group will share analysis responsibilities and results gathered. Results of all groups will be shared with all students in the class. Alternatively, students may work individually, instead of in groups of two. In this case, each student is given a food sample and is expected to complete the analysis that is assigned to the group in the current procedure. Students should review previous chapters for sample preparation, dilution, plating, colony counting, and determination of population counts.

Food Handling Precautions

In addition to the information shared in Chapter 2, the following precautions should be observed in *all laboratory exercises*:

- Foods should be kept in the laboratory refrigerator until just prior to laboratory start.
- The food package should be inspected visually for integrity or abnormalities.
- The package label should be read, and label information (e.g., weight or volume, brand, ingredients) should be recorded appropriately and used in results interpretation. Sell-by date or other pertinent information may be found stamped on the label or other parts of the package. Alternatively, a picture of the package, showing package information, may be taken and shared with the class.
- Tools used in sample preparation (particularly cutting boards, knives, blender jars, food containers, spatulas, tongs, scissors, mortar and pestles, and cheese shredders) should be clean and ideally sterile. If this requirement cannot be achieved, these items need to be cleaned appropriately and sanitized with ethanol immediately before use.

  Caution: To prevent fire hazard, sanitizing utensils or gloves with alcohol SHOULD NOT be done in proximity to a lit Bunsen burner.

- The contents of the food's retail package should be transferred aseptically into a suitable plastic container (with cover) and mixed thoroughly with a sterile spatula before sampling. Some foods require grinding or shredding before samples can be taken for analysis. Such food preparations should be done under aseptic conditions using sterilized (or sanitized) tools and surfaces.
- In most dilution schemes, 10 g ($\pm$ 0.1 g) of food sample are used for analysis. To weigh a food sample, a tared sterile Petri plate may be used, and its contents transferred to a sterile stomacher bag. Preferably, the sample may be weighed directly into a stomacher bag. The bottom of the stomacher bag may be spread into a cup shape to support the bag in an upright position on the balance while weighing the sample directly into the bag. Care should be taken to not touch the inside of the stomacher bag. A stomacher bag holder, which is available commercially, is useful in holding the bag open above a balance during sample weighing.
- Note that it is crucial that sample contamination is minimized or eliminated by proper and prompt handling and weighing. Students who spend extra time weighing 10 g with high accuracy (e.g., 10 ± 0.01 g) are more likely to contaminate their samples than those measuring the sample more rapidly at lesser accuracy (e.g., 10 ± 0.1 g).

Examination of Colonies

It is sometimes beneficial to gather information about the types of colonies resulting from plating the food sample. Note that in the case of pour plating, a single species may produce more than one type of colony morphology, depending on the position of the colony in the agar medium. Surface colonies tend to be round, whereas those embedded in the agar may show an oval or star shape. Differences in morphology of colonies of a given microorganism are minimal in the case of spread plating. In this laboratory exercise, spread plating will be used; therefore, colony morphology should be relatively easy to interpret.

Microscopic examination provides additional information about the morphology of the cells that formed these colonies. If colonies of a given morphology are predominant, this raises the curiosity about the type of contaminants in the food. In this case, the analyst or the food processor may become interested in identifying the microorganism(s) representing the prevailing morphology. In fact, examining the colony and cell morphology is an essential preliminary step in identifying isolates from food. For identification of a microorganism, the isolate should be tested for biochemical, physiological, serological, or genetic traits (see later chapters on identification of pathogens in foods). It is important to realize that morphological testing, as conducted in this exercise, is not sufficient to identify the microorganism represented by the examined colonies; however, it may provide important information to support further microbiological tests.

If identity of isolated colonies must be determined, it is recommended to use genetic techniques. An isolated bacterium may be identified by extracting its DNA, amplifying the 16S rDNA gene by PCR, sequencing the amplified DNA fragment, and comparing the resulting sequence with published databases. This approach helps identify isolates at the genus, and sometimes at the species, level. Alternatively, whole genome sequencing of the isolate may be obtained using commercially available sequencing facilities. This type of analysis would help identify an isolate at the subspecies level and assist in pathogen tracking during disease outbreaks. Discussion of these identification methods is presented in later chapters.

SELECTED REFERENCES

Maturin, L., and Peepler, J.T. (2001). Aerobic plate count. *Bacteriological Analytical Manual.* U.S. Food and Drug Administration. https://www.fda.gov/Food/FoodScienceResearch/LaboratoryMethods/ucm063346.htm

Pothakos, V., Samapundo, S., and Devlieghere, F. (2012). Total mesophilic counts underestimate in many cases the contamination levels of psychrotrophic lactic acid bacteria (LAB) in chilled-stored food products at the end of their shelf-life. *Food Microbiology,* 32: 437–443.

Ryser, E.T. and Schuman, J.D. (2015). Mesophilic aerobic plate count. In: *Compendium of Methods for the Microbiological Examination of Foods*, 5e. (ed. Y. Salfinger and M.L. Tortorella), 95–101. Washington, DC: APHA Press.

SESSION 1: SAMPLING, HOMOGENIZATION, DILUTION, PLATING, AND INCUBATION

In this laboratory session, recovery of microorganisms from the food matrix will be performed by stomaching. Homogenized food samples will be spread plated on an agar medium (Figure 5.2). Counts from different foods will be compared in subsequent laboratory sessions.

MATERIALS AND EQUIPMENT

Per Two Students (i.e., one group)

- Food sample
- One bottle of peptone water (90 ml)
- Three peptone water tubes (9 ml each)
- Six TSA plates (agar medium pre-poured 24–48 h before use)
- Sterile spatula (held in a closed sterile test tube)
- Stomacher bag holder
- Stomacher bag
- Vortex mixer
- Pipette pump
- Individually-wrapped sterile 1-ml pipettes (disposable)

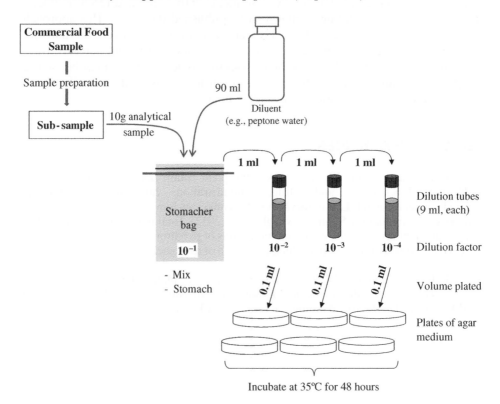

Figure 5.2 Suggested dilution scheme for aerobic mesophilic plate count exercise.

- Pipetters: 1000 µl capacity, and 200 µl (or 100 µl) capacity
- Sterile pipette tips (1000 and 200 µl)
- Cell spreader (e.g., triangular stainless steel spreader, or bent glass rod spreader)
- Alcohol jar for decontaminating spreader
- Other common laboratory supplies (e.g., labeling and masking tapes)

Sample Preparation Utensils (as needed)

- Alcohol squeeze-bottle for sanitizing utensils
- Plastic container for holding and mixing chopped sample
- Cutting board
- Knife
- Tongs

Class-Shared

- Common laboratory equipment (e.g., Bunsen burner and incubator)

PROCEDURE

Labeling

1. Using labeling tape, label peptone water (diluent) tubes with appropriate dilution factors.
2. Label agar plates using the permanent marker. Duplicate plates are needed for each dilution. Plate should be labeled on the periphery of the bottom (agar side) surface. Include analyst initials, date, food sample analyzed, dilution plated, and volume of dilution dispensed.

Sampling

For this analysis, a retail food package of 100–500 g may be considered a laboratory sample. Samples should be prepared for subsampling and analysis as described in Chapter 2. Sterilized utensils are ideally used in sample preparation. If this condition is not met, these utensils should be appropriately cleaned and then sanitized with alcohol (ethanol).

Caution: To prevent fire hazard, sanitizing utensils or gloves with alcohol SHOULD NOT be done in proximity to a lit Bunsen burner.

1. **Package information.** Record pertinent information from the package of the food sample.
2. **Preparing bulky packaged food for subsampling.** In the case of bulky food (e.g., meat strips, lettuce head, etc.), follow this procedure:
   a. Prepare clean and alcohol-sanitized (i) wooden or plastic cutting board, (ii) knife, and (iii) sample container with a cover (sanitizing utensils is needed if these have not been autoclaved before use). Wear disposable gloves and sanitize gloves with alcohol; replace or re-sanitize as needed.

b. Open the package aseptically, and transfer the entire contents to the cutting board. Dispose of the empty package in the appropriate waste receptacle.

c. Cut the sample with the knife into pieces as small as practically feasible. The smaller the pieces, the better the subsample represents the retail sample. However, the longer the time of preparation, the greater the chances of sample contamination.

d. Transfer the chopped sample into the sterilized (or sanitized) container. Close the container and shake it to mix its contents thoroughly.

e. Place the stomacher bag into provided holder and place the holder on the balance. Take care to not touch the inside of the stomacher bag. Tare the weight of the bag and holder.

f. Transfer 10 g (± 0.1) of the chopped food into the stomacher bag aseptically. Use the provided sterile spatula or a pair of sanitized tongs, depending on the food.

g. If other groups are using the same sample, close the container to protect the sample from contamination. If not, dispose of the remainder of the sample appropriately. It is advisable that open food packages and samples are disposed of in biohazard boxes.

3. **Preparing chopped packaged food for subsampling.** In the case of chopped food (e.g., ground meat, bagged fresh-cut salad), follow this procedure:

a. Prepare clean and sterilized (or sanitized) sample container with a cover. Wear disposable gloves and sanitize with alcohol; replace or re-sanitize as needed.

b. Open the food package aseptically, and transfer the contents to the sample container. Close the container and dispose of the empty package appropriately. If the size of the original container allows for mixing without displacing the sample, it is acceptable not to transfer the food to the plastic container.

c. Mix the food thoroughly in the sample container. Use the spatula (e.g., ground meat) or simply shake the container (e.g., chopped salad).

d. Place the stomacher bag into the provided holder and place the holder on the balance. Take care to not touch the inside of the stomacher bag. Tare the weight of the bag and holder.

e. Transfer 10 g (± 0.1) of the chopped food into the stomacher bag aseptically. Use the provided sterile spatula or a pair of sanitized tongs, depending on the food.

f. If other groups are using the same sample, close the container to protect the sample from contamination. If not, dispose of the remainder of the sample appropriately. It is advisable that open food packages and samples are disposed of in biohazard boxes.

Homogenization

1. Add 90 ml sterile peptone water to the stomacher bag. Carefully push most of the air out of the bag and close the bag with its built-in fastener (if applicable) before stomaching.

2. Homogenize the sample by stomaching for 2 min. The contents of the stomacher bag (i.e., the homogenized sample) is the 1/10 (i.e., the 10^{-1}) dilution of the original analytical sample.

Preparing Dilutions

All dilution preparation work should be done in proximity to a lit Bunsen burner. When opening the stomacher bag or tube, care should be taken to avoid contamination of its contents. Use appropriate technique to open tubes and sanitize (briefly flame) the opening before and after pipetting. Never set tube lids down on the bench. Accurate setting of the pipette and pipetting are essential.

1. Using a disposable 1-ml pipette, transfer 1 ml (i.e., 1000 µl) of the 10^{-1} dilution (stomacher bag contents) into one labeled peptone water tube. This is the 10^{-2} dilution.
2. Vortex the contents of the 10^{-2} dilution tube for 10 seconds.
3. Repeat this process (transferring from the most recently prepared dilution tube) to prepare 10^{-3} and 10^{-4} dilutions.

Spread-Plating

Use the following dilutions for spread-plating on tryptic soy agar: 10^{-2}, 10^{-3}, and 10^{-4}. All spreading work should be done in proximity to a lit Bunsen burner. During dispensing and spreading of diluted sample, the agar plate should be only partially opened, and the opening should be facing the Bunsen burner. The spreader should be sanitized before and after each use, by dipping in an alcohol jar and flaming off excess alcohol.

1. Use the appropriate pipetter to transfer and dispense 100 µl of sample dilutions onto the agar surface.
2. Vortex the contents of the 10^{-2} dilution tube for 10 seconds and transfer 0.1 ml (i.e., 100 µl) into each of the appropriately labeled pair of agar plates. Using a freshly sanitized (or disposable) cell spreader, carefully and evenly spread the inoculum onto the surface of the agar. Rotate the plate during spreading. Two complete turns of the agar plate are often sufficient to spread the inoculum evenly.
3. Repeat for the 10^{-3} and 10^{-4} dilution tubes.

Plate Incubation

1. Tape the plates together using masking tape. Label the stack of plates appropriately.
2. Incubate the plates inverted at 35°C for 48 h.

Note: The dilution scheme described is based on the expected microbial load in raw foods collected from a regional retail market within the Midwest United States. For foods analyzed in different regions, or in different countries, the number of dilutions needed should be adjusted based on expected microbial load of local foods.

SESSION 2: COLONY COUNTING AND ISOLATION

In this laboratory session, incubated plates will be examined visually to become acquainted with the diversity of bacteria recovered. Examples of the incubated plates resulting from the analysis foods by aerobic mesophilic plate count method is shown in Figure 5.3. In addition to visual examination, counts of microbial population in the food sample (CFU/g) will be determined by following the colony counting rules discussed in Chapter 3. Selected colonies will be streaked on fresh agar medium for morphological examination.

MATERIALS AND EQUIPMENT

Per Group of Two

- Incubated plates
- Colony counter
- Four TSA plates
- Other common laboratory supplies (e.g., permanent marker)

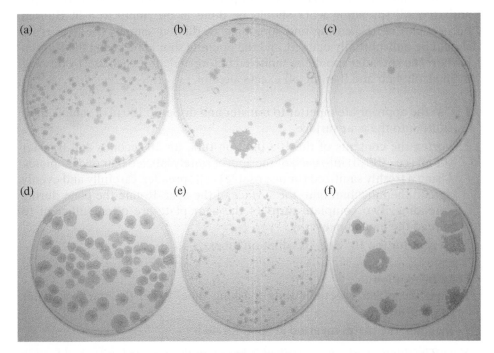

Figure 5.3   A sample of results of market food analyzed using the mesophilic aerobic plate count method (as described in this chapter) by students taking the Food Microbiology Laboratory course at the Ohio State University, spring semester, 2020. (a–c): Plates receiving 10^{-2}, 10^{-3}, and 10^{-4} dilutions of a sample of bagged spinach; (d–f): Plates receiving 10^{-2} dilutions of mint, ground beef, and raw beef sausage, respectively.

Class-Shared

- Common laboratory equipment (e.g., Bunsen burner and incubator)

PROCEDURE

Colony Counting

1. Remove the tape from incubated plates and lay the plates on the bench in order of the dilution series.
2. Examine the plates as follows:
   a. Observe the diverse morphology of colonies appearing on these plates. This diversity will be more obvious on the plates receiving lower dilutions. It is likely that colonies with very similar in morphology belong to the same species. Based on these similarities, how many phenotypes can you visually detect? Is there a predominant phenotype?
   b. Confirm that the dilution series produced an approximate 10-fold difference, that pairs of plates from the same dilution are similar in count, and that there is no obvious contamination. If inconsistencies are noted, ask an instructor for feedback before proceeding.
3. Do a preliminary estimate of the number of colonies on all of the plates, regardless of whether the counts would be used in calculations. If plates are estimated to contain colonies in the 20-200 range, use these to determine colony and population counts; disregard those from other dilutions.
4. Record the colony counts (from two plates) in Table 5.2, using only the dilution most suitable for determining population counts (i.e., countable plates).
5. Determine the AMPC in sampled food as described for sample population calculation in Chapter 3.

$$\text{Population count}\,(\text{CFU}/\text{g}) =$$

$$\frac{\text{Average colony count}\,(\text{on 2 countable plates receiving the same dilution})}{\text{Volume plated}\,(\text{ml}) \times \text{dilution factor for dilution tube used in plating}}$$

$$(5.1)$$

6. If incubated plates do not contain colony counts in the range described, use the decision tree (Chapter 3) to determine the rules applicable to this situation.

TABLE 5.2 {Add a descriptive title for this data set, including food sample used, media, recovery method, and what data were obtained}

Food Sample	Dilution	Colony count	Population count(CFU[a]/g)
Plate 1			Calculation detail:
Plate 2			

[a] Colony-forming units
Include additional footnotes as needed. Feel free to modify the footnote above if needed.

7. Record the AMPC as CFU/g in Table 5.2, and in the "Class Data" table provided by the laboratory instructor. Population counts should be presented in scientific notations, with two significant digits reported.

Streak Selected Colonies for Isolation

1. Using the provided four TSA plates, appropriately label each with a permanent marker on the bottom of the Petri plate.
2. Select four morphologically distinct colonies from the incubated plates and 3-phase streak each for isolation on the corresponding TSA plate. Colonies do not all have to be from the same original plate, or from one of the plates that were counted. Colonies may differ by color, size, consistency, margin, or other characteristics. Try to choose colonies that are as different from each other as possible.
3. Tape streaked plates together using masking tape, invert the stack of plates and incubate plates at 35°C for 24–48 h.
   Notes: Shorter incubation (24 h) of plates produces colonies most suitable for Gram staining. Longer incubation of some Gram-positive species may result in the thinning of the peptidoglycan layer. This thinning reduces the ability of the crystal violet-iodine complex to adhere to the cells and can lead to inconclusive (Gram-variable) or incorrect (Gram-negative) determinations. On the contrary, slow-growing isolates need 48 h, or longer, to produce colonies with cell mass sufficient for isolation or staining.

SESSION 3: EXAMINING COLONY AND CELL MORPHOLOGIES

In this laboratory session, isolates on the incubated plates will be examined for colony and cell morphologies.

MATERIALS AND EQUIPMENT

Per Group of Two

- Incubated streaked plates
- Light microscope (bright-field setting)
- Gram stain kit
- Staining rack
- Sink or waste bin
- Microscope slides (2 per group)
- Wax pencil
- Inoculating loop
- Bunsen burner
- Bibulous paper
- Immersion oil
- Lens cleaner
- Lens paper
- Other common laboratory supplies

Class-Shared

- Colony counters
- Dissecting microscopes (stereo-microscope)
- Other common laboratory equipment

PROCEDURE

Colony Morphology

1. Examine the colony morphology of the four isolates (on the incubated streak plates) using a dissecting microscope (stereo-microscope).
2. Observe colonies for differences in morphology; e.g., colony color, size, consistency, elevation, and margin.
3. Record these observations in Table 5.3.

Cell Morphology

Use isolated colonies from the streaked, incubated agar plates to prepare smears on glass slides and perform Gram staining as described in Chapter 4, and summarized in this section. Bacteria often appear Gram-positive or Gram-negative. If an isolate happens to be a yeast or mold, cells will appear large (e.g., 10 times larger than many bacterial cells) when examined micro-

TABLE 5.3 {Add a descriptive title for this data, including food sample used, media, recovery method, and what data were obtained}

Colony Number	Colony Morphology	Gram Reaction[a]	Cell Morphology	Relevant Microorganism[b]

[a] Gram reaction is not relevant to fungi.
[b] Data gathered in this laboratory are not sufficient to identify any of the isolates. These data, however, may be sufficient to make guesses about potential microbial groups (genera) in the food.
Note: These footnotes are provided as examples for instruction purpose. Do not copy these footnotes; instead, develop your own.

scopically, and result of staining should not be interpreted as Gram-positive or Gram-negative.

1. **Prepare a smear from each of the four isolates (two colonies per microscope slide) as follows:**
   a. Identify the colonies to be transferred and number their locations on the base of the agar plates.
   b. Label each end of a microscope glass slide with the corresponding colony number, using a wax pencil.
   c. Place a small drop of water near the end of the first slide.
   d. Transfer a portion of the first colony, using sterile loop, to the water drop and emulsify.
   e. Spread the emulsified colony into a smear, flame the loop and set it aside.
   f. Repeat steps c–e for the second colony using the other end of the slide.
   g. Repeat steps b–f for the third and fourth colonies; this should be completed simultaneously by the laboratory partner.
   h. Let the smears air dry, and fix by heat (pass through flame 2–3 times).
2. **Gram stain each smear as follows:**
   a. Place each slide on a staining rack.
   b. Cover the smears with crystal violet (primary stain) and let sit undisturbed for 60 seconds then rinse with water.
   c. Cover the smears with Gram's iodine. After 60 seconds, rinse with water as described earlier. Lightly blot the slides with bibulous paper to remove excess water; do not let the slides dry during blotting.
   d. Apply acetone-alcohol decolorizer for 3–5 seconds; immediately rinse with water to prevent further decolorization. Drain off excess water.
   e. Cover the smears with safranin for 60 seconds then rinse with water. Carefully blot the slides dry using bibulous paper.

3. **Examine the smears under a bright-field light microscope**. Remember to start with the low magnification objective lens (10×), then the intermediate magnification lens (e.g., 40×) and lastly with the 100× oil immersion lens. Determine the Gram reaction of the isolate.

4. **Record observed cell morphology in Table 5.3.**

5. **Clean and store the microscope.**

QUESTIONS

1. Were the dilutions prepared and plated suitable for aerobic mesophilic plate count in the food you analyzed? If these were not suitable, what dilutions do you suggest if this exercise is to be repeated?

2. Was the microbiota of the food analyzed by your group relatively homogenous or diverse? How do you know? Include colony and cell morphology descriptions in your answer.

3. What are the most likely sources of the genera you proposed for the isolates examined under the microscope? *Hint: Considering that raw foods were analyzed, it is important to realize that, in nature, different microorganisms thrive in different habitats. Consider also that foodborne pathogens are exceedingly rare in commercial samples.*

CHAPTER 6

MESOPHILIC SPOREFORMING BACTERIA

AEROBIC AND ANAEROBIC INCUBATION. DIFFERENTIAL
STAINING. PHASE-CONTRAST MICROSCOPY

INTRODUCTION

Some bacterial species within the phylum *Firmicutes* enter a highly dormant
state by forming endospores. These bacteria are described as sporeformers, and
their endospores are simply referred to as spores. Spore formation is often triggered by high cell density, as is the case when the cell population enters the stationary phase of growth, or by environmental stresses such as starvation. The
spore structure protects the bacterial genetic material in environments that can
be detrimental to the cell. Formation of spores can be detected microscopically;
for example, spores appear refractile when observed with a phase-contrast
microscope. Sporeforming genera of considerable impact on food quality and
safety include *Bacillus*, *Paenibacillus*, *Alicyclobacillus*, and *Clostridium*. Cells of
a young sporeforming culture are characteristically Gram-positive, but older
cells of some species appear Gram-negative or Gram-variable. While *Clostridium*
spp. are obligate or aerotolerant anaerobes, species of *Bacillus*, *Paenibacillus*,
and *Alicyclobacillus* are aerobes or facultative anaerobes. It is important to note
that presence or lack of oxygen does not affect the dormant spores. On the contrary, the metabolic activity, multiplication, or even viability of sporeforming
cells is affected by oxygen level. *Clostridium* spp. are catalase-negative, whereas
most aerobic sporeformers are catalase-positive. Considering the abundance of
published literature on *Bacillus* spp., their spores are often perceived as the ideal
model for all bacterial spores. Generalizations presented in this chapter are
mostly applicable to the spores of this genus.

Analytical Food Microbiology: A Laboratory Manual, Second Edition. Ahmed E. Yousef,
Joy G. Waite-Cusic, and Jennifer J. Perry.
© 2022 John Wiley & Sons, Inc. Published 2022 by John Wiley & Sons, Inc.

Thermal Requirements for Growth and Inactivation

Bacteria, including sporeformers, vary considerably in optimum growth temperatures (see introduction to Part II). Larkin and Stokes (1966) isolated psychrophilic sporeformers from soil; these isolates grew well at 0°C. Mesophilic sporeformers isolated by these authors had a growth minimum of 10°C and a maximum of ~50°C. Mesophilic sporeforming bacteria described in this chapter are those having their most optimum growth in the range of 30 to 40°C. Thermophilic sporeformers typically have minimum, optimum, and maximum growth temperatures of 45, 50 to 60, and 70°C, respectively. It should be obvious that the temperatures needed for optimum bacterial growth are lower than those required to inactivate these bacteria. Furthermore, temperatures required to inactivate cells of sporeforming species are lower than those required to inactive their spores. Inactivating bacterial spores generally requires heating at temperatures higher than 90°C, but thermal resistance varies considerably, depending on the species. Warth (1978) compared the heat resistance of cells and spores of different *Bacillus* spp. To achieve the same degree of lethality in cells and spores, the researcher found that spores needed to be treated at 41 to 54°C higher than that needed for cells of the sporeformers.

Prevalence and Impact on Food

Sporeforming bacteria are commonly found in soil and thus are considered natural contaminants of most foods. Some *Clostridium* spp. are present in the intestinal tract of animals and appear as contaminants in food of animal origin. Because of their widespread nature and tolerance to dryness and other lethal factors, bacterial spores are commonly found in spices, cereal grains, dried fruits, flour, starches, and dried milk. Sporeforming bacteria are common causes of food spoilage (Table 6.1). *Bacillus* spp. cause rapid deterioration of pasteurized milk, softening and stickiness in the center of bread loaves (ropy bread spoilage), and flat sour spoilage (acid, but no gas production) in containers of thermally processed low-acid (pH > 4.6) foods. *Alicyclobacillus* spp. are particularly problematic in pasteurized fruit juices and may cause spoilage of these products. *Clostridium* spp. produce gas in cheese blocks during the late stage of ripening (late gassiness defect) and spoil canned products such as sweet corn and spaghetti with tomato sauce. When some *Bacillus* and *Clostridium* spp. grow in food, they produce toxins that cause foodborne diseases. *B. cereus* and *C. botulinum* are notable for causing food intoxications. When *C. perfringens* is ingested in large numbers with food, the bacterium may sporulate in the intestine. A toxin is released during the sporulation process, causing a toxicoinfection.

Presence of high counts of bacterial spores in food indicates heavy contamination of the production or processing environment, poor handling and storage, or lack of processing treatments that eliminate spores. Commercial sterilization of canned foods (e.g., retorting) is designed to eliminate pathogenic sporeformers. Gamma radiation is used to decrease spore load of the spices. Bacterial spores in food, however, survive processing treatments that are less severe than that achieved by commercial sterilization. Faulty thermal processing also results in food containers with surviving bacterial spores. In this laboratory exercise,

TABLE. 6.1 Sporeforming bacteria of importance in food.

Group	Bacteria	Cell characteristics	Spore characteristics	Importance in foods
Mesophilic aerobes	*Bacillus anthracis*	Catalase positive. Facultative anaerobe. Large cell. No hemolysis. Non-motile. Hydrolyze gelatin, starch, casein. Optimum growth: 30 to 40°C. Growth in 7% NaCl.	Ellipsoidal, central to subterminal. No swelling of sporangium.	Foodborne infection. Linked to ingestion of undercooked infected meat.
	Bacillus cereus	Catalase positive. Facultative anaerobe. Large cell. Hemolysis. Hydrolyze gelatin, starch. Grows at 5 to 50°C (Optimum: 28–35°C). Growth at pH 4.9 to 9.3.	Ellipsoidal, central to subterminal. No swelling of sporangium. $D_{100°C} = 3$ to 200 min	Causes food intoxication. Disease outbreaks linked with starchy foods (cooked rice, pasta), cooked meats, vegetables, soups, salads, puddings.
	Bacillus licheniformis	Catalase positive. Facultative anaerobe. Grows at 30 to 55°C. Growth in 7% NaCl.	Oval, central to subterminal. No swelling of sporangium. $D_{100°C} = 13.5$ min	Spoilage of cooked meat (softening, discoloration). Opportunistic pathogen.
	Bacillus megaterium	Catalase positive aerobe. Large cell. Grows in the range of 5°C to 40°C (Optimum: ~30°C).	Oval, round, elongated central to subterminal. No swelling of sporangium. $D_{100°C} = 1$ min	Spoilage: coagulation of canned evaporated milk.
	Bacillus subtilis	Catalase positive. Obligate aerobe. Grows in the range of 10°C to 50°C. Growth at pH 5.5 to 8.5	Oval, central to subterminal. No swelling of sporangium. $D_{100°C} = 7$–70 min	Spoilage: "Ropy" bakery product; softening of pickles. Opportunistic pathogen.
	Paenibacillus polymyxa	Catalase positive. Facultative anaerobe. Grows in the range of 5°C to 40°C. Growth at pH 5.	Oval endospore. Swelling of sporangium. $D_{100°C} = 0.1$–0.5 min	Spoilage of canned vegetables (swelling)

Group	Organism	Characteristics	Spore	Spoilage/Food implications
Thermophilic aerobes	*Alicyclobacillus* spp.	Catalase positive. Hydrolyze starch. No growth in 5% NaCl. Acidophile: grows at pH 2.5 to 5.5 (Optimum: 3.5 to 4.0). Grows in the range of 20°C to 70°C (Optimum: 42°C to 60°C).	Oval, subterminal to terminal. Swelling of sporangium. $D_{95°C} = 2.2–8.7$ min	Flat sour spoilage. Spoilage of fruit juices and iced tea.
	Bacillus coagulans	Catalase positive. Facultative anaerobe. No growth in 5% NaCl. Grows in the range of 30°C to 65°C (Optimum: > 50°C). Aciduric: grows at pH 4.5 (Optimum: pH 6).	Oval endospore. Swelling of sporangium. High D-value: $D_{100°C} = 20–300$ min	Flat sour spoilage. Spoilage of canned tomato juice and other acid foods. Coagulation of canned condensed milk.
	Geobacillus stearothermophilus	Facultative anaerobe. Grows in the range of 40°C to 70°C (Optimum: > 50°C). No growth in 7% NaCl. Limited tolerance to acid.	Oval endospore. Swelling of sporangium. High D-value: $D_{100°C} = 100–1600$ min	Flat sour spoilage. Spoilage of low-acid canned vegetables. Coagulation of canned condensed milk.
Mesophilic anaerobes	Proteolytic *Clostridium botulinum*	Catalase negative. Strict anaerobe. Produce botulinum neurotoxin types A, B, F. Digest gelatin, milk, meat. Grows in the range of 10°C to 48°C (Optimum: 35°C to 40°C). Minimum growth pH: > 4.6. No growth in 6.5% NaCl or at pH 8.5.	Oval, subterminal endospore. Swelling of sporangium. $D_{100°C} = 15–25$ min	Food intoxication. Food implicated: home-canned foods; foods subjected to faulty commercial processing or temperature abuse; vegetables, particularly those in contact with soil.
	Non-proteolytic *Clostridium botulinum*	Catalase negative. Strict anaerobe. Produce botulinum neurotoxin types B, E, and F. Saccharolytic. Digest gelatin. Grows in the range of 3°C to 45°C (Optimum: 25 to 37°C). Minimum growth pH: > 5.0. No growth in 6.5% NaCl or at pH 8.5.	Oval, subterminal endospore. Swelling of sporangium. $D_{100°C} < 0.1$ min	Food intoxication. Food implicated: fermented marine products, dried fish, and vacuum-packed fish.

(Continued)

92

TABLE.6.1 (*Continued*)

Group	Bacteria	Cell characteristics	Spore characteristics	Importance in foods
Mesophilic anaerobes (*Continued*)	*Clostridium butyricum*	Catalase negative. Strict anaerobe. Some strains produce toxin type E. Growth temperature: Minimum: > 10 to 15°C; Optimum: 30°C to 37°C. Minimum growth pH: > 4.0–5.2. No growth in 6.5% NaCl.	Oval, central to subterminal. No swelling of sporangium. $D_{100°C} < 1$–5 min	Spoilage of canned tomatoes, peas, olives, cucumbers (swelling and butyric odor).
	Clostridium perfringens	Catalase negative. Air-tolerant anaerobe. Rapid growth. Stormy fermentation of lactose in milk. Grows in the range of 20°C to 50°C (Optimum: 43°C to 45°C). Growth pH: 5.5 to 8.5. No growth in 6.5% NaCl.	Large, oval, central or subterminal. Swelling of sporangium. No exosporium. $D_{100°C} = 0.3$ to 18 min	Foodborne non-invasive infection. Foods implicated: meat and poultry products. Outbreaks usually attributed to temperature abuse during food storage. Cause of several non-food transmitted diseases.
	Clostridium sporogenes	Catalase negative. Putrefactive anaerobe. Grows in the range of 25°C to 45°C (Optimum: 30°C to 40°C). Good growth in 100% CO_2. Growth in 6.5% NaCl, or at pH 8.5.	Oval, subterminal. Swelling of sporangium. $D_{100°C} = 80$–100 min	Spoilage of canned vegetables (swelling and putrid odor). Putrefaction of cured bacon. Off-odor of Swiss cheese.
	Clostridium tyrobutyricum	Catalase negative. Some strains produce toxin type E. Grows in the range of 25°C to 45°C (Optimum: 30°C to 37°C). No growth in 6.5% NaCl.	Oval, subterminal endospore. Swelling of sporangium.	Gas formation ("blowing") of cheeses during ripening.
Thermophilic anaerobes	*Clostridium thermosaccharolyticum*	Catalase negative. Optimum growth at 55°C to 62°C. Some strains grow slowly at 37°C. Few grow poorly at 30°C. No growth at 70°C	Oval/round, terminal. Swelling of sporangium. High D-value: $D_{100°C} = 400$ min	Spoilage of canned vegetables (swelling and sour, butyric odor).
	Desulfotomaculum nigrificans	Catalase negative. Obligate anaerobes. Minimum growth temperature: ≥ 43°C; optimum: 55°C. Growth pH: 5.6 to 78 (Optimum: 6.8–73)	Oval/round, terminal to subterminal. Slight swelling of sporangium. High D-value: $D_{100°C} < 480$ min, $D_{120°C} = 2.0$–3.0 min	Hydrogen sulfide spoilage of low-acid canned vegetables (blackened appearance and rotten egg odor). No swelling of cans.

Source: Table compiled by Y.-K. Chung.

spores of aerobic and anaerobic mesophilic sporeforming bacteria will be counted in selected foods. The exercise is designed to enumerate spore population after excluding cells.

Life-Cycle of a Sporeforming Bacterium

The complete life cycle of a sporeforming bacterium includes multiplication of the cell by binary fission (i.e., growth), cell sporulation, and spore-to-cell transformation (Fig. 6.1). Cell growth has been addressed in other chapters, and the following is a brief description of the two other stages, sporulation and spore-to-cell transformation; the latter is often referred to as germination.

Spore formation

Spore formation is frequently observed when cells reach high density as the population enters the stationary phase of growth. Nutritional and environmental stresses such as starvation are known triggers of sporulation. Response to these stresses leads to the initiation of sporulation. This process begins with the duplication of bacterial genome and formation two unequal cell divisions. The smaller division (forespore) is engulfed by the larger division (mother cell), which contributes to spore development. Subsequently, a think later of peptidoglycan (cortex) is added to the forespores. The spore formation is completed by the

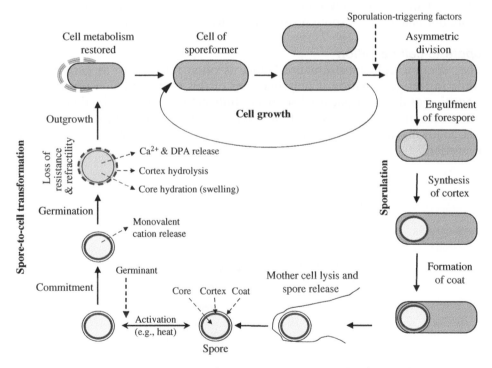

Figure 6.1 Life cycle of a spore-forming bacterium. Dark shades (e.g., cell cytoplasm) indicate phase-dark and lighter shades (spore core) indicate phase-bright when seen under phase-contrast microscope. DPA: Dipicolinic acid.

deposition of a proteinaceous layer called spore coat. The spore may remain in the remnants of mother cell or released as a free spore after the degradation of these remnants.

Spore structure and resistance

The structure of the bacterial spore contributes to its resistance. Bacterial spore has several distinct structures (Fig. 6.2). Many spore structures, including the coats and cortex, have no counterparts in the cell. Spores may have a delicate outer layer known as exosporium. Underlying the exosporium are spore coats that protect the spore cortex from attack by lytic enzymes. The coats also may provide an initial barrier to chemicals such as oxidizing agents. However, the coats seem to provide no protection to the spore against physical biocidal factors such as heat or irradiation. The coats contain mainly spore proteins. These proteins are rich in hydrophobic amino acids and contain a high level of cysteine. Hydrophobicity and disulfide bridges in these outer layers may contribute to the dormancy of the spore. Below the spore coat is the cortex, which is largely made of peptidoglycan, and thus it is structurally similar to the bacterial cell wall. However, when compared with cell wall peptidoglycan, cortex peptidoglycan is loosely cross-linked and lacks amino acid cross-links between adjacent peptide chains. The germ cell wall and inner membrane are structurally similar to their counterparts in the cell. The innermost region, the core, contains DNA, RNA, ribosomes, enzymes, small acid-soluble proteins, as well as dipicolinic acid (DPA) and divalent cations. Low water content in the core may play a major role in spore dormancy and in spore resistance to physical biocidal agents.

Spore-to-cell transition

The "spore-to-cell transformation" stage is often referred to as germination, while in fact germination is only one step of this transformation process. The transition of dormant bacterial spores to fully active forms can be divided into four sequential processes: activation, commitment, germination, and outgrowth. The following are the important events that take place during this cell-to-spore transition.

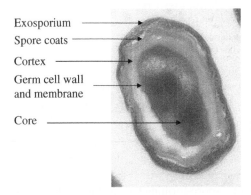

Exosporium
Spore coats
Cortex
Germ cell wall and membrane
Core

Figure 6.2 Transmission electron micrograph of *Bacillus subtilis* spores (Courtesy of M. Khadre).

Activation

Activation of spores or the breaking of dormancy is commonly achieved by heating spores in an aqueous medium. The temperature and duration of optimal heating vary among different species and even among different spore preparations of the same strain. Commonly used heat treatments fall within the range of 75°C to 80°C for 15–30 minutes.

Commitment

The spore ends its dormancy in response to the exposure to germinants. The germinant could be a small-molecule nutrient (e.g., L-alanine, L-valine, and L-asparagine), a non-nutrient chemical (e.g., calcium dipicolinate), or a physical factor, particularly ultra-high pressures in the range of 100 to 350 MPa. Nutrient germinants are known to initiate the germination by binding to germinant receptors. These receptors are protein complexes embedded in the spore inner membrane. This binding activates the receptors and starts the cascade of the germinative events. It is apparent that the germinants have to cross spore coats and cortex before they bind to and activate the deeply embedded germinant receptors. Shortly after the application of a germinant, the spore begins its transformation process by releasing monovalent cations such as H^+, K^+, and Na^+. Optimum temperature for this stage seems to match that for the growth of the corresponding cells. Once committed, the spore transformation process progresses irreversibly.

Germination step

The germination step is defined as a series of degradative events triggered by specific germinants that lead to the loss of typical spore characteristics (Figure 6.1). Hydrolysis of the cortex, core swelling, and release of dipicolinic acid and calcium ions are some of the changes detected during the germination step. Tangible changes observed during the germination step include the decrease in light absorbency of the spore suspension, loss of light reflactility, and loss of heat resistance. Note that dormant bacterial spores are typically refractile (shiny) when examined by phase-contrast microscope and thus described as "phase-bright." The refractility is lost during the germination step and thus the geminated spore appears phase-dark.

Outgrowth

Outgrowth refers to the emergence of bacterial cells from the germinating spore. These cells obviously are more sensitive than spores to heat and other deleterious factors.

OBJECTIVES

1. Determine the counts of spores in selected foods.
2. Compare the aerobic and anaerobic spore load in food.
3. Visually examine the morphology of bacterial endospores using a bright-field microscope, following staining with malachite green or other strains, and by phase-contrast microscope.

MEDIA AND INCUBATION CONDITIONS

Thioglycolate Agar

Thioglycolate medium is traditionally used to assess the oxygen requirement of microorganisms. Thioglycolate is a reducing agent that slows the diffusion of oxygen through the medium. In this exercise, it is used as an overlay on a growth medium in anaerobically incubated plates to ensure an anaerobic environment. Thioglycolate agar may be made of thioglycolate broth after supplementation with agar to the level of 1.5%. It is advisable that the medium used has been freshly prepared.

Tryptone Glucose Extract (TGE) Agar

Tryptone glucose extract (TGE) agar (also known as yeast dextrose agar) is a non-selective medium. This medium has been used historically for detection and enumeration of bacteria in dairy products. Because there are numerous amino acids and carbon/energy sources in this medium, TGE agar is also useful for promoting the germination of bacterial endospores.

Anaerobic Incubation

Anaerobic jars are airtight containers that have oxygen removed or reduced to a desired level. Jars of different sizes normally hold 10 to 20 Petri plates. Chemical reactions that consume oxygen are initiated using oxygen-scavenging packets (gas packs). The opened packet is placed in the jar, along with an anaerobic indicator, and the jar is sealed. Anaerobic jars will be used in this exercise. Unfortunately, this method sometimes results in conditions that are not truly anaerobic, leading to underestimation of strict anaerobic populations. Anaerobic chambers are more useful in ensuring adequate anaerobic conditions. These chambers also are particularly useful when a large number of plates need to be incubated anaerobically. In this case, the oxygen in the chamber is replaced with another gas (commonly nitrogen) or gas mixtures from an external source, e.g., a nitrogen or gas mixture tank. The jar-and-candle is the oldest method to achieve anaerobic conditions for incubated plates. A candle is lit next to the Petri plate stack and both are covered with a bell-shaped glass jar. The lit candle consumes the oxygen in the confined space under the jar and the flame is auto-extinguished. Anaerobic conditions are maintained during the incubation of the plates, provided that the setup is airtight.

MICROSCOPY METHODS

Brightfield and phase-contrast microscopes will be used to detect endospores and free spores of mesophilic sporeformers recovered from food samples. Both will also be used to estimate the degree of sporulation of these isolates.

Spore Staining and Examination by Brightfield Microscope

Malachite green reliably stains bacterial spores. Smears of bacterial colonies likely to contain spores are stained with malachite green while being steamed.

Steaming is necessary for malachite green penetration into spore coats. The slide is then rinsed thoroughly with water and counterstained with safranin to strain sporangia (singular sporangium; remnants of mother cell, surrounding the endospore) and cells. Malachite green binds poorly to cells, therefore a decolorizer is not needed in this staining procedure. Spores should appear green, whereas cells and sporangia appear pink to red. The ability to distinguish between cells and spores using a combination of malachite green and safranin is a type of differential staining.

Note: Malachite green is a suspected carcinogen and should be handled with care. Additionally, the stain is difficult to remove from skin; therefore, wearing disposable gloves should be emphasized during this procedure. It is recommended that the malachite green staining is completed in a chemical hood.

Alternatively, smears may be stained with Gram's crystal violet without application of steam. This causes cells to appear blue to violet in color. The sporangium picks up this stain, but the spore itself is not stained. The contrast between stained and unstained portions of the endospore makes it easy to recognize spores under a brightfield microscope.

Phase-Contrast Microscope

Phase-contrast microscopes make it possible to detect spores due to their inherently different refractive index when compared to cells and surroundings. No stain is necessary for visualization of spores using this microscope. To produce the desired optical effect, specific condenser settings and phase objectives are selected to produce a phase shift of approximately one-quarter of a wavelength of the illuminating light. Therefore, the analyst adjusts the condenser turret to match the label on the phase objective being used (For example on Nikon microscopes, phase 3 objective needs the condenser setting to be 3). The specimen will be visualized after preparing a simple wet mount on a microscope slide. Free spores and spores entrapped in parent cells (endospores) appear refractile (i.e., phase-bright), whereas cells and germinated spores appear phase-dark. Using the 60× objective lens may be sufficient for the examination of spore refractility by phase-contrast microscope, otherwise the 100× oil immersion lens will be used. Note that a phase-contrast microscope often has a brightfield setting so that it can be used as brightfield microscope.

PROCEDURE OVERVIEW

The procedure used in this exercise is illustrated in Figure 6.3. Briefly, the food is sampled, the sample is homogenized in a suitable diluent (e.g., peptone water), and dilutions are made. One-milliliter aliquots of selected dilutions are mixed with molten agar medium in test tubes and the mixtures are promptly heated in a water bath at 80°C for 20 min. The medium used in this laboratory exercise is TGE agar.

Note: Delay in heating may cause spores to begin germination and lose heat resistance; this leads to an underestimation of spore count in the sample.

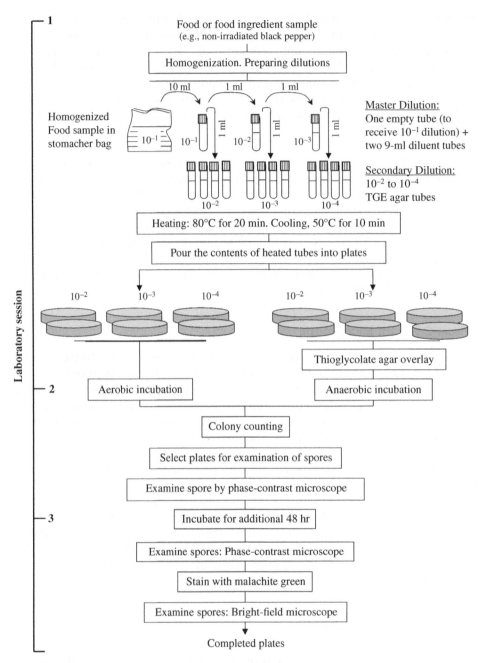

Figure 6.3 Outline of procedure for counting aerobic and anaerobic mesophilic sporeformers.

The heat treatment kills the cells in the sample homogenate but not the bacterial spores. It also activates highly dormant spores and enhances their ability to germinate. The heat treatment, therefore, selects for bacterial spores. The heated mixtures are cooled in a water bath set at ~50°C and then poured into Petri plates. The contents of one set of plates are overlaid with thioglycolate agar, incubated anaerobically, and used for determination of the anaerobic mesophilic spore count. The other set of plates is incubated aerobically to determine the aerobic mesophilic spore count. All plates are incubated at 35°C for 48 hours and colonies are counted. Selected plates are incubated for an additional 48 hours after counting to allow for further sporulation. Sporulation of cells in selected colonies will be observed using a phase-contrast microscope and a spore-staining technique.

Unprocessed spices (e.g., organic black pepper) and other dry products are analyzed in this exercise; these food ingredients may contain high spore population (e.g., 10^5 spores/g). Adjustments to the current procedure may be required if different spore loads are expected or if other foods or food ingredients are tested. When the tested product has low spore count, a greater portion of the sample is plated to increase the test's detection limit. In this case, multiple TGE agar tubes, e.g., five tubes, should be prepared for each dilution, and the colony counts on all plates receiving a given dilution are tallied and reported. When testing foods expected to contain relatively high spore loads, as done in this laboratory exercise, only duplicate TGE agar tubes are needed.

Organization

Food ingredients most suitable for counting mesophilic aerobic sporeformers include spices, flour, milk powder, and similar dry goods. Meat products, seafood, and vacuum-packaged mildly processed low-acid foods may be suitable for counting mesophilic anaerobic sporeformers. One food sample will be analyzed by a pair of students. Both members of the group will cooperate to analyze the sample.

SELECTED REFERENCES

Higgins, D. and Dworkin, J. (2012). Recent progress in *Bacillus subtilis* sporulation. *FEMS Microbiology Reviews* 36: 131–148.

Larkin, J.M. and Stokes, J.L. (1966). Isolation of psychrophilic species of *Bacillus*. *Journal of Bacteriology* 91: 1667–1671.

Stevenson, K.E. and Lembke, F. (2015). Mesophilic aerobic endo sporeforming bacilli. In: *Compendium of Methods for the Microbiological Examination of Foods*, 5e (ed. Y. Salfinger and M.L. Tortorello), 299–304. Washington, DC: American Public Health Association Press.

Tortorello, S. and Anderson, J.E. (2015). Mesophilic anaerobic sporeformers. In: *Compendium of Methods for the Microbiological Examination of Foods*, 5e (ed. Y. Salfinger and M.L. Tortorello), 305–318. W Washington, DC: American Public Health Association Press.

Warth, A.D. (1978). Relationship between the heat resistance of spores and the optimum and maximum growth temperatures of *Bacillus* species. *Journal of Bacteriology* 134: 699–705.

SESSION 1: SAMPLE PREPARATION, PLATING, AND INCUBATION

In this session, food samples will be diluted, transferred to the molten growth media (TGE), and heat-treated to select for spores. Dilutions will be pour-plated for aerobic and anaerobic incubation.

MATERIALS AND EQUIPMENT

Per Pair of Students

- Food sample
- One 90-ml bottle of 0.1% peptone water
- One empty sterile tube (to receive 10^{-1} dilution) and two 9-ml 0.1% peptone water tubes
- Twelve 9-mL molten TGE agar tubes in the 50°C water bath (2 incubation conditions × 3 dilutions × 2 replicates)
- Six 5-ml thioglycolate agar tubes for overlay (in the 50°C water bath)
- Twelve empty, sterile Petri plates
- One net-lined stomacher bag
- Micropipetters
- Sterile pipette tips (1000 µl)
- Other common laboratory supplies (e.g., permanent markers and disposable gloves)

Class-Shared

- Stomacher
- Anaerobic jars
- Gas packs for anaerobic jars
- Water bath set at 50°C
- Water bath set at 80°C
- Other common laboratory equipment (e.g., balance and incubator)

PROCEDURE

Notes:
- *The TGE agar tubes are held in the 50°C water bath to remain molten. Do not remove the tubes from the water bath until ready to inoculate them with diluted food homogenate. When asked, label these tubes while in the water bath and stick the labels on top of the tubes. When asked to remove these tubes from the water bath, work with no more than four tubes at a time to prevent premature solidification of the medium.*
- *In every water bath, the temperature should be monitored externally using a thermometer, and temperature uniformity at different locations in the bath should be ensured. A circulating water bath is ideal for maintaining temperature uniformity.*

1. **Food Homogenization**

   a. Weigh 10 g of food to be analyzed and transfer the sample into a net-lined stomacher bag, making sure that all food is contained on one side of the bag. Add the 90-ml diluent.

   b. Homogenize the sample-diluent mixture in a stomacher for 2 min; the homogenized sample represents the 10^{-1} dilution.

2. **Preparation of Master Dilution**

   a. Label the empty tube as 10^{-1} and the two diluent tubes as 10^{-2} and 10^{-3} dilutions.

   b. Using 10-ml sterile pipette, transfer 10 ml of stomacher bag contents into the empty tube (10^{-1}). The pipetted volume should be taken from the filtered compartment of the bag.

   c. Vortex the contents of the 10^{-1} tube and transfer 1 ml into the diluent tube labeled 10^{-2}.

   d. Vortex the contents of the 10^{-2} tube and transfer 1 ml into the diluent tube labeled 10^{-3}. Vortex the contents of the tube.

3. **Inoculation of TGE Agar Tubes (Secondary Dilutions):**

   a. Labeling: Label molten TGE agar tubes: Labeling should be done with consideration that four molten TGE agar tubes (two for aerobic and two for anaerobic incubation) are needed for each of the three dilutions, i.e., a total of 12 tubes. The first set of four tubes will be receiving 1 ml each from the 10^{-1} homogenized sample; therefore, these tubes will be labeled as dilution 10^{-2}. The second set of four tubes will be receiving 1 ml each from the 10^{-2} dilution tube (of the master dilution); therefore, these tubes will be labeled as dilution 10^{-3}. The third set of four tubes will be receiving 1 ml each from the 10^{-3} dilution tube (of the master dilution); therefore, these tubes will be labeled as dilution 10^{-4}. Label the top of tubes with labeling tape; label information should include dilution, tube number, and group identification. Placing labels on the lids of tubes prevents the tape from falling off in the water bath.

   b. Pipette 1 ml aliquots of the 10^{-1} dilution tube (of the master dilution), after vortexing tube contents, into the four TGE agar tubes labeled 10^{-2}. Mix the tube contents very gently by hand. Do not vortex these tubes to avoid trapping air in the molten agar or splashing the sample on tube walls; some areas of the wall may not receive the same heat treatment delivered to the submerged portion of the tube.

   Caution: Place inoculated tubes without delay into the 80°C water bath.

   c. Repeat the previous step with the 10^{-2} and 10^{-3} master dilutions to prepare the 10^{-3} and 10^{-4} TGE agar tubes; place inoculated tubes without delay into the 80°C water bath.

4. **Heat Treatment**

   a. Heat the agar tubes at 80°C for 20 min in the water bath.

   Notes:

   - *To minimize confusion, it is advisable that a student group keeps its prepared tubes together in one test-tube rack and that two or three groups may share a rack in the water bath.*

- *To ensure heating uniformity, a circulating water bath should be used and it should remain covered with its lid during heating.*
- *Water level in the water bath should be higher than the level of the mixture in TGE agar tubes.*
- *Water bath temperature should be checked (and adjusted if needed) periodically.*

  b. After the heating time is met, transfer the heated tubes to a water bath set at 50°C and allow to cool for ~10 min. To prevent solidification of inoculated agar in the tubes, leave the tubes in the 50°C water bath until ready to plate.

5. **Plating**

  a. Label Petri plates with the dilutions 10^{-2}–10^{-4} for aerobic incubation. A total of six plates (two for each dilution) are needed. Repeat this step for the anaerobic plates.

  b. Pour the contents of the heated food-agar media into the appropriately labeled Petri plates and swirl plates gently to cover the bottom of the dish. Hold plates at room temperature until agar solidifies.

  c. Plates prepared for anaerobic incubation receive an additional 5 mL overlay of thioglycolate agar. After the overlay solidifies, invert the plates and incubate in an anaerobic jar at 35°C for 48 hr.

  d. Invert the plates for aerobic incubation and incubate aerobically at 35°C for 48 hr.

  e. Make a note of the appearance of agar in plates. If food particles were incorporated into dilutions, observe their appearance in the agar plates. This observation may help in differentiating these particles from bacterial colonies when the plates are inspected in the subsequent session.

SESSION 2: SPORE ENUMERATION AND EXAMINING SPORULATION

In this session, colonies will be counted and the degree of sporulation in selected aerobic colonies will be estimated using phase-contrast microscope. At the end of the session, plates will be re-incubated to promote further sporulation before spore staining in Session 3.

MATERIALS AND EQUIPMENT

Per Pair of Students

- Incubated aerobic and anaerobic TGE agar plates
- Colony counter
- Phase-contrast microscope
- Other common laboratory supplies (e.g., permanent marker)

Class-Shared

- Common laboratory equipment (e.g., Bunsen burner and incubator)

PROCEDURE

1. **Enumeration of Aerobic and Anaerobic Spores**
   a. Count colonies on the aerobic and anaerobic plates using a colony coun-
      ter. Note that colonies will be suspended throughout the agar, not just
      on the surface. For safety, make sure the plates remain shut while count-
      ing colonies. If excessive water condensation is observed on the interior
      of agar plate lids, replace these using sterile Petri plate lids.
   b. Record the colony counts in Table 6.2.
   c. Calculate the aerobic and anaerobic spore populations in food using the
      counting rules provided in a Chapter 3. The following equation can be
      used in the calculation:

$$\text{Spore CFU} / \text{g} = \frac{\text{Average colony count from the duplicate plates}}{\text{Volume plated (ml)} \times \text{Dilution factor in plated tubes}}$$

   Important note: *To apply this equation, notice that the volume plated is*
   the whole content of a secondary dilution tube, which is 10 ml, and the
   dilution factor describes the content of this plated tube.

TABLE 6.2 *{add a descriptive title and footnotes for this data, including food sample*
used, media, recovery method, and what data were obtained}

Aerobic incubation			Anaerobic incubation		
Dilution Factor	Plate number	Colony count	Dilution Factor	Plate Number	Colony count
	1			1	
	2			2	
Aerobic spores CFU/g:			Anaerobic spores CFU/g:		

   d. Select two *aerobically* incubated plates with well-isolated colonies for microscopic examination of sporulation as described in the next section. Discard the remaining plates.

   e. Upon completion of microscopic evaluation (see next section), incubate these selected plates aerobically for an additional 48 hr at 30°C.

2. Spore Examination by Phase-Contrast Microscope (48-hr incubation)

   a. Select, mark, and label four colonies from the selected aerobic plates; using the inoculation loop, transfer part of each colony to droplets of water on microscope slides; two colonies per slide.

   Note: These colonies will be re-examined during Session 3 of this exercise; therefore, be careful to not use the entire colony mass for this slide preparation.

   b. Mix the cell mass with the water droplet and place carefully slide cover slip on top; do not dry or heat-fix. Make sure no air bubbles exist between the slide and the cover slip.

   c. Examine the slide using the phase-contrast microscope. Use the 40× or the 60× objective initially and, if needed, use the 100× objective with oil.

   Note: Each phase objective has a corresponding position on the rotating turret of the condenser. Make sure that the phase objective and the condenser turret are matched.

   d. Observe the presence of refractile free spores and endospores. For endospores, determine the placement of the spore in the mother cell. If possible, capture an image of the field using a camera-equipped microscope. A suitable camera-equipped handheld device (e.g., mobile phone) may be used to take the picture through the eye piece if this can be done safely and the permission of the instructor is granted.

   Note: Changing light intensity (that illuminates the slide) by changing the diaphragm and condenser settings may be needed to produce the desirable contrast between phase-bright spores and phase-dark cells.

   e. Roughly estimate the cell-to-spore ratio. This estimate will be compared with cell-to-spore ratio following additional incubation (Session 3).

   f. Draw and describe the observed spores; record these observations in Table 6.3.

TABLE 6.3 *{add a descriptive title and footnotes for this data, including food sample used, media, recovery method and what data were obtained}*

Microscope	Isolate #	Description	Drawing
Phase-contrast	1		
	2		
	3		
	4		
Brightfield (after malachite green staining)	1		
	2		
	3		
	4		

SESSION 3: MICROSCOPIC EXAMINATION OF SPORES

In this session, spores in selected colonies will be re-examined by phase-contrast microscope and also by brightfield microscope after malachite green staining.

MATERIALS AND EQUIPMENT

Per Student Pair

- TGE agar plates after extended incubation
- Phase-contrast microscope
- Microscope slides and cover slip
- Malachite green stain
- Safranin stain
- Other common laboratory supplies

Class-Shared

- Boiling water bath with a staining tray + slide rack mounted over the bath (or a suitable alternative setup).
- Other common laboratory equipment

PROCEDURE

1. **Spore Examination by Phase-Contrast Microscope (96-hr incubation)**
   a. Identify the four colonies that have been marked in Session 3 and transfer a loopful of each colony to a droplet of water on a microscope slide.
      Notes:
      - *The four colonies examined in session two ideally are re-examined in this session. If the remaining cell mass is not enough for this examination, choose other colonies with similar morphology.*
      - *Two colonies can be examined using one slide.*
   b. Prepare smears and examine by phase-contrast microscope as described in the previous session under the title "Spore Examination by Phase-Contrast Microscope (48-hr incubation)."
   c. Draw and describe the observed spores; record these observations in Table 6.3.
2. **Spore Staining and Examination by Brightfield Microscope**
   a. Prepare smears of the four colonies examined by the phase-contrast microscope.
      Notes:
      - *The four colonies examined in Session 2 ideally are stained in this session. If the remaining cell mass is not enough for this examination, choose other colonies with similar morphology.*
      - *Two colonies can be examined using one slide.*

b. Air-dry and heat-fix the smears as described in Chapter 4.

c. Place the slides containing the heat-fixed smears (with the smear-side up) on the staining rack of the water bath.

Caution: The slides on the water bath rack should be handled carefully to avoid any injuries or burns.

d. Add malachite green stain to the slides and stain for 15 min.

Notes:

• *The stain may be added periodically to the slide to compensate for water evaporation, which can adversely lead to smear drying and dye crystallization.*

• *Handle the slides using clothespins and wearing gloves to avoid the exposure to the stain.*

e. Using water, rinse the stain off the slide.

f. Add safranin to the slide and stain for 60 seconds (steaming is not needed in this step).

g. Rinse off the stain with water and blot dry using bibulous paper.

h. Examine the stained smears using a brightfield microscope with 100× oil immersion lens. The brightfield setting of the phase-contrast microscope can be used to examine the slide.

i. Notice the green-stained spores and the red-stained cells.

j. Observe the presence of spores within parent cells or in a free state.

k. Determine the spore placement in the parent cell, i.e., central, subterminal, or terminal, and observe whether the sporangium is swollen at the location of the spore.

l. Record these observations in Table 6.3.

QUESTIONS

1. Construct a table (Table I) for class data based on the aerobic spore counts compiled during this exercise. Group the data by food and calculate the average Log spores/g and standard deviation for spore population of each food. Include an appropriate descriptive title and footnotes. Mark your own group data in the table. Show an example of the calculations using your own data.

2. Table I. _____

Food	Group #	Spores/g	Log spores/g	Average Log spores/g	Standard deviation
Black pepper	1				
	13				

3. Use information in Table I to construct a bar-graph (Figure I) displaying aerobic spore populations for each food analyzed in the class. Include standard deviations as error bars. Add an appropriate descriptive title and footnotes. Label axes correctly.

   a. Briefly describe and compare results shown in Figure I in text form (one paragraph).

   b. Describe your microscopy observations for *two* of the sporeforming isolates. Include information from both phase-contrast microscopy and spore staining. Where were the spores placed in parent cells? What was the approximate cell-to-spore ratio?

   c. Why is spread-plating using a cell spreader not recommended for plating sporeformers?

CHAPTER 7

Pseudomonas SPECIES AND OTHER SPOILAGE PSYCHROTROPHS

SELECTION BY INCUBATION TEMPERATURE; DETECTION OF PROTEOLYTIC AND LIPOLYTIC ACTIVITY

INTRODUCTION

The production environment contributes significantly to food microbiota. Fresh produce is expected to harbor microorganisms originating from soil, dairy products routinely contain bacteria from bovine skin, and seafood carries cold-tolerant microbiota (psychrotrophs) originating from aquatic environments. As food travels through the supply chain, original microbiota may increase to a level that causes product spoilage. Food processing, by design, is expected to decrease or eliminate microbial contaminants, but many products are only mildly or minimally processed. Additionally, some food products are kept in the raw state throughout the supply chain and many of these are susceptible to rapid spoilage (i.e., perishable). Perishable raw and minimally processed products are commonly stored at refrigeration temperatures to decrease the risk of spoilage, thus increasing product shelf life. These low temperatures decrease the metabolic activity and multiplication of microbial contaminants. However, when the contaminants are mainly psychrotrophic microorganisms, usefulness of refrigeration is diminished. This chapter addresses aspects of food spoilage pertaining to psychrotrophic bacteria, particularly *Pseudomonas* spp.

Analytical Food Microbiology: A Laboratory Manual, Second Edition. Ahmed E. Yousef,
Joy G. Waite-Cusic, and Jennifer J. Perry.
© 2022 John Wiley & Sons, Inc. Published 2022 by John Wiley & Sons, Inc.

Microbial Spoilage of Food

The phenomenon of food spoilage is generally defined as the manifestation of observable negative sensory characteristics sufficient to render a food inedible in the opinion of the end consumer. Microbial spoilage results from the metabolic activity of large microbial populations in the food. It is generally presumed that the population of bacteria should reach 10^6–10^7 CFU/g before spoilage becomes apparent. At such a high population, the type and degree of spoilage depend on the metabolic activity of members of that population. The ability to secrete degradative enzymes, such as those causing proteolysis and lipolysis, is the primary cause of food spoilage by these bacteria. Signs of spoilage include off-odors, discoloration, slime production, or other unacceptable sensory attributes. The majority of spoilage microorganisms are not pathogenic, and these are believed to play a role in preventing the proliferation of pathogens by competitive inhibition. Many factors affect the rate of microbial spoilage, but temperature is a major contributor. It is generally understood that shelf life and storage temperature are negatively correlated; that is, the lower the storage temperature, the longer the shelf life.

Psychrotrophic Bacteria in Food

The average temperature of Earth's surface air is approximately 15°C, a temperature conducive to the proliferation of psychrophilic and psychrotrophic microorganisms. Based on this fact, it is presumed that psychrotrophic microorganisms are abundant in the environment. If the environmental contaminants in food are psychrotrophic microorganisms, the food is further enriched in these microorganisms during cold storage. As defined in the introduction to Part II of this book, psychrotrophs are typically capable of growing at refrigeration temperatures (0 to 7°C). Their optimum growth temperatures range between 20 and 30°C, and the maxima exceed 30°C. The overlap of this growth range with those for psychrophiles and mesophiles allows psychrotrophs to grow in environments where the temperature fluctuates.

Incubation of food at refrigeration temperatures can select for the psychrotrophs among product microbiota. This approach is used in this exercise by incubating agar plates, which have been inoculated with food homogenates, at a refrigeration temperature. A typical condition for this incubation is 7°C ± 1 for 7–10 days. Once psychrotrophic organisms are isolated from incubated plates, these isolates can be grown relatively rapidly at temperatures within the group's optimum range (e.g., 25°C for 48 hours).

Many psychrotrophic organisms can be isolated from food. Yeast and filamentous fungi are common psychrotrophs in food; these are described in a Chapter 9. Foodborne bacterial genera known to comprise psychrotrophic species include *Achromobacter*, *Acinetobacter*, *Aeromonas*, *Alcaligenes*, *Enterobacter*, *Flavobacterium*, *Moraxella*, *Pseudomonas*, *Serratia*, *Shewanella*, *Vibrio*, and *Yersinia* (Gram-negatives); and *Bacillus*, *Brochothrix*, *Clostridium*, *Carnobacterium*, *Lactobacillus*, *Lactococcus*, *Leuconostoc*, *Listeria*, *Microbacterium*, *Micrococcus* and *Paenibacillus* (Gram-positives). Among these genera, *Pseudomonas* is the most studied in relation to food spoilage.

Production of enzymes that cause lipolysis, proteolysis, or both is considered an indicator of an isolate's ability to spoil food. *Acinetobacter*, *Pseudomonas*, and *Microbacterium* are psychrotrophic genera known to be strong producers of these degradative enzymes. Minimal amounts of these enzymes are produced by the psychrotrophic lactic acid bacteria *Lactococcus, Lactobacillus,* and *Leuconostoc.* If these enzymes are heat-stable, they may survive food pasteurization and similar thermal treatments, and cause spoilage of processed foods. Although these thermal treatments may eliminate food's psychrotrophic cell populations, psychrotrophic sporeformers (e.g., some strains of *Bacillus cereus*) produce spores that are not eliminated by such treatments.

Pseudomonas spp.

Pseudomonas spp. are Gram-negative, non sporeforming, rod-shaped bacteria that are motile by one or several polar flagella. They are catalase positive and oxidase positive or negative. Members of this genus are aerobic with strict respiratory carbohydrate metabolism. In their respiration, they use oxygen as the terminal electron acceptor (aerobic respiration), but in some cases, nitrate can serve an alternate electron acceptor, allowing these microorganisms to grow and metabolize carbohydrates anaerobically (anaerobic respiration). *Pseudomonas* spp. are incapable of performing carbohydrate fermentation. Some *Pseudomonas* spp. have exceptional ability to utilize a large variety of molecules as carbon and energy sources and therefore have been exploited to detoxify a number of environmental pollutants. The majority of *Pseudomonas* spp. fail to grow at pH 4.5 or lower.

Pseudomonas spp. are widely distributed in nature and can be found in soil, water, and on plant surfaces. Some pseudomonads are plant, animal, and human pathogens. Many *Pseudomonas* spp. grow optimally at 28°C and the psychrotrophic members of the genus also grow considerably at 4°C. Most *Pseudomonas* spp. species grow without the need for organic growth factors. They generally have low nutritional needs and may grow in minimal media. *Pseudomonas* spp. are capable of forming biofilms on or around food processing equipment, thus they are persistent in the processing environment. The same traits that contribute to the formation of biofilms allow environmental pseudomonads to adhere to the surface of muscle foods at higher rates than other spoilage organisms.

The ability of *Pseudomonas* spp. to spoil refrigerated food is related not only to their ability to grow, but also to excrete appreciable amounts of extracellular degradative enzymes at refrigeration temperatures. Their lipolytic and proteolytic enzymes hydrolyze lipids and proteins of dairy and muscle foods, giving rise to off-odors. Production of pectic lyase enzymes cause soft rots of plant-derived foods. Species commonly isolated from food include *P. fluorescens, P. fragi, P. putida, P. lundensis,* and *P. brennerii.*

OBJECTIVES

1. To demonstrate the effect of storage temperature on the proliferation of psychrotrophic bacteria in seafood.
2. To determine the contribution of *Pseudomonas* spp. to the psychrotrophic population of the food.
3. To assay for isolates' proteolytic and lipolytic activity as traits that contribute to food spoilage.

MEDIA AND TESTS

Milk Agar

Milk agar is used in testing the ability of *Pseudomonas* isolates to produce proteolytic enzymes. The medium contains peptone and yeast extract, which serve as essential nutrients for the growth of many microorganisms. The medium also contains skim milk powder, which is added as a source of casein. Proteolytic *Pseudomonas* spp. produce enzymes that hydrolyze the casein, producing a zone of clearing around the colonies, whereas the rest of the medium appears turbid.
Note: Milk agar can be made from other commonly used media such as plate count agar after supplementation with skim milk powder.

Pseudomonas Isolation Agar (PIA)

PIA (*Pseudomonas* isolation agar) is a selective medium for isolation of pseudomonads. The medium is somewhat minimal, using a carbon source (glycerol) not commonly utilized by bacteria. The medium contains peptone, which serves as a nitrogen source. Selectivity is afforded by inclusion of Irgasan, a broad-spectrum antimicrobial agent that is not active against *Pseudomonas* spp. In addition, the inclusion of magnesium chloride and potassium sulfate in the medium enhances the production of pyocyanin. Pyocyanin is a blue to blue-green pigment produced by *Pseudomonas aeruginosa* that usually diffuses into the growth medium. Pyoveridin (historically, but incorrectly referred to as fluorescein) is a yellow to yellow-green, fluorescent pigment that may be produced by many *Pseudomonas* spp. Other pigments have been described in the literature. Some of the pigments produced by *Pseudomonas* spp. are known to act as siderophores (iron transporters). Pigment production and color may be used to differentiate pseudomonads phenotypically.

Tributyrin Agar

Tributyrin agar is used to test for the ability of *Pseudomonas* isolates to produce lipolytic enzymes. The medium contains peptone and yeast extract that serve as nutrient sources. The medium also contains 1% tributyrin, a simple liquid triglyceride that can be found naturally in oils and fats. Addition of tributyrin gives the medium a uniform turbid appearance. Hydrolysis of tirbutyrin by lipolytic bacteria results in clearing around the colonies.
Note: Tributyrin agar can be made from other commonly used media such as plate count agar, to which 10 ml of tributyrin is added and mixed in a liter of dissolved medium before autoclaving.

Tryptic Soy Agar (TSA)

This is a rich general-purpose medium that will be used to enumerate aerobic psychrotrophs. Description of this medium has been included in Chapter 5.

PROCEDURE OVERVIEW

Two seafood samples will be analyzed, representing one that has been stored frozen for seven days and one that has been refrigerated for the same period of time. Whole fish or fish fillet may be tested. During the analysis, counts of psychotrophic microbiota and *Pseudomonas* spp. will be determined. Additionally, selected *Pseudomonas* isolates will be tested for their ability to produce proteolytic and lipolytic enzymes; these enzymes are major causes of food spoilage by these bacteria.

The exercise requires four laboratory sessions to complete (Figure 7.1). In the first session, analytical samples will be taken, sample homogenate will be

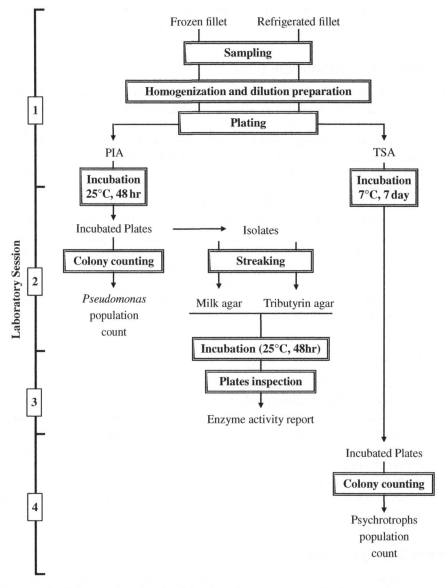

Figure 7.1 Scheme of testing food for *Pseudomonas* spp. and other psychrotrophs.

prepared and dilutions made, and appropriate dilutions will be plated on non-selective (TSA) and selective (PIA) media. Selectivity for psychrotrophs on TSA will be achieved by an extended incubation under refrigeration conditions. During the second session, incubated PIA plates will be inspected and populations of *Pseudomonas* spp. will be determined. Additionally, selected *Pseudomonas* isolates, from the two sample types, will be streaked on milk agar and tributyrin agar. After incubation of these plates, the proteolytic and lipolytic activity of the isolates will be observed during the third session. Finally, the fourth session includes counting the colonies of psychrotrophs that grew on TSA plates during incubation.

Organization

Students will work in pairs to analyze two samples. Therefore, each student will be responsible for sampling, diluting, and plating one sample. The two members of the groups will work cooperatively on determining the proteolytic and lipolytic activities of selected *Pseudomonas* isolates.

SELECTED REFERENCES

Ercolini, D., Russo, F., Nasi, A., et al. (2009). Mesophilic and psychrotrophic bacteria from meat and their spoilage potential *in vitro* and in beef. *Applied and Environmental Microbiology* 75: 1990–2001.

Moyer, C.L., Collins, R.E., and Morita, R.Y. (2017). Psychrophiles and psychrotrophs. *Reference Module in Life Sciences*. Elsevier Inc. doi.org/10.1016/B978-0-12-809633-8.02282-2

Rawat, R. (2015). Food spoilage: microorganisms and their prevention. *Asian Journal of Plant and Science Research* 5: 47–56.

Vasavada, P.C. and Critzer, F.J. (2015). Psychrotrophic microorganisms. In: *Compendium of Methods for the Microbiological Examination of Foods*, 5e. (ed. Y. Salfinger and M.L. Tortorello), 175–189. Washington, DC: American Public Health Association Press.

SESSION 1: SAMPLING AND PLATING

Raw whole fish or fish fillets that have been stored frozen or refrigerated for one week prior to this session will be analyzed to determine the counts of psychotrophic microbiota and *Pseudomonas* spp. In this session, analytical samples will be taken, sample homogenate will be prepared and dilutions made, and appropriate dilutions will be plated on a non-selective (TSA) and selective (PIA) media. In groups of two, one student will analyze the refrigerated and the other will analyze the frozen sample.

MATERIALS

Per Student

- Seafood sample:
  - One fish fillet (or whole fish) that has been refrigerated for 7 days following purchase
  - One fish fillet (or whole fish) that has been frozen for 7 days following purchase
- Diluent bottle: 90-ml bottle of peptone water
- Diluent tubes: Three 9-ml peptone water tubes
- Six TSA plates (for plating 3 dilutions in duplicates)
- Seven PIA plates (6 for plating 3 dilutions in duplicates, and 1 for positive control)
- Culture of *Pseudomonas* sp. (positive control)
- Other common laboratory supplies

Class-Shared

- Common laboratory equipment (stomachers, incubators, etc.)

PROCEDURE

Dilutions

1. Prepare the whole fish or fish fillet for sampling to ensure that a representative sample is analyzed (see sampling details in Chapter 2).

Caution: When whole fish is tested, care should be taken to avoid including bones in the analytical sample. Fish bones can puncture stomacher bags and cause sample leakage. Additionally, sharp bones should be handled carefully to avoid analyst's injury.

2. Weigh 10 g of seafood directly into a stomacher bag, then add 90 ml peptone water.
3. Stomach for 2 min.
4. Prepare 10^{-2}, 10^{-3}, and 10^{-4} dilutions using the peptone water diluent tubes (see dilution procedure in Chapter 4).

Plating

1. Properly label six plates each of PIA and TSA per sample.
2. Spread 0.1 ml of the 10^{-2}, 10^{-3}, and 10^{-4} dilution tubes onto appropriate duplicate plates of PIA and TSA.
3. Streak the positive-control culture for isolation (3-phase streaking) onto one labeled PIA plate (see streaking procedure in Chapter 4).
4. Incubate the PIA plates aerobically at 25°C for 48 hours.
5. Incubate the TSA plates aerobically at 7°C for 7–10 days.

SESSION 2: ENUMERATION OF *Pseudomonas* SPECIES AND TESTING FOR ENZYMATIC ACTIVITY

During this session, incubated PIA plates will be inspected, population of *Pseudomonas* spp. will be determined, and results of the two samples will be shared by the two members of each group. Additionally, selected *Pseudomonas* isolates, from the two food samples, will be streaked on milk agar and tributyrin agar to demonstrate the proteolytic and lipolytic activity of these isolates.

MATERIALS

Per Student

- Incubated PIA plates (Session 1)
- Goggles for eye protection against ultraviolet radiation
- Plate of milk agar medium
- Plate of tributyrin agar medium
- Other common laboratory supplies

Class-Shared

- Ultraviolet (UV) lamps
- Common laboratory equipment (e.g., colony counters)

PROCEDURE

Inspection and Enumeration of *Pseudomonas* spp.

1. Inspect the PIA plates visually and observe the similarity or variation in the morphology of colonies.
2. Count *Pseudomonas* colonies. Assume that all colonies growing on the plate are indeed *Pseudomonas* spp. Calculate the population count of *Pseudomonas* spp. in the food analyzed. Record the colony and population counts in Table 7.1.

TABLE 7.1 *[Add descriptive title]*

Sample storage temperature	Number of colonies on PIA[a]			*Pseudomonas* spp. (CFU[b]/g)
	Dilution factor	Plate 1	Plate 2	

[a] Pseudomonas isolation agar
[b] Colony forming units

3. Take the plates into a darkened room to observe them using ultraviolet light. UV will cause the pigments (e.g., pyocyanin) to glow with characteristic color. Do this under the supervision of a laboratory instructor. Eye protection (UV-absorbing goggles or eyeglasses) and gloves are required. Describe the appearance and relative abundance of those colonies that fluoresced.

Testing Enzymatic Activity of Isolates

1. Label the milk agar plate and draw two perpendicular lines on the plate base to divide the area of each plate into four quadrants. Repeat using the tributyrin agar plate.
2. Select four well-isolated colonies of *Pseudomonas* spp. from the incubated PIA plates; number these colonies.
3. Streak the colonies onto the quadrants of milk agar; one colony per quadrant. Repeat using the same four colonies to streak the quadrants of the tributyrin agar plates.
4. Incubate the plates at 25°C for 48 hours.

SESSION 3: DETECTING ENZYMATIC ACTIVITY

In this session, incubated milk agar and tributyrin agar plates will be inspected for signs of proteolytic and lipolytic activities.

MATERIALS

Per Pair of Students

- Incubated milk agar and tributyrin agar plates (Session 2)

PROCEDURE

1. Inspect the milk agar plate for clearing around the growth of streaked isolates. Presence of clearing indicates isolate's ability to produce proteolytic enzymes.
2. Inspect the tributyrin agar plate for clearing around the streaks. Presence of clearing indicates isolate's ability to produce lipolytic enzymes.
3. Determine the number of colonies (out of 4 total) producing proteolytic activity, lipolytic activity, or both.
4. Record the results in Table 7.2.

TABLE 7.2 Proteolytic and lipolytic activity of *Pseudomonas* spp. isolated from refrigerated and frozen fish.

Isolate #	Sources	Proteolytic	Lipolytic	Remarks

SESSION 4: ENUMERATION OF AEROBIC PSYCHROTROPHS

In Session 4, TSA plates that were incubated at 7°C for 7–10 days will be inspected and the psychrotrophic population counted.

MATERIALS

Per Student

- Incubated TSA plates (Session 1)

Class-Shared

- Colony counters

PROCEDURE

Psychrotrophic Count

1. Inspect TSA plates to determine the dilution that has produced the most countable numbers.
2. Count all colonies and record counts in Table 7.3.
3. Calculate the aerobic psychrotroph population (CFU/g) and record results in Table 7.3.

QUESTIONS

1. Compare the counts of the *Pseudomonas* spp. obtained from the two fish samples analyzed by your group. What do you conclude from this comparison?
2. Compare the counts of the *Pseudomonas* spp. and aerobic psychrotrophs obtained from each of the two fish samples analyzed by your group. Are there discrepancies between the counts of these two populations? If so, provide an explanation.

TABLE 7.3 *{Add descriptive title}*

Sample storage temperature	Number of colonies on TSA[a]			Aerobic psychrotrophs (CFU[b]/g)
	Dilution factor	Plate 1	Plate 2	

[a] Tryptic soy agar
[b] Colony forming units

3. Did you notice any signs of spoilage for the two samples analyzed by your group? If so, did the *Pseudomonas* and psychrotrophic counts support this observation?

4. If you were asked to run additional experiments to gather more information about the isolates you obtained, what experiments would you suggest? Why?

5. There is an interesting saying: "*Milk does not really spoil, it just changes from one product to another.*" Based on your knowledge in microbiology and food science, critique this statement. If there are any merits to this saying, would it also be applicable to seafood?

CHAPTER 8

DETECTION AND ENUMERATION OF *Enterobacteriaceae* IN FOOD

USE OF SELECTIVE-DIFFERENTIAL MEDIA. MOST PROBABLE NUMBER TECHNIQUE.

INTRODUCTION

Enterobacteriaceae is a family of Gram-negative, rod-shaped, motile bacteria. It belongs to the order *Enterobacteriales,* phylum *Proteobacteria*, and the class *Gammaproteobacteria* (see the introduction to Part II: Food Microbiota). Members of this family have both respiratory and fermentative metabolism and thus are considered facultative anaerobic bacteria. Most members are catalase-positive, oxidase-negative, and capable of reducing nitrate to nitrite. Members of *Enterobacteriaceae* are substantially heterogeneous in ecology and host range; some are human, animal, or plant pathogens. As the family name implies, these organisms are common inhabitants of the intestinal tract. In addition, *Enterobacteriaceae* are found readily in soil, water, and vegetation. These bacteria are often found in raw or even processed foods and their presence is indicative of poor satiation. Some members of *Enterobacteriaceae* cause food spoilage and other are responsible for food-transmitted diseases; hence, food processors strive to minimize the risk of contamination of processed products with these bacteria. Genera of *Enterobacteriaceae* that are of interest to food processors include *Citrobacter, Enterobacter, Erwinia, Escherichia, Hafnia, Klebsiella, Proteus, Salmonella, Shigella,* and *Yersinia.* Additional information on *Escherichia coli* and *Salmonella* are found in chapters 12 and 13 of this book.

Detection and selective enumeration of *Enterobacteriaceae* are based on their ability to grow in the presence of bile salt and crystal violet and to ferment glucose, producing acidic end products. These conditions are well-met in violet-red bile glucose (VRBG) agar, a selective-differential medium. Bile salts and crystal violet

Analytical Food Microbiology: A Laboratory Manual, Second Edition. Ahmed E. Yousef, Joy G. Waite-Cusic, and Jennifer J. Perry.
© 2022 John Wiley & Sons, Inc. Published 2022 by John Wiley & Sons, Inc.

in this medium inhibit Gram-positive cocci such as *Staphylococcus* spp. and *Enterococcus* spp. *Enterobacteriaceae* ferment glucose in VRBG agar and the resulting acids are detected by the pH indicator, neutral red, thus colonies appear red with red-purple halos. Selectivity for *Enterobacteriaceae* on this medium is improved further by providing semi-anaerobic conditions (using agar overlay) or by anaerobic incubation; this excludes non-fermentative Gram-negative bacteria such as *Pseudomonas* spp. The colonies on VRBG agar need to be confirmed as members of *Enterobacteriaceae* by conducting an oxidase test. Negative oxidase test results provide additional evidence that these colonies belong to the *Enterobacteriaceae*.

Despite the numerous advantages of VRBG agar, this medium can't be used directly to enumerate the small *Enterobacteriaceae* populations ($< 10^2$ CFU/g) commonly encountered in food. Additionally, bacterial cells that have been stressed or injured during food production or processing may not be recovered adequately on this strongly selective medium. To address these issues, an enrichment can be included at the start of the analysis to help in resuscitating injured cells and increase the ability of VRBG agar to detect *Enterobacteriaceae*. However, such an enrichment would not allow the analyst to quantify *Enterobacteriaceae* population in food. The approach used in the current exercise allows for detecting and enumerating small *Enterobacteriaceae* populations in food while overcoming some of the drawbacks just described. *Enterobacteriaceae* enrichment broth (EEB) will be used to enrich and enumerate *Enterobacteriaceae* simultaneously; enumeration will be accomplished using the most-probable number (MPN) technique. This will be followed by confirmation of the identity of bacteria in positive MPN tubes using VRBG agar and similar media. Note that EEB is a highly buffered medium. This buffering minimizes the deleterious effect of acids which cause growth inhibition at the early stage and auto-inhibition at the end of the enrichment.

Enterobacteriaceae in Food

Enterobacteriaceae is of great importance in food since the family includes spoilage and pathogenic organisms. Members of *Enterobacteriaceae* are commonly found in food production environments such as fields, fishing water, barns, and milking parlors. These organisms also may be found in processing environments (e.g., slaughterhouses), transportation vehicles, storage facilities, and retail outlets. The presence of *Enterobacteriaceae* in ready-to-eat food is often indicative of contamination during production, processing, or handling, hence the poor microbiological quality of the product.

OBJECTIVES

1. Detect and enumerate *Enterobacteriaceae* in foods that may contain a small population of these bacteria.
2. Apply the most-probable number technique as an approach for enumerating small microbial populations.
3. Develop familiarity with selective-differential media.

MEDIA AND TESTS

Enterobacteriaceae Enrichment Broth (EEB)

This is a highly-buffered selective medium used to recover *Enterobacteriaceae* from food. The medium contains brilliant green and Oxgall (bile salts), which are selective against non-enterics. Glucose is included as the sole carbohydrate source. Incubation conditions may vary depending on selection criteria. In this laboratory exercise, the inoculated EEB will be incubated at 35°C for 18–24 hr. Growth of foodborne microorganisms in this medium will be judged by the turbidity of the tube contents; these will be considered EEB-positive tubes. Streaking the contents of EEB-positive tubes on the selective differential medium, VRBG agar, will help in concluding that the growth is presumptive for *Enterobacteriaceae*.

Glucose Agar

This medium is used for testing glucose fermentation by *Enterobacteriaceae*, and it is prepared as agar tubes suitable for stab inoculation. The medium contains tryptone and yeast extract as growth-supporting nutrients. Glucose is the sole carbohydrate source and bromothymol blue serves as a pH indicator. Glucose metabolism to acidic end products will cause the pH indicator to change color from blue-green to yellow. Glucose agar tubes are stabbed once in the center using inocula from turbid EEB tubes. Inoculated glucose agar tubes will be overlaid with sterile mineral oil at a depth of 0.5–1.0 cm to create anaerobic conditions for determining the fermentation of glucose. Inoculated tubes will be incubated at 35°C for 18–24 hr. All *Enterobacteriaceae* should produce a positive reaction with color change.

Tryptic Soy Agar (TSA)

This is a rich, general-purpose medium that will be used to provide colonies for testing the Gram reaction and oxidase production. Description of this medium was provided in Chapter 5.

Violet-Red Bile Glucose (VRBG) Agar

This is a selective-differential medium that is useful in the presumptive stage of *Enterobacteriaceae* detection or enumeration in food samples. Bile salts and crystal violet in the medium select against Gram-positive and non-enteric organisms. Organisms that are capable of fermenting glucose to produce acid end products cause a color change of the pH indicator (neutral red). Colonies capable of glucose fermentation will appear red-purple in color. In addition, *Enterobacteriaceae* will precipitate bile acids from bile salts in the medium. This results in an opaque, red-purple halo of precipitate surrounding the colonies. The inoculated VRBG agar plates will be incubated at 35°C for 18–24 hr.

Oxidase Test

The oxidase test is used to determine whether a particular organism possesses the enzyme cytochrome c oxidase. Cytochrome c is an electron carrier involved in the electron transport chain in some bacteria. Cytochrome c oxidase is responsible for the oxidation of cytochrome c in the electron transport chain to aid in the reduction of oxygen, the final electron acceptor. To perform this test, a portion of a colony is picked from an agar plate using a wooden stick or toothpick (metal implements may lead to erroneous results) and transferred to a piece of filter paper. It is often necessary to repeat the transfer until a visible portion of the colony is placed on the filter paper. A single drop of oxidase reagent (tetramethyl-p-phenylenediamine, TMPD) is added to the smear. The oxidase reagent should be freshly prepared and appear clear and colorless; if it is purple or blue, it makes results difficult to interpret. If the smear changes color from clear to bluish-purple within 20 seconds, the colony is considered oxidase positive. Changes occurring after 30 seconds are not valid. Commercially available oxidase test cards and strips, which only require the transfer of a portion of the colony, without the addition of reagent, can be used instead of the traditional test method just described. All members of *Enterobacteriaceae* are oxidase negative. This test is particularly useful in differentiating between members of *Enterobacteriaceae* and *Pseudomonadaceae*; the latter are often oxidase positive.

PROCEDURE OVERVIEW

An overview of the procedure used in this exercise is shown in Figure 8.1. Raw, minimally processed, and ready-to-eat food will be analyzed to detect and enumerate *Enterobacteriaceae*. The MPN technique, which is described in more detail in Chapter 3, will be used for the enumeration; this will be followed by a number of confirmatory tests. Briefly, enumeration by MPN technique consists of preparing dilutions of the food homogenate and transferring portions of three consecutive dilutions to nine tubes (three tubes per dilution) of a selective enrichment medium; EEB will be used in this exercise. Although EEB favors *Enterobacteriaceae,* other bacteria may grow during incubation of these enrichment tubes.

Tubes of incubated EEB showing turbidity will be streaked on VRBG agar before presumed to contain *Enterobacteriaceae.* The VRBG agar plates will be incubated at 35°C for 18–24 hours and colonies that are red-purple with a red-purple halo will be considered presumptive *Enterobacteriaceae.* The tubes from which these presumptive colonies are obtained will be tallied and the "presumptive MPN/g food" will be deduced from the MPN table (See Chapter 3). Note that calculation of MPN is based on sample weight dispensed into each tube, and calculation adjustment may be needed depending on the dilution scheme utilized in the analysis.

Enterobacteriaceae-presumptive isolates should be confirmed as *Enterobacteriaceae.* This will be accomplished by (i) confirming acid production from glucose fermentation in glucose agar stabs; (ii) demonstrating the absence of cytochrome c oxidase

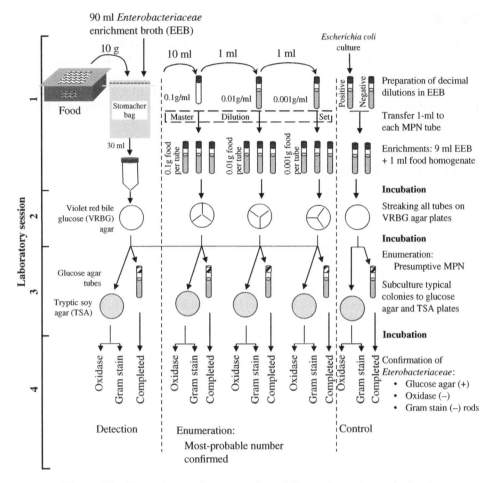

Figure 8.1   Detection and enumeration of *Enterobacteriaceae* in food.

using the oxidase test; and (iii) examining cellular morphology to confirm that isolates are Gram-negative rods. Enrichment tubes confirmed to contain
Enterobacteriaceae will be tallied and the confirmed MPN of these bacteria in food
will be determined using the MPN table.

Based on this series of tests, Gram-negative, oxidase-negative, glucose-
fermenting, rod-shaped bacteria that grow in the presence of bile salts and crystal violet are confirmed as *Enterobacteriaceae*.

*Note: If experimental procedure cannot be performed on a sequential daily basis,
the inoculated media should be held refrigerated and then moved to the incubator
only 18–24 hours before the subsequent laboratory session. On the contrary, it is
not recommended that inoculated media are incubated first then held refrigerated
until examined in the subsequent laboratory session as this approach may produce
inconsistent results.*

Organization

Students will work in groups of two. Each pair of students will test one food sample. Food samples to be analyzed may include various fresh produce (e.g., mushrooms, sprouts, and parsley), raw meat (e.g., raw sausages), and raw poultry products (e.g., chicken parts). It is advisable that students of a given group analyze a food product different to what they already analyzed in previous laboratory exercises.

SELECTED REFERENCES

Health Protection Agency. (2005). Detection and enumeration of *Entrobacteriaceae*. National Public Health of Wales, Reference no: F 18i1

ISO. (2017). Microbiology of the food chain-horizontal method for the detection and enumeration of *Enterobacteriaceae* – Part 1: Detection of *Enterobacteriaceae*. The International Organization for Standardization (ISO) 21528:2017.

Kornacki, J.L., Gurtler, J.B., and Stawick, B.A. (2015). *Enterobacteriaceae*, coliforms, and *Escherichia coli* as quality and safety indicators, In: *Compendium of Methods for the Microbiological Examination of* Foods, 5e. (ed. Y. Salfinger and M.L. Tortorella), 103–120. Washington DC: APHA Press.

Paulsen, P., Borgetti, C., Schopf, E., and Smulders, F.J.M. (2008). Enumeration of *Enterobacteriaceae* in various foods with a new automated most-probable-number method compared with Petrifilm and international organization for standardization procedures. *Journal of Food Protection* 71: 376–379.

SESSION 1: PREPARATION OF MOST PROBABLE NUMBER TUBES

During this session, food will be sampled, and the sample will be homogenized. A master dilution set will be prepared and used to inoculate a 3×3 MPN tube set.

MATERIALS AND EQUIPMENT

Per Pair of Students

- Sample of food
- One bottle containing 90 ml of EEB
- One empty sterile test tube (for receiving homogenized sample; the 10^{-1} tube)
- One empty sterile 50-ml tube (for receiving 30 ml of homogenized sample)
- Thirteen 9 ml EEB tubes (Two for making dilutions, nine for MPN tubes, and two for controls)
- One 10-ml sterile pipette and the corresponding pipette-aid
- Positive control: overnight culture of *Escherichia coli*
- Micropipetters (200- and 1000-μl capacity)
- Sterile pipette tips (for the 200- and 1000-μl micropipetters)
- Other common laboratory supplies (e.g., disposable gloves and permanent marker)

Class-Shared

- Sample preparation utensils (Cutting brads, knives, pair of tongs, alcohol bottles, Styrofoam clam-shell containers or reusable sanitizable containers, etc.)
- Common laboratory equipment

PROCEDURE

Dilutions

1. Prepare the food for sampling to ensure that a representative sample is analyzed (see sampling details in Chapter 2).
2. Weigh 10 g of food (analytical sample) directly into a stomacher bag, then add 90 ml EEB as diluent.
3. Homogenize food-diluent mixture in the stomacher for 2 min.
4. Prepare the 10^{-1} tube by transferring 10 ml of the homogenized sample into an empty sterile test tube.
5. Prepare 10^{-2} and 10^{-3} dilution tubes using two 9-ml EEB tubes. The three dilution tubes (10^{-1}, 10^{-2}, and 10^{-3}) will be referred to as the **master dilution** set. The amounts of food in these tubes are **1, 0.1, and 0.01 g/tube**, respectively.

Inoculating EEB – MPN Method

1. **Control**
   a. Label one EEB tube as a positive control. Inoculate the tube with 0.1 ml of overnight culture *E. coli*.
   b. Label one additional tube as a negative control; do not inoculate this tube, but incubate as normal.
   c. Vortex tubes and incubate at 35°C for 48 hr.

2. **Detection Food Sample**
   a. Label 50-ml tube as a "detection sample."
   b. Using the 10-ml disposable serological pipette, transfer 30 ml of the homogenized food (from the stomacher bag) into the tube.
   c. Incubate the tube at 35°C for 48 hr.

3. **Enumeration Food Sample**
   a. **Labeling:** Label nine tubes of EEB, three corresponding to each dilution of the master dilution set. The first three tubes will be receiving 1 ml each, from the 10^{-1} master dilution tube; these will be labeled as 0.1 g. The next three EEB tubes will receive 1 ml each, from the 10^{-2} master dilution tube; these will be labeled as 0.01 g. The last three tubes will receive 1 ml each, from the 10^{-3} master dilution tube; therefore, these will be labeled as 0.001 g.
   b. Transfer 1-ml aliquot from the 10^{-1} master dilution tube into each of the three tubes labeled 0.1 g.
   c. Transfer 1-ml aliquot from the 10^{-2} master dilution tube into each of the three tubes labeled 0.01 g.
   d. Transfer 1-ml aliquot from the 10^{-3} master dilution tube into each of the three tubes labeled 0.001 g.
   e. Vortex the nine inoculated EEB tubes and incubate at 35°C for 48 hr.

SESSION 2: PRESUMPTIVE MPN DETERMINATION

In this session, MPN tubes will be tested to determine if they are presumptively positive for *Enterobacteriaceae*.

MATERIALS AND EQUIPMENT

Per Pair of Students

- Incubated EEB tubes (from previous session)
- Incubated "detection sample" tube
- Five VRBG agar plates (Three for the nine MPN tubes, one for the detection sample, and one for the positive control)
- Other common laboratory supplies (e.g., inoculation loop)

Class-Shared

- Common laboratory equipment (e.g., incubator)

PROCEDURE

1. **Control**
   a. Examine the EEB tube inoculated with *E. coli* (positive control). The tube should be obviously turbid. Compare this tube to the uninoculated (negative control) tube. Record your observations in Table 8.1.
   b. Label one VRBG agar plate with information about the positive control culture.
   c. Streak for isolation (three-phase streaking) from the positive control tube onto the VRBG agar plate.
   d. Incubate the plate at 35°C for 18–24 hr.
2. **Detection Food Sample**
   a. Examine the tube containing the detection sample. The tube should be obviously turbid. Compare this tube to the uninoculated (negative control) tube. Record your observations in Table 8.1.
   b. Label one VRBG agar plate with information about the detection sample.
   c. Streak for isolation (three-phase streaking; see Chapter 4) from the incubated detection sample tube onto the VRBG agar plate. This step is completed regardless the presence or absence of turbidity in the incubated tube.
   d. Incubate the plate at 35°C for 18–24 hr.
3. **Enumeration Food Sample**
   a. Examine the incubated EEB tubes containing food. Identify tubes showing turbidity, which may be caused by the growth of *Enterobacteriaceae* originally present in food. Record your observations in Table 8.1. An example of these incubated EEB tubes is shown in Figure 8.2.

TABLE 8.1 *[add a descriptive title and footnotes for this data, including food sample used, media, recovery method, and what data were obtained]*

Tube (dilution / tube #)	Tube turbidity (+/-)	Isolate morphology on VRBG agar	Presumptive *Enterobcateriaceae* (+/-)	Gram reaction and cellular morphology	Glucose fermentation (+/-)	Oxidase (+/-)	Confirmed *Enterobcateriaceae* (+/-)
0.1g/1							
0.1g/2							
0.1g/3							
0.01g/1							
0.01g/2							
0.01g/3							
0.001g/1							
0.001g/2							
0.001g/3							
Control (*Enterobacteriacea-positive*)							
Detection sample							

[Add any relevant footnotes]

130

Figure 8.2 Most-probable number (MPN) *Enterobacteriaceae* enrichment broth (EEB) tubes that have been inoculated with dilutions of homogenized food and incubated before scored for presumptive MPN. The tubes from left to right correspond to following amounts of food: 0.1 g (tubes 1–3), 0.01 g (tubes 4–6), 0.001 g (tubes 7–9), and 0 g (tube 10, which serves as negative control). Note that all nine tubes were considered positive even though the degree of turbidity or color change was different.

    b. Label three VRBG agar plate with information about the enumeration sample. Draw lines that divide each plate into three equal segments. Number these segments to correspond to the numbers on the nine MPN tubes.

    c. Apply three-phase streaking to the contents of all MPN tubes, whether they are turbid or not, onto the divided VRBG agar plates; three tubes/ plate.

    d. Incubate inoculated VRBG agar plates at 35°C for 18–24 hr.

SESSION 3: PRESUMPTIVE MPN DETERMINATION AND CONFIRMATION

In this session, presumptive MPN *Enterobacteriaceae*/g food will be determined based on the results obtained from the VRBG agar plates. For further confirmation, typical colonies from VRBG agar will be inoculated in glucose agar to test for anaerobic glucose utilization and production of acidic end products. Additionally, these colonies will be streaked onto TSA plates in preparation for additional testing in Session 4.

MATERIALS AND EQUIPMENT

Per Pair of Students

- Incubated VRBG agar plates
- Up to five TSA plates for streaking isolates from VRGB plates (three for the MPN, one for the detection sample and one for the positive control); the number of TSA plates needed depends on the number of MPN tubes that were positive on VRBG agar plates.
- Up to 11 glucose agar tubes for stabbing isolates from VRGB plates (nine for MPN, one for detection sample, and one for the positive control); the number of glucose agar tubes needed will depend on the number of MPN tubes that were positive on VRBG agar plates.
- Sterile mineral oil (~12 ml)
- 1-ml sterile pipettes to transfer mineral oil
- Other common laboratory supplies (e.g., inoculation loop)

Class-Shared

- Common laboratory equipment (e.g., incubator)

PROCEDURE

1. **Control**
   a. Examine the incubated VRBG agar plate representing the positive control. Check for typical *Enterobacteriaceae* colonies. The colonies should be red-purple in color with a red-purple halo. Record your observations in Table 8.1.
   b. Label a TSA plate and complete a three-phase streak of a typical *Enterobacteriaceae* isolate from the incubated VRBG agar plate of the positive control.
      *Note: The **same** colony must be used to inoculate TSA plates and glucose agar tubes, so it is critical not to transfer the entire colony with this first assay.*
   c. Label a glucose agar tube with positive control culture's information.

    d. Transfer the same isolated typical colony from the VRBG agar plate that was used for the TSA streak into the glucose agar tube by stabbing once with an inoculating needle.

    e. Overlay the glucose agar tube with sterile mineral oil to a depth of 0.5–1 cm to achieve anaerobic conditions. Use a pipette to add the oil layer.

    f. Incubate the TSA plate and the glucose agar tube at 35°C for 18–24 hr.

2. Detection Food Sample

    a. Examine the incubated VRBG agar plate representing the detection sample. Check for typical *Enterobacteriaceae* colonies. The colonies should be red-purple in color with a red-purple halo. Record your observations in Table 8.1. If typical *Enterobacteriaceae* colonies are found, proceed to next step. If no typical *Enterobacteriaceae* colonies are found, stop analyzing the detection sample and consider the food sample negative for *Enterobacteriaceae*.

    b. Label a TSA plate and complete a three-phase streak of a typical *Enterobacteriaceae* isolate from the incubated VRBG agar plate.

    c. Label a glucose agar tube, and transfer the same isolated typical colony from the VRBG agar plate into the glucose agar tube by stabbing once with an inoculating needle.

    d. Overlay the glucose agar tube with sterile mineral oil to a depth of 0.5–1 cm to achieve anaerobic conditions. Use a pipette to add the oil layer.

    e. Incubate the TSA plate and the glucose agar tube at 35°C for 18–24 hr.

3. Enumeration Food Sample

    a. Determination of presumptive MPN in food

        i. Examine the incubated VRBG agar plates. Check the plates for typical *Enterobacteriaceae* colonies. The colonies should be red-purple in color with a red-purple halo. Any straw-colored colonies should be ignored, as these are not *Enterobacteriaceae*. Record your observations in Table 8.1.

        ii. Using results from the VRBG agar plates, originating from the food sample, determine the MPN tubes that are presumptively positive for *Enterobacteriaceae*. MPN tubes that resulted in typical colonies on VRBG agar are counted as positive. Report the number of positive tubes receiving each of the three dilutions (e.g., 3, 1, 0). Follow the instructions for determining MPN provided in Chapter 3 and calculate the presumptive *Enterobacteriaceae* MPN/g food, using the MPN table in that chapter. Report the results in Table 8.2.

    b. Streaking presumptive *Enterobacteriaceae* isolates on TSA plates for confirmation

        i. Using the VRBG agar plates representing the food sample, identify colonies with typical *Enterobacteriaceae* morphology (red-purple colony with red-purple halo). Select and mark a typical colony representing each MPN tube from each dilution (0.1, 0.01, and 0.001 g food), if available, to be further analyzed.

TABLE 8.2 *{Add a descriptive title and footnotes for this data, including food sample used, media, recovery method and what data were obtained}*

Food weight/tube	MPN[a] tube presumptively positive			Presumptive *Enterobacteriaceae* count
	1	2	3	
g				
g				
g				MPN/g
	MPN tubes confirmed *Enterobacteriaceae*			Confirmed *Enterobacteriaceae* count
g				
g				
g				MPN/g

[a] Most probable number
{Add any relevant footnotes}

    ii. Using an inoculating loop, transfer a portion of each of the selected colonies from VRBG agar plates to TSA plates and streak for isolation. If necessary, split TSA plates into three sections to accommodate all MPN tube isolates.

    iii. Incubate the inoculated TSA plates at 35°C for 18–24 hr.

c. Stabbing presumptive *Enterobacteriaceae* isolates in glucose agar for confirmation

    i. Label up to 11 glucose agar tubes appropriately (up to 9 for MPN, 1 for detection sample, and 1 for the positive control). Make sure that you can match the colony streaked on TSA to the colony inoculated in glucose agar, and these should be linked to the original MPN tube (see Table 8.1).

    ii. Inoculate the presumptive *Enterobacteriaceae* isolates (from the VRBG agar plates) into the glucose agar tubes by stabbing.

    iii. Overlay each glucose agar tube, originating from an isolate, with sterile mineral oil to a depth of 0.5–1 cm to achieve anaerobic conditions. Use a pipette to add the oil layer.

    iv. Incubate the stabbed glucose agar tubes at 35°C for 18–24 hr.

SESSION 4: CONFIRMATION

In this session, confirmation of isolate identity as *Enterobacteriaceae* will be completed by observing glucose agar tube results, conducting Gram staining and the oxidase test.

MATERIALS AND EQUIPMENT

Per Pair of Students

- Incubated TSA plates
- Incubated glucose agar stabs
- Sterile toothpicks
- Filter paper
- Wax pencil
- Other common laboratory supplies

Class-Shared

- Oxidase test reagent (freshly prepared)
- Oxidase-positive colonies on a suitable agar medium (e.g., *Pseudomonas* spp.)
- Pasteur pipettes/droppers
- Common laboratory equipment

PROCEDURE

1. **Control (*Enterobacteriaceae*-Positive Control)**
   a. Examine incubated TSA plates inoculated with the positive control cultures, and observe the colony morphology.
   b. Using an isolated colony on TSA, complete Gram staining following previously described procedure (see Chapter 4). Observe Gram reaction and cell morphology of the isolate and record the results in Table 8.1. *Enterobacteriaceae* are Gram-negative rods.
   c. Run oxidase test on colonies isolated from the TSA plate, as follows:
      i. Mark an isolated colony from the TSA plate.
      ii. Using a sterile toothpick (not the metal loop), transfer a portion of the colony to supplied filter paper; multiple colonies can be tested on one piece of filter paper.
      iii. Repeat the previous two steps using the provided oxidase-positive control plate.
      iv. Using provided Pasteur pipette, add one drop of the oxidase reagent onto the smears.
      v. Wait 20–30 sec. for a color change. *Enterobacteriaceae* are oxidase negative, so there should be no color change within 30 sec. (color

changes after this time are common, but do not indicate a positive result). The smear from the oxidase-positive control culture should produce blue color within 30 sec of the application of the oxidase reagent.

vi. Record oxidase result in Table 8.1.

d. Examine glucose agar stabs for *E.coli* control culture. In a positive reaction, the tubes change color to yellow. Record the result in Table 8.1.

2. **Detection Food Sample**

a. Examine incubated TSA plates, inoculated with detection sample, for colony morphology.

b. Using an isolated colony on TSA, complete Gram staining following previously described procedure (see Chapter 4). Observe Gram reaction and cell morphology of the isolate and record the results in Table 8.1. *Enterobacteriaceae* are Gram-negative rods.

c. Run oxidase test on a colony from the TSA plate, as describe above for the *Enterobacteriaceae*-positive control. Also, run a side-by-side oxidase test on the provided oxidase-positive culture. Compare the oxidase results observed in the two smears.

d. Record oxidase result in Table 8.1.

e. Examine glucose agar stab for the detection sample. In a positive reaction, the tubes should develop a yellow color. Record the result in Table 8.1.

3. **Enumeration Food Sample**

a. Examine incubated TSA plates corresponding to the food sample and observe the colony morphology of isolates.

b. Using isolated colonies on TSA, representing each presumptively positive MPN tube, complete Gram staining following previously described procedure (see Chapter 4). Observe Gram reaction and cell morphology of the isolates and record the results in Table 8.1. *Enterobacteriaceae* are Gram-negative rods.

c. Using isolated colonies on TSA, representing each presumptively positive MPN tube, run the oxidase test, as describe above for the *Enterobacteriaceae*-positive control. Also, run a side-by-side oxidase test on the provided oxidase-positive culture. Compare the oxidase results observed in the two smears. Record results in Table 8.1.

d. Examine the glucose agar stabs. In a positive reaction, the tubes will have yellow color. Record results in Table 8.1.

e. Using results in Table 8.1, determine the MPN tubes that are confirmed positive for *Enterobacteriaceae*. Report the number of positive tubes receiving each of the three dilutions (e.g., 3, 1, 0). Follow the instructions for determining MPN provided in Chapter 3 and calculate the confirmed *Enterobacteriaceae* MPN/g food, using the MPN table in that chapter. Report the results in Table 8.2.

DATA INTERPRETATION

Results obtained in this laboratory exercise can be interpreted as follows:

- Presumptive MPN counts for *Enterobacteriaceae* will be determined after examining VRBG agar plates and determining the number of EEB tubes presumptively positive for *Enterobacteriaceae*.
- Confirmed MPN counts for *Enterobacteriaceae* will be determined after examining results of glucose agar stabs, Gram reaction, and oxidase test. These will help us determine the number of EEB tubes confirmed positive for *Enterobacteriaceae*.

QUESTIONS

1. An incubated "detection sample" was analyzed in this exercise. Answer these related questions:
   a. Did you detect *Enterobacteriaceae* in the food you analyzed? What is the theoretical minimum detection level (MPN/g food) that is measurable in the detection sample?
   b. If *Enterobacteriaceae* were detected in the food you analyzed, how did this food become contaminated with these bacteria?
2. This exercise allowed you to enumerate *Enterobacteriaceae* in the food you analyzed. Answer these related questions:
   a. Based on class data, would you consider the turbidity of incubated MPN tubes a good indication of the presence of *Enterobacteriaceae*?
   b. What is the theoretical minimum detection level (MPN/g food) that is measurable in the MPN tubes?
3. Suggest a rapid method that can be used to detect and enumerate *Enterobacteriaceae* in food. In addition to time savings, what are the other advantages or disadvantages of this method in comparison with the one used in this laboratory exercise?

CHAPTER 9

EXAMINATION AND ENUMERATION OF FOODBORNE FUNGI

FUNGI MORPHOLOGY AS IDENTIFICATION TOOL. SELECTIVE MEDIA COMPARISON.

INTRODUCTION

Fungi (singular, fungus) are eukaryotic organisms, possessing membrane-enclosed nuclei (singular, nucleus). The fungal nucleus carries multiple chromosomes; for example, *Saccharomyces cerevisiae* cells contain 16 chromosomes. In contrast, bacteria are prokaryotes with a cell that typically contains a single chromosome but without a well-defined nucleus. The walls of fungal cells usually contain chitin, a polymer of N-acetylglucosamine, and glucans.

Anatomically, fungi form unicellular, filamentous, or complex structures. The commonly recognized fungal groups – yeasts, molds, and mushrooms – represent these three structures, respectively. Yeasts are unicellular organisms and their colonies on agar media cannot be readily distinguished from those of bacteria. Upon microscopic examination, yeast cells are recognizably much larger than those of bacteria. Molds are filamentous fungi, made of thread-like, branching formations known as hyphae (singular, hypha). Fungal hyphae may contain cross walls (septate hyphae) or lack these walls (non-septate or coenocytic hyphae). Growth and branching of hyphae produce a structure called mycelium (plural, mycelia). Some fungi are unicellular (i.e., yeast-like) under one set of conditions, but produce pseudohyphae (i.e., mold-like) or true hyphae when these conditions change; these are described as dimorphic fungi. Presence of molds on surfaces is often easy to detect visually, but the structure of the mycelium is recognizable only under the microscope.

Analytical Food Microbiology: A Laboratory Manual, Second Edition. Ahmed E. Yousef,
Joy G. Waite-Cusic, and Jennifer J. Perry.
© 2022 John Wiley & Sons, Inc. Published 2022 by John Wiley & Sons, Inc.

Fungi may reproduce through budding, hyphal elongation and fragmentation, or spore formation. Fungal mycelia often differentiate to produce spores. Fungi produce asexual and sexual spores. The asexual spore develops from a haploid cell through mitosis, i.e., division producing identical cells. For sexually reproducing fungi, two haploid cells (or hyphal structures) of different mating types fuse together, forming a cell with a diploid nucleus. The diploid cells represent a new form of the fungus known as the teleomorphic phase, compared to the anamorphic (haploid) form. The diploid cells reproduce asexually through mitotic division, but may undergo meiosis (i.e., reduction division), producing haploid sexual spores. Therefore, sexually reproducing fungi develop cells (or mycelia) that are haploid, diploid, or mixed. Fungal sexual spores are often resistant to heat, dry conditions, acidity, and antimicrobial agents.

Fungi are ubiquitous microorganisms with diverse habitats including soil, water environments, decaying vegetation, tissues of living plants, and stored crops. Many plant diseases and a few human diseases are caused by fungi. Fungi are heterotrophic organisms, i.e., lack the ability to synthesize cell components from inorganic sources such as carbon dioxide. Instead, heterotrophic organisms use pre-formed organic substrates including glucose as carbon and energy sources. Fungi are generally aerobic microorganisms, but most yeasts also grow anaerobically (i.e., facultative anaerobes). Compared to bacteria, fungi are generally slow growers.

Classification

Members of the fungi kingdom are classified into phyla (singular, phylum) based on the type of sexual and asexual spores produced and whether they produce septate or non-septate hyphae. Phyla are further classified into classes, orders, families, genera, and species based mainly on phenotypic traits. The taxonomic standing for some food-related fungi is shown in Figure 9.1. Additionally, descriptions of food spoilage fungi are shown in Tables 9.1 and 9.2.

Foodborne Fungi

Fungi include yeasts, molds, and mushrooms, but foodborne fungi refer only to the first two of these three groups. Foodborne fungi grow well in the psychrotrophic temperature range and tend to tolerate high osmotic and acid conditions. Xerophilic fungi (e.g., *Zygosaccharomyces rouxii*) can grow on food with water activity ≤ 0.85. Some fungi produce heat-resistant spores (e.g., *Byssochlamys* spp.) that can survive the thermal pasteurization of food. Preservative-resistant fungi (e.g., *Zygosaccharomyces bailii*) are problematic for the beverage, bakery, and dairy industries. The ubiquity of fungi and their adaptability to different environments make these organisms a major cause of food spoilage. Consequently, contamination of food with certain fungi may cause considerable economic losses to the food industry.

Some foodborne molds produce toxins. Consumption of the toxin-contaminated food may lead to disease (i.e., intoxication). Symptoms of these diseases vary from simple gastrointestinal disturbances to hemorrhages of internal organs and cancer. Mold toxins, commonly referred to as mycotoxins, vary

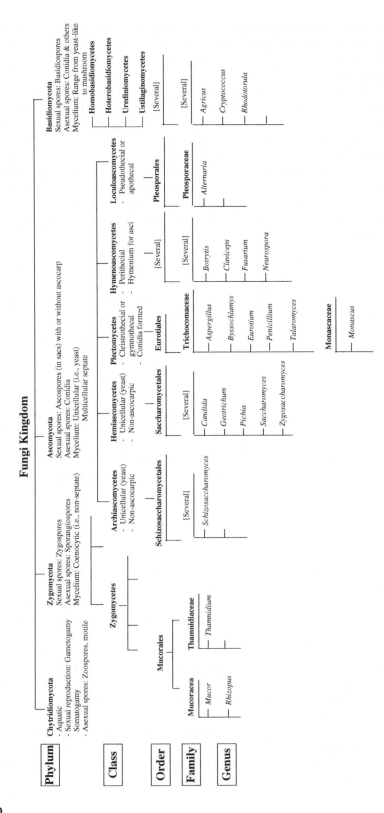

Figure 9.1 Classification of fungi associated with food. Some genera are grouped together, yet each group belongs to multiple orders or families; these orders or families are reported in this tree as [Several].

TABLE 9.1 Characteristics of foodborne molds. The anamorph-teleomorph relationship is not emphasized in this list.

Phylum	Genus	Identifying characteristics	Morphology	Characteristics related to food quality and safety
Ascomycota	*Alternaria*	• Large, brown club-shaped conidia • Conidia with longitudinal and traverse septa • Conidia formed in chains on simple conidiophores		• Brown to black rots in apple, figs, citrus fruits • Found on red meats • Field fungus: grows on barley & wheat grow indoors, cause hay fever
	Aspergillus	• Asexual spore produced on aspergillum • Black, brownish black, purple-brown conidiophore • Yellow to green conidia • Dark sclerotia		• *A. niger*: black rot on fruits & vegetables • Yellow, green to black on large number of foods • Some produce mycotoxins • Some used in food fermentations
	Botrytis	• Known as grey mold • Conidiophore stout and branching		• Infect vegetables and small berries • Infect strawberries and cause "noble rot" in grapes • *B. cinerea* is most common in food
	Byssochlamys	• *B. fulva*: tannish yellow to olive green colonies • *B. nivea*: white to cream colonies • Absence of ascocarp • Asci in open clusters		• Heat-resistant ascospores • Spoil canned & bottled fruit juice • Some produce the mycotoxin patulin

(Continued)

TABLE 9.1 (*Continued*)

Phylum	Genus	Identifying characteristics	Morphology	Characteristics related to food quality and safety
Ascomycota	*Cladosporium*	• Form thick, velvety colonies • Green, olive green, dark-blue, black, or brown colonies • Some lemon-shaped conidia • Variously branched		• Black spot on meat, beef • Some spoil butter, margarine • Field fungus: grows on barley & wheat
	Eurotium	• Bright yellow cleistothecia • Pale yellow, oblate (may have ridges) ascospores • Gray, green conidia		• Xerophilic • Spoilage in grape jam & jelly • Grow in dry, residential building environments
	Fusarium	• Cottony, pink, red, purple, brown colonies • Fusiform to crescent-shaped septate conidia (macroconidia) • Extensive mycelium		• Soft rot of figs • Brown rot of citrus fruits, pineapples • Field fungus: grows on barley & wheat • Bacon, refrigerated meat spoilage, pickle softening • Produce the mycotoxins trichothecenes, fumonisins, and zearalenone
	Geotrichum	• Dimorphic • Arthroconidia formation • White colonies • Colorless conidia		• Machinery mold • Soft rot of citrus fruits, peaches • Common in dairy products • Some have strong odors

142

Ascomycota

Genus	Characteristics	Properties
Monilinia (Anamorph: *Monilia*)	• Pink, gray, or tan conidia	• Red bread mold • Brown rot of stone fruits
Neosartorya	• White, cottony, fluffy colony • White cleistothecia • Colorless ascospores • Gray, green conidia • Teleomorph of some *Aspergillus* spp.	• Produce heat-resistant ascospores • Spoil canned & bottled fruit juice • Not xerophilic
Penicillium	• Produce brush-like conidiophores • *P. digitarum*: yellow-green conidia • *P. italicum* & *P. expansum*: blue-green conidia • *P. camemberti*: grey conidia	• Blue/green rots of citrus fruits • Soft rots of apple, pear, peaches • Some produce mycotoxins • Some used in cheesemaking
Trichothecium	• Form small pinkish colonies • Conidia form V-formation on tips of long simple hyphae • Ellipsoidal to pyriform conidia, with single lateral septum • Common species: *T. roseum*	• Pink mold grows on fruits • Common in grains including barley, wheat, and maize • Produce the mycotoxins trichothecenes

(Continued)

143

TABLE 9.1 (*Continued*)

Phylum	Genus	Identifying characteristics	Morphology	Characteristics related to food quality and safety
Zygomycota	*Mucor*	• Non-septate hyphae • Cottony colony • Smooth, non-striated sporangiospore • Produce no rhizoids		• Grow on refrigerated meat, so may cause defect referred as "whiskers" • Black spot on frozen mutton • Very common on bread
	Rhizopus	• Nonseptate hyphae • Stolons, rhizoids • Umbrella-shape columellae • Large sporangiospore with striated wall • Dark sporangia containing dark to pale spores		• Bread mold • Watery soft rot of fruits • Black spot on beef, bacon, frozen mutton
	Thamnidium	• Nonseptate hyphae • Sporangia on highly branched structure • Clusters of smaller sacs (sporangioles) present as well as sporangia (often on the same stipe)		• "Whiskers" of beef

Source: (compiled originally by Y-K Chung).

TABLE 9.2 Characteristics of foodborne yeasts. The anamorph-telemorph relationship is not emphasized in this list.

Phylum	Genus	Identifying characteristics	Morphology	Characteristics related to food quality
Ascomycota	*Brettanomyces*	• Multilateral budding • Formation of cells pointed at one end, rounded at the other. • Acidogenic yeast, producing acetic acid from glucose under anaerobic conditions		• Spoil beer, soft drinks, wine, pickles • *B. intermidius*: Prevalent species, grows at pH as low as 1.8 • *B. bruxellensis*: Off-odors in beer, cider, soft drinks
	Candida	• Cells are spheroidal, cylindrical, ovoid, or elongated • Pseudomycelium formation		• Ubiquitous: Found on plants, in water and other habitats • Common in fresh ground beef and poultry • *C. krusei*: Preservative resistant, forms film on pickle, olives, sauces • *C. parapsilosis*: Spoils cheese, margarine, dairy, and fruit products
	Debaryomyces	• Multilateral budding • Spherical or oval ascospore • Produce pseudomycelium		• Prevalent in dairy products • Forms slime on frankfurters • Grow in cheese brine • *D. hansenii*: High salt tolerance (grows at a_w 0.65), spoils yogurt, orange juice concentrate, etc.

(Continued)

TABLE 9.2 (*Continued*)

Phylum	Genus	Identifying characteristics	Morphology	Characteristics related to food quality
Ascomycota	*Hanseniaspora*	• Bipolar budding • Lemon-shaped (apiculate) cell • Ascospores: 2–4 per ascus		• Found on figs, tomatoes, citrus fruits, strawberries • Grow in fruit juices
	Kluyveromyces	• Multilateral budding • Spherical spores • Vigorous fermenter of sugars		• Prevalent yeasts in dairy products • *K. marxianus*: cheese spoilage
	Pichia	• Multilateral budding • Ascospores: Often hat-shaped, 1–4 per ascus • Few produce pseudomycelium and arthroconidia		• Found on fresh fish, shrimp. • Grow in olive brine • *P. membranifaciens*: Preservative resistant, film formation on fermented olives, pickles, sauces
	Saccharomyces	• White or cream colonies • Multilateral budding • Produce 1–4 globose ascoscopes in asci • Do not ferment lactose • Typical yeasty odor		*S. cerevisiae:* • Ubiquitous contaminant, sometimes fermentative • Spoilage of soft drinks • Some strains are preservative resistant • Used in baking and brewing

Ascomycota

Zygosaccharomyces
- Multilateral budding
- "Dumbbell" shape asci
- 1–4 ascospores/ascus
- Bean-shaped ascospores
- Strong fermenter of sugars

- Cause spoilage of tomato sauce, mayonnaise, salad dressing, soft drinks, fruit juices etc
- Resistant to preservatives
- Xerophile: grow down to a$_w$ 0.8 or lower
- Some are acidophiles
- Heat-resistant ascospores
- Examples: *Z. bailii* & *Z. rouxii*

Basidiomycota

Cryptococcus
- Ubiquitous: Found in all climate zones
- Multilateral budding
- Nonfermenter of sugars

- Found on plants and in soil
- Found on strawberries, other fruits, marine fish, shrimp, fresh ground beef

Rhodotorula
- Orange or salmon pink color pigmentation
- Often mucoid colonies
- Multilateral budding
- Nonfermenter of sugars

- Found on fresh poultry, shrimp, fish, beef
- Some grow on surface of butter
- *R. mucilaginosa*: Spoils dairy products, occasional spoilage of fresh fruits

Source: (compiled originally by Y-K Chung).

greatly in structure, properties, toxicity, and carcinogenicity. Mycotoxins include aflatoxin B1, which is highly carcinogenic; T-2 toxin, which causes alimentary toxic aleukia; and ochratoxin, which leads to kidney failure.

Despite the many problems associated with fungi, some have beneficial applications in food. These include baker's yeast (*Saccharomyces cerevisiae*), which is used in bread making, and some *Penicillium* spp., which serve as starter cultures in making mold-ripened cheeses (e.g., Camembert and blue cheeses). Additionally, *Saccharomyces* spp. are used in beer and wine fermentation. Other beneficial molds produce valuable food ingredients. *Aspergillus niger*, for example, is used to produce microbial rennet. Rennet is a food ingredient containing a proteolytic enzyme that is used as a coagulant in cheese manufacture.

Considerable presence of yeasts and molds is undesirable in food, except for mold-ripened cheese and some other fermented products. Presence of a large population of yeasts and molds in food may indicate poor sanitation and handling, temperature abuse, inadequate processing, or post-processing contamination, and ultimately result in food spoilage. Fungi growing in acidic food may consume acidic ingredients and thus raise the pH of the product. This may lead to conditions that support the growth of hazardous bacteria that were inhibited initially by the food's acidity. Food ingredients that were exposed to mold growth (e.g., flour made from moldy corn) may contain hazardous mycotoxins. Processing a food made from contaminated ingredients may decrease the fungi count substantially, but not the level of the mycotoxin. In this case, analysis for mycotoxins is needed to assess the safety and quality of these foods.

Quantifying Foodborne Fungi

A commonly used procedure to quantify the population of yeasts and molds in food involves homogenizing the test sample in peptone water, using a stomacher or a blender, preparing additional dilutions of the homogenate, and spread-plating selected dilutions (0.1 ml each) onto the surface of a suitable agar medium. For enumeration of fungi in food, it is recommended that plates are incubated aerobically at 25°C for 4–5 days. Reliability of this procedure in estimating food's fungal population depends on many factors including the type of fungal contaminants, selection of the agar medium, and choice of incubation conditions.

1. **Type of fungal contaminants**
   - When the fungal contaminants in food are mainly unicellular yeasts, enumerating this population with reasonable reliability is possible using the dilution plating technique just described. The enumeration process, in this case, is similar to that used for quantifying a bacterial population.
   - In the likelihood that the fungi in the food are mainly in the form of spores, each spore potentially produces a colony on agar plates, and thus the count reflects the population of fungal spores. Since molds tend to produce a very large number of spores, counting spores does not reflect the mass of fungi that is actively involved the in the spoilage of food.
   - When a food is minimally or mildly contaminated with mold, it is sometimes difficult to reliably quantify the fungal population in that food. If

the food contaminant is mainly mold mycelium, it is not clear what is being enumerated by dilution plating. A likely scenario is that during homogenization of the test sample (e.g., in a blender), the mycelium in the food is chopped into pieces, and upon plating, each piece potentially becomes a colony forming unit. If this is the case, a greater degree of homogenization theoretically produces larger colony counts on the incubated plates. In case of unreliability of the enumeration technique, it may be replaced with a simple qualitative test as follows. The analyst inspects the undisturbed test sample visually and then examines it under a stereoscope (dissecting microscope); if a fungus is detected on the food sample, a specimen of the mycelium may be examined using a light microscope. The degree of food contamination may be estimated subjectively.

2. **Agar media for recovery of fungi from food**

Several selective agar media have been developed for enumerating fungi in food. These media support colony formation by cells and spores of yeasts and molds and suppress or prevent bacteria from forming colonies. It is also desirable that selective media reduce the colony diameter of fast-growing fungi so that colonies on agar plates can be counted accurately. Selective agents used in these media include acids (e.g., tartaric in acidified potato dextrose agar), dehydrating agents (e.g., glycerol in dichloran 18% glycerol agar), antibiotics (e.g., chloramphenicol in antibiotic plate count agar [APCA]), and other antimicrobial agents (e.g., Rose Bengal in dichloran Rose Bengal chloramphenicol [DRBC] agar). These media vary considerably in their ability to recover fungi from various foods. For example, media with high water activity are suitable for recovering fungi from vegetables or meat, whereas those formulated with water-binding agents (e.g., high concentrations of glycerol or sugar) are more suitable for recovering xerophilic fungi from dry foods.

3. **Incubation conditions.** For enumeration of fungi in food, it is recommended that plates are incubated aerobically at 25°C for 4–5 days due to the slow-growing nature of most fungal isolates. Incubation in the dark may be necessary with the use of some selective media (e.g., Rose Bengal chloramphenicol agar). Fungi that grow optimally at different temperatures are not well-represented in this count.

4. **Important considerations**

   a. Fungi grow well aerobically and for their enumeration, surface plating is preferred over pour plating. Compared to pour plating, in which 1 ml of homogenized sample is plated, surface plating decreases the method's detection limit, since only 0.1 ml homogenized sample is spread on the plate.

   b. Inoculated Petri plates should always be incubated in the upright position for optimal fungal growth. This also will minimize fungal spore shedding in the environment, which can lead to persistent contamination issues in a laboratory setting and increases likelihood of allergic reaction. When handled, examined, or stored, the incubated plates should remain in the upright position with lids in place. If incubated plates must be uncovered, e.g., during slide preparation for microscopic examination, this should be done in a biological hood.

c. Proper safety procedures should be exercised while working with fungi.

OBJECTIVES

1. Assessing the mycological quality of food through visual inspection and microscopic examination.
2. Enumerating foodborne fungi (i.e., yeasts and molds) in different foods using different selective agar media.
3. Becoming familiar with the macroscopic and microscopic appearance of diverse foodborne fungi.

MEDIA

Two of these media can be used to complete this laboratory exercise.

Antibiotic Plate Count Agar (APCA)

Plate count agar (PCA) is a general-purpose medium typically used for enumeration of organisms in food. Presence of tryptone, glucose (also called dextrose), and yeast extract in PCA makes this medium suitable for growing a wide variety of microbes. The addition of appropriate antibiotics (i.e., tetracycline and chloramphenicol) makes this medium selective against bacteria. Antibiotic PCA medium is prepared by mixing 2 ml of a mixed antibiotic stock solution with 100 ml of molten (~48°C) sterile PCA. The stock solution is prepared as follows: 500 mg each of chlortetracycline HCl and chloramphenicol are dissolved in 100 ml of ethanol, filter sterilized, and the solution is stored in the refrigerator until use in media preparation.

Dichloran Rose Bengal Chloramphenicol (DRBC) Agar

Selectivity of DRBC agar for yeasts and molds is conferred by two antibacterial ingredients: Rose Bengal and chloramphenicol. Dichloran (2,6-dichloro-4-nitroaniline) has antifungal activity and is included to prevent the spread of fungal colonies, resulting in less crowded plates. This medium is suitable for enumerating fungi in high-moisture foods including fruits, vegetables, meats, and most dairy products. Protection of this medium from light is essential for its proper functionality.

Dichloran 18% Glycerol (DG18) Agar

The addition of appropriate antibacterial agents (i.e., chloramphenicol) makes this medium selective for yeasts and molds. The medium contains 18% glycerol, which results in water activity (a_w) of 0.955. This relatively low water activity minimizes the interference of bacteria and rapidly growing molds.

Dichloran also is included to aid in production of countable plates. The medium is suitable for enumerating fungi in foods with reduced water activity (< 0.95 a_w).

Yeast and Mold Count Petrifilm™

A Petrifilm™ plate is made of a dehydrated medium on a thin plastic film with a flexible transparent covering. Adding 1 ml of homogenized sample to the Petrifilm™ rehydrates the medium. After incubation, yeasts are typically indicated by small, blue-green colonies with defined edges. Molds are indicated by large, variably colored colonies with diffuse edges. A built-in grid facilitates counting colonies.

PROCEDURE OVERVIEW

Consistent with the objectives just described, the laboratory exercise encompasses three main tasks:

1. Visual and microscopic assessment of food moldiness. This involves the following tests: (a) inspecting undisturbed food visually; (b) examining a food specimen under a dissecting microscope for the presence of molds; and (c) preparing a slide culture of the predominant mold, incubating the culture, and examining the resulting colony.
2. Enumeration of fungi population in food, by applying the counting rules presented previously in Chapter 3.
3. Predicting the identity of food isolates by comparison with (a) pre-mounted microscope specimens of known fungal genera; (b) fungi listed in Tables 9.1 and 9.2; and (c) fungi reported in various scientific publications.

The activities related to these tasks will be completed in three sessions according to the following schedule (Figure 9.2):

Session 1: Visual and microscopic assessment of food moldiness (Task 1), and food sample preparation, dilution, plating, and incubation of inoculated plates.
Session 2: Microscopic examination of pre-mounted fungal specimens, representing genera commonly found in food.
Session 3: Counting colonies, determining population count, and examining colony and cell morphology of food isolates for predicting their identities.
Note that fungi populations will be recovered on two selective media and resulting counts will be compared.

Foods Analyzed

Fresh produce, particularly fruits, often support the growth of fungi; spoilage of these commodities by yeasts and molds is common. Therefore, fresh and "stored"

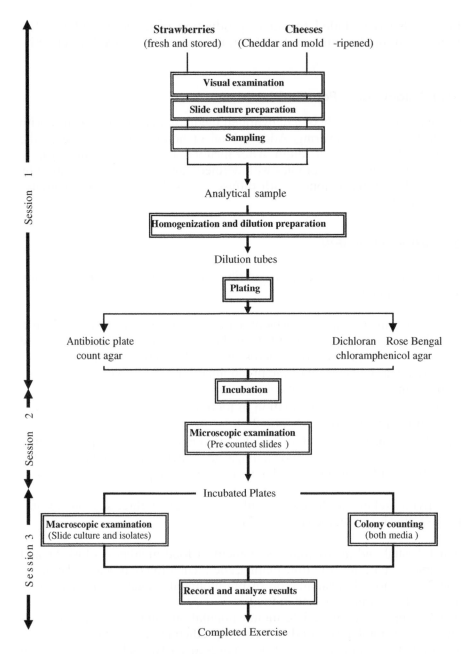

Figure 9.2 Scheme of testing food for fungi.

fruits (purposefully subjected to temperature abuse) will be analyzed in this laboratory exercise. Additionally, cheeses will be examined mycologically. A typical hard cheese, such as Cheddar, has low fungi count, yet mold growth is the common cause of spoilage of this type of cheese. Mold-ripened cheeses such as blue, Roquefort, and Camembert naturally contain a large mold population.

Therefore, presence of a large population of fungi is expected in mold-ripened cheeses, but such a population would cause the spoilage of other cheese types. In addition to fruits and cheese, nuts could be analyzed. Examples of nuts suitable for analysis in this laboratory include peanuts, walnuts, and chestnuts. Ideally, these are raw, in-shell nuts that have been shelled and ground aseptically immediately prior to analysis.

Sample Preparation

Preparing a sample for counting its microbial population depends on the nature of food. Using a blender is suitable for releasing organisms embedded in a hard food matrix, whereas the stomacher is more appropriate for the release of surface contaminants. For delicate food (e.g., strawberries) sample-diluent mixture is manipulated gently by hand so that surface fungi are released. This gentle homogenization alleviates the release of sample components (e.g., acid), which may have antimicrobial characteristics. Additionally, breaking the matrix of a surface-contaminated product into fine particles may conceal some of the microorganisms that are recoverable by milder homogenization techniques.

Cheese should be shredded, and nuts may be ground before an analytical sample is withdrawn. These sample preparations should be done aseptically. Autoclavable or sanitizable hand-held cheese shredders and porcelain mortar and pestle sets can be used in these sample preparations. Homogenization should maximize the recovery of fungal spores, yeast cells, and mold mycelia from solid food samples. A warm emulsifying diluent (e.g., 2% sodium citrate solution at 40°C) is suitable for emulsifying cheese during homogenization in a blender or a stomacher. Proper homogenization in the presence of this diluent transforms the cheese into a milky slurry and releases a considerable portion of the microbial population of the sample.

SELECTED REFERENCES

Pitt, J.I. and Hocking, A.D. (2009). *Fungi and Food Spoilage*, 3e. New York, NY: Springer.

Ryu, D. and Wolf-Hall, C. (2015). Yeasts and molds. In: *Compendium of Methods for the Microbiological Examination of Foods*, 5e (ed. Y. Salfinger and M.L. Tortorella), 277–298. Washington, DC: American Public Health Association Press.

Webster, J. and Weber, R. (2007). *Introduction to Fungi*. Cambridge, UK: Cambridge University Press.

SESSION 1: EXAMINATION OF FOOD FOR MOLDINESS AND PREPARING SAMPLES FOR FUNGI ENUMERATION

The following three tests will be completed in this laboratory session:
1. Visual inspection: Foods will be evaluated visually for the degree of moldiness.
2. Slide culture preparation: A moldy piece of the food sample will be examined under a stereoscope (dissecting microscope), and a small portion of the mycelium of the predominant fungus will be used inoculate the agar on the slide. The slide assembly will be incubated, and the resulting mold colony will be examined in the second session.
3. Sample preparation for fungi enumeration: This involves sampling, homogenization, dilution, plating, and incubation. Populations will be recovered on two selective media (antibiotic-PCA and DRBC agar) and resulting counts will be compared during the second session.

FOODS

Foods with and without signs of moldiness are suitable for analysis in this laboratory exercise. Analysis of these two food types provides quantitative data that can be used to assess the mycological quality of the product.

Note: To protect the analysts against the hazard of inhaling mold spores or exposure to mycotoxins, excessively moldy food should not be analyzed in this exercise.

Samples of the following two food pairs will be analyzed during this exercise:

- **Fresh and Stored Strawberries**

  Defect-free strawberries sold in retail stores, which are referred to as "fresh" herein, usually contain very small populations of fungi. The product is often sold in containers that allow air exchange and minimize condensation of moisture. Fresh strawberries will be analyzed in this exercise as an example of consumer-acceptable product that is liable to fungal spoilage under abuse conditions. In addition to the fresh strawberries, storage-abused, mildly spoiled product ("stored strawberries") also will be analyzed. To produce the storage-abused strawberries, samples of fresh strawberries are placed in plastic bags that are sealed and held for 2–4 days in a refrigerator or at room temperature (approximately 22°C) before analysis. Sealing the bags keeps the environment moist, but sufficient headspace (i.e., air) over the strawberries should be allowed. This procedure often encourages visible moldiness in the product. The stored product may contain not only the visible mold, but also yeast, which may not be readily visible.

  Note: Do not analyze stored strawberries that have become excessively moldy. These should be considered a biohazard and discarded appropriately.

- **Hard and Mold-Ripened Cheeses**
  A hard cheese (e.g., Cheddar or Parmesan) will be tested as a product that should not contain countable fungal population. For this exercise, it is preferable to use an intact rather than a factory-shredded cheese sample. The intact sample should be shredded manually in the laboratory before analysis. A consumer-size package (200–500 g) is suitable for this purpose. Factory-shredded cheese may contain antifungal agents added by the manufacturer to protect it against mold spoilage during storage. Presence of antifungal agents would interfere with this test.

  The mold-ripened cheese (e.g., blue or Camembert) contains beneficial mold that would be suitable for examination and enumeration in this exercise. Such a cheese may also support the growth of yeast. Most mold-ripened cheeses are too soft to be shredded with mechanical or hand-held shredders. These samples can be crumbled and mixed well using sterile spatulas before the analytical sample is withdrawn.

  Note: Considering the beneficial role of molds in mold-fermented foods, describing such a food as moldy does not mean it is poor in quality.

ORGANIZATION

Students, in groups of two, will cooperate to analyze a pair of related foods: fresh and stored (moldy) strawberries, or hard and mold-ripened cheeses. One member of the group will analyze the moldy sample and the other will analyze the non-moldy sample of the same food type. Both members of the group will cooperate to inspect visually their pair of foods for signs of moldiness and to prepare the slide culture.

Session I. Visual inspection of food for moldiness

MATERIALS

Per Pair of Students

A pair of food samples: Fresh and stored strawberries or Cheddar and blue cheeses

PROCEDURE

1. The two members of each group will cooperate to inspect their pair of foods for the following:
   a. Presence or absence of visible mold
   b. Distribution of mold on the sample (e.g., surface, or embedded)
   c. Diversity of the food mycobiota
   d. Morphology of the predominant mold
2. On a scale of 1–5, report the degree of moldiness; 1 being mold-free and 5 being very moldy.
3. Record these observations in Table 9.3.

TABLE 9.3 {Provide a descriptive title}

Characteristic	Food 1 ( )	Food 2 ( )
Presence of mold		
Distribution		
Diversity		
Morphology		
Degree (scale 1–5)		

II. Preparation of slide culture

Slide cultures will be prepared and incubated; these will be examined micro-scopically during a subsequent laboratory session.

MATERIALS

Per Pair of Students

- Empty sterile Petri dish
- Microscope slides and cover slips
- Slide culture assembly: Sterile glass Petri dish containing a glass rod for elevating a microscope slide (Figure 9.3)
- Tube of sterile water (~2 ml)
- Individually wrapped 1-ml sterile serological pipette and pipettor bulb (or pipette-aid).
- Food samples: Stored strawberries or mold-ripened cheese (from part I).

Class-Shared

- Stereoscopes
- Dissecting needles
- Molten sterile plate count agar (PCA), held in a 50°C water bath
- Other common laboratory supplies and equipment

PROCEDURE

Microscope Examination of Mold on Food

1. Transfer a piece of food with visible mold to the empty Petri dish.
2. Place the Petri dish under the stereoscope and adjust the lighting and focus.
3. Identify a spot on the food having a mass of mold mycelium or spores, suitable for preparing the slide culture.

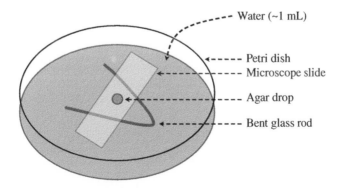

Figure 9.3   Slide culture assembly (courtesy of Matthew Mezydlo).

Preparation of Slide Culture

1. Collect the slide culture assembly (Figure 9.3).
2. Dispense 1 ml of sterile water into the bottom of the glass Petri dish, to provide a moist environment during incubation of the slide culture.
3. Sanitize the microscope slide with alcohol.
4. Using the sterile 1-ml pipette (or sterile Pasteur pipette), transfer a drop of molten PCA medium onto the slide in the glass Petri dish and allow the agar medium to solidify.
5. Pick some of the fungal mycelium (or spores), from the food spot identified under the stereoscope, using a dissecting needle or sterile swab, and inoculate the agar on the slide. Touching the outer edges of the agar drop lightly with the needle (or swab) serves as an inoculum. Cover the inoculated agar with an alcohol-sanitized slide cover slip.

   Note: allow alcohol to evaporate; do not pass cover slip through a flame.
6. Cover the Petri dish with its lid, tape closed with labeling tape, label with your group's initials, and incubate at 25°C for 4–5 days.

III. Sample preparation for fungi enumeration

This test involves sampling, sample homogenization, dilution, plating, and incubation. For homogenization, groups examining strawberries will only massage the sample-diluent mixture by hand. For groups analyzing shredded cheese, use the stomacher to homogenize the sample. Dilutions of the food sample are prepared and plated on two selective agar media, antibiotic-PCA and DRBC agar. Plates are incubated right-side up (i.e., agar-side down) at 25°C for 4–5 days. Colonies on incubated plates will be counted during the second session.

Note: The number of dilutions to be prepared and those to be plated depend on the size of the fungal population in the tested food. The dilution scheme presented in this exercise is based on the experience of the instructors with foods from their localities. This scheme should be adjusted when tested foods are expected to contain different fungal populations.

MATERIALS

Per Pair of Students

- **Foods**
  Each group of two will analyze one of these food pairs (i.e., one food per student):
  - Hard cheese (e.g., a retail small package of Cheddar cheese) and mold-ripened cheese (e.g., a retail small package of blue cheese)
  - Recently purchased (fresh) strawberries and stored strawberries; see food description presented previously.
- **Media**
  - 12 APCA plates; 6 plates/student
  - 12 DRBC agar plates; 6 plates/student
- **Supplies for analyzing fresh strawberries**
  - One 90-ml bottle of peptone
  - Two 9-ml sterile peptone water tubes (diluent)
  - Sterile 100-ml graduated cylinder, covered with aluminum foil
- **Supplies for analyzing stored strawberries**
  - One 90 ml bottle of peptone
  - Four 9-ml sterile peptone water tubes (diluent)
  - Sterile 100-ml graduated cylinder, covered with aluminum foil
- **Supplies for analyzing Cheddar cheese**
  - A sanitizable or autoclavable hand-held cheese shredder
  - One 90-ml bottle of citrate solution (2% trisodium citrate) in a 40°C water bath
  - Two 9-ml sterile peptone water tubes (diluent)

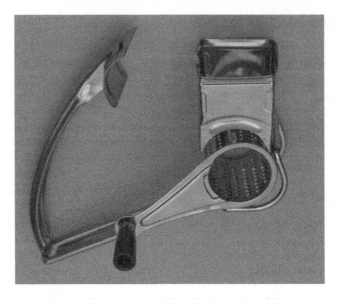

Figure 9.4   Hand-held hard cheese shredder.

- **Supplies for analyzing blue cheese**
  - One 90-ml bottle of citrate solution (2% trisodium citrate) in a 40°C water bath
  - Four 9-ml sterile peptone water tubes (diluent)

Class-Shared

- Sample preparation utensils, as needed. These include cheese shredder (Figure 9.4), cutting brads, knives, pair of tongs, alcohol bottles, Styrofoam clam-shell containers, or reusable sanitizable containers, etc.
- Common laboratory equipment (e.g., incubators)

PROCEDURE

Sample Preparation

Strawberries (fresh and stored)
Considering that fungi on strawberries are mainly surface contaminants, these will be recovered by rinsing the sample in equal amount of the diluent. Therefore, the rinse liquid represents the non-diluted sample (i.e., 10^0 dilution). In other words, one ml of this liquid represents one gram of strawberries.

1. Weigh three strawberries into a sterile stomacher bag.
2. Measure a volume of sterile peptone water that is equivalent to the weight of strawberries.
3. Combine the sample with the peptone water in the stomacher bag.
4. Hand massage the sample-diluent mixture gently for two minutes.

Cheddar cheese

Cheddar and harder cheeses require shredding before homogenization. The following procedure is used with this type of cheese, assuming package size of 200–500 g.

1. Remove the cheese aseptically from the commercial package and place it on a pre-sanitized cutting board.
2. Cut the cheese into strips or cubes using the sanitized knife.
3. Transfer the cheese pieces aseptically to the hopper of a sterile or sanitized cheese shredder (Figure 9.4). Alternatively, if this type of shredder is unavailable, shred using a sanitized box grater (in this case, do not remove the cheese from overwrap and do not cut).
4. Collect shredded cheese in a suitable pre-sanitized container and mix container contents well.
5. Weigh a 10 g analytical sample into a sterile Petri dish and transfer the sample to a sterile stomacher bag. Alternatively, weigh the analytical sample directly into the stomacher bag.
6. Dispense the 90 ml warm sodium citrate diluent into stomacher bag. Warm sodium citrate solution is an ideal diluent considering its ability to emulsify the shredded cheese.
7. Homogenize the mixture in a stomacher for two minutes; the sample homogenate represents the 10^{-1} dilution of the analytical sample.

Blue cheese

Compared to Cheddar, blue cheese is much softer. The procedure used to prepare Cheddar cheese is used with blue cheese except that the sample is divided into small pieces with the knife instead of the shredder. The following is the procedure as applied to blue cheese:

1. Remove the cheese aseptically from the commercial package and place it on a pre-sanitized cutting board.
2. Cut the cheese into small pieces (3–5 mm in dimeter, approximately) using the sanitized knife.
3. Collect the cut cheese in a suitable pre-sanitized container and mix container contents well.
4. Weigh a 10 g analytical sample into a sterile Petri dish and transfer the sample to a sterile stomacher bag. Alternatively, weigh the analytical sample directly into the stomacher bag.
5. Dispense the 90 ml warm sodium citrate diluent into stomacher bag.
6. Homogenize the mixture in a stomacher for two minutes; the sample homogenate represents the 10^{-1} dilution of the analytical sample.

Dilution and Plating

Fresh strawberries

1. Transfer 1 ml of strawberry rinse solution into a 9-ml diluent tube to pre-pare the 10^{-1} dilution.
2. Vortex the contents of the 10^{-1} dilution tube, and transfer 1 ml to another 9-ml diluent tube to prepare the 10^{-2} dilution.
3. Dispense 0.1 ml of the 10^0 dilution (rinse solution from the stomacher bag) onto two plates of each selective medium (antibiotic PCA and DRBC agar). Repeat for the 10^{-1} and 10^{-2} dilutions.
4. Spread the inocula evenly on the plates using a cell spreader. Sanitize the spreader in alcohol and flame the spreader before and after each use.
5. Tape the plates together using a masking tape.
6. Incubate plates, right side up, at 25°C for 4–5 days. The plates should remain undisturbed until counted and should not be exposed to light.

Stored strawberries

1. Transfer 1 ml of strawberry rinse solution into a 9-ml diluent tube to pre-pare the 10^{-1} dilution.
2. Vortex the contents of the 10^{-1} dilution tube, and transfer 1 ml to another 9-ml diluent tube to prepare the 10^{-2} dilution. Repeat to prepare the 10^{-3} and 10^{-4} dilutions.
3. Dispense 0.1 ml of the 10^{-2} dilution onto two plates of each selective medium (antibiotic PCA and DRBC agar). Repeat for the 10^{-3} and 10^{-4} dilutions.
4. Spread the inocula evenly on the plates using a cell spreader. Sanitize the spreader in alcohol and flame the spreader before and after each use.
5. Tape the plates together.
6. Incubate plates, right side up, at 25°C for 4–5 days. The plates should remain undisturbed until counted and should not be exposed to light.

Cheddar cheese

1. Using the homogenized cheese (i.e., 10^{-1} dilution), prepare the 10^{-2} and 10^{-3} dilutions.
2. Dispense 0.1 ml of the 10^{-1} dilution onto two plates of each selective medium (antibiotic-PCA and DRBC agar). Repeat for the 10^{-2} and 10^{-3} dilutions.
3. Spread the inocula evenly on the plates using a cell spreader. Sanitize the spreader in alcohol and flame the spreader before and after each use.
4. Tape the plates together using a masking tape.
5. Incubate plates, right side up, at 25°C for 4–5 days. The plates should remain undisturbed until counted and should not be exposed to light.

Blue cheese

1. Using the homogenized cheese (i.e., the 10^{-1} dilution), prepare 10^{-2}–10^{-5} dilutions.

2. Dispense 0.1 ml of the 10^{-3} dilution onto two plates of each selective medium (antibiotic PCA and DRBC agar). Repeat for the 10^{-4} and 10^{-5} dilutions.

3. Spread the inocula evenly on the plates using a cell spreader. Sanitize the spreader in alcohol and flame the spreader before and after each use.

4. Tape the plates together.

5. Incubate plates, right side up, at 25°C for 4–5 days. The plates should remain undisturbed until counted and should not be exposed to light.

   Note: If inoculated DRBC agar plates are incubated in a lighted room, these plates should be covered appropriately to protect the medium from the deleterious effect of light.

SESSION 2: EXAMINATION OF PRE-MOUNTED FUNGAL SPECIMENS

Cultures on plates of agar media and pre-mounted microscope slides, for genera commonly found in food, will be examined. These plates and slides will be prepared in advance by course instructors. This examination may help in recognizing the genera of food isolates, which will be examined in the next laboratory session.

MATERIALS AND EQUIPMENT

Class-Shared

- Brightfield microscopes
- Five side demonstrations (agar plates under stereoscopes and slides under phase-bright microscopes) for the following:
  a. Molds:
     - *Rhizopus* sp.
     - *Penicillium* sp.
     - *Aspergillus* sp.
  b. Yeast
     - *Saccharomyces* sp.
  c. Bacterium (for comparison with fungi)
     - *Escherichia coli*

PROCEDURE

Demonstrated Yeast and Bacteria

1. With the stereoscope, examine the colony morphology of the yeast and the bacterium on the corresponding agar plates. Look at colony characteristics such as color, margin, shape, and size. Record the observations in Table 9.4.
2. Use the brightfield microscopes to examine the cell morphology of the yeast and the bacterium. Record the observations in Table 9.4.

Demonstrated Mold

1. Using the stereoscope, inspect the morphology of the mold colony on the agar plates and record the observations in Table 9.4.
2. Use the brightfield microscope to examine the molds and describe the mycelium structures (e.g., septate, or non-septate, and types and arrangements of spores). Record the observations in Table 9.4.

TABLE 9.4 {add a descriptive title for this data. Add also appropriate footnotes}

Sample	Colony morphology	Cell morphology
Saccharomyces sp. [Demonstrated]		
Rhizopus sp. [Demonstrated]		
Penicillium sp. [Demonstrated]		
Aspergillus sp. [Demonstrated]		
Escherichia coli [Demonstrated]		
Yeast (from sample: _____)		
	1	
	2	
	3	
Mold (from sample: _____)		
	1	
	2	
	3	
Slide culture		

SESSION 3: DETERMINING FUNGI POPULATION AND ISOLATE MORPHOLOGY

During this session, incubated plates of the two selective media (antibiotic PCA and DRBC agar) will be examined and fungi populations will be counted and compared. Additionally, students will be asked to recognize some of the counted fungi based on visual and microscopic examination. Compared to bacteria, fungi under the microscope have much larger cells and easier-to-recognize cell arrangements. Simple staining is often sufficient to complete the microscopic examination. Crystal violet stain will be used for staining yeast cells and lactophenol cotton blue for staining mold mycelia. Instead of the brightfield microscope, a phase-contrast microscope may be used to examine fungal cells or mycelia. When examined under the phase-contrast microscope, specimen staining is not needed.

Note: Exposure to some fungi may pose health hazards. Some fungi cause allergies or other ailments. Therefore, care should be exercised when handling plates containing fungi colonies or when taking samples for microscopic examination. Handling these plates is ideally carried out in a biological safety cabinet.

MATERIALS AND EQUIPMENT

Per Pair of Students

- Incubated plates (antibiotic PCA and DRBC agar)
- Crystal violet stain
- Lactophenol cotton blue stain
- Microscope, slides, and cover slips
- Incubated slide culture
- Dissecting needle
- Tweezers
- Other common laboratory supplies

Class-Shared

- Common laboratory equipment (e.g., colony counter)

PROCEDURE

Colony Counting and Calculating Population Count

1. Examine the incubated plates for the two media, antibiotic PCA and DRBC agar. Determine which dilutions produced countable plates. The countable range for fungi is 15–150 colonies per plate. Discard the plates that fall outside the countable range into the biohazard disposal container.

TABLE 9.5 Fungi colony and population counts in foods analyzed by _____ and _____

Food analyzed	Agar medium	Replicate plate	Volume plated	Dilution factor	Colony count	Population count (CFU/g)
		1				
	Antibiotic PCA	2				
	DRBC agar	1				
		2				

{add at least two footnotes clarifying the presentation of the data and abbreviations}

2. In preparation for colony counting, tape the lid to the base of each plate, and place the plate on a lighted colony counter, with the base facing up.

3. Count fungi colonies (molds and yeasts, together) on the plates identified as countable. Mark the colonies with a marker as they are counted to avoid double counting.

   Notes:

   • *Yeast colonies often are similar in appearance and size to bacterial colonies, but mold colonies are typically larger and fuzzier.*

   • *Overgrowth of mold may cause difficulties in counting.*

   • *Use reasonable judgment in counting; if mycelium appears to originate from two different overlapping colonies, it probably does.*

4. Calculate the fungi CFU/g food, on each of the two media, and record results in Table 9.5. Use appropriate equation to determine the population count (see data analysis description at the end of this chapter).

5. Compare the population counts determined on the two media.

Microscopic Examination

Yeast isolates

1. Using the counted plates, choose uncrowded plates and mark up to three colonies suspected to be yeasts. If possible, choose colonies with different morphologies. Report the observed colony morphologies in Table 9.5.

2. Using inoculation loop, transfer a portion of each colony to a small drop (i.e., droplet) of water on microscope slides.

3. Emulsify and spread the mixtures and heat-fix the resulting smears.

4. Stain using a simple stain (i.e., crystal violet) for 60 seconds. Rinse the smears with water and blot-dry.

TABLE 9.6 Colony and cell morphology of fungi isolated from _____ {food analyzed} and their likely genera[a].

Source	Observation	
	Morphology	
Yeast isolate (Incubated plates)	Drawing	
	Likely genus	
	Morphology	
Mold isolate (Incubated plates)	Drawing	
	Likely genus	
	Morphology	
Mold grown onslide culture	Drawing	
	Likely genus	

[a] *Table should be expanded if more isolates were obtained.*

5. Examine the cells under the microscope using two objective lenses; the 40× and the oil immersion (100×). This microscopic examination will aid in distinguishing between colonies of yeast and bacteria.

6. Report the yeasts' cellular morphology and their likely genera in Table 9.6.

Mold isolates

1. Using the counted plates, choose uncrowded plates and mark up to three colonies having the appearance of molds. If possible, choose colonies with different morphologies. Report the observed colony morphologies in Table 9.5.

2. Dispense drops of the lactophenol cotton blue stain on glass slides.

3. Using inoculation loop or dissecting needle, cut portions of the selected colonies, transfer to the slides, and mix with the stain.

4. Cover the specimens with slide cover slips and wait for 5 minutes to allow mycelium staining.

5. Examine the mycelium under the microscope using a 40× or 60× objective lens. You may need to view different fields to see all structures present on the slide.

   Note: do not attempt to use the oil immersion lens in the presence of a coverslip.

6. Report molds' cellular and micellular morphologies, and their likely genera, in Table 9.6.

Slide culture

1. Remove the incubated slide from the glass Petri dish.
2. Keep the cover slip on the slide and place on the stage of bright-field microscope. Alternatively, use phase-contrast microscope to examine the slide culture.
3. Examine the slide under the microscope using the 40× or 60× objective. Observe fungal structures. Record your observations in Table 9.6.

DATA ANALYSIS AND QUESTIONS

Group Data

1. For each food analyzed, calculate fungi population counts, for both media, and report the results in Table 9.5. As described Chapter 3, population count in a food sample is determined as follows:

$$\text{Population count}(\text{CUF}/\text{g}) = \frac{(\text{Colony count on plate }1 + \text{Colony count on plate }2)/2}{\text{Volume plated}(\text{ml}) \times \text{Dilution factor}(\text{of the dilution plated})}$$

2. Enter these data in the class computer for sharing with other groups.

Class Data

3. Review class data and correct any calculation errors. Present the pooled class data in a table like the following (include appropriate title and footnotes):

Group initials {e.g., AX/BZ}	Colony forming units per gram (CFU/g)							
	Fresh strawberries		Stored strawberries		Cheddar cheese		Blue cheese	
	APCA	DRBC	APCA	DRBC	APCA	DRBC	APCA	DRBC

4. Construct a table for class data, similar to the one just created, but based on $\log_{10}$ CFU/g, instead of CFU/g. Use these data to calculate the average population and standard deviation for each food type on each medium.

   Note: By convention, log population (log10 CFU/g) data points can be averaged and used to determine population standard deviation.

5. Plot class data in a bar graph. Cluster appropriately samples of the same type. Use a bar to represent the population in a food, determined on a given medium. Add the standard deviation to each bar.

Questions

1. Was the visual assessment of the degree of moldiness of food, in the first session, consistent with the quantitative determination completed in the second session?

2. Considering data from your group of two, did the two selective media recover similar fungal populations from the two foods analyzed? If population recovery was different, which of the two media resulted in greater recovery? Was this trend noticeable when pooled class data were compared?

3. What were the morphological traits of the predominant organism in the fungi population of the two samples analyzed by the group? Based on this knowledge, predict the genus of the predominant fungus.

4. Summarize, in one statement, the new knowledge one may gain from completing this exercise.

PART III

FOODBORNE PATHOGENS

Part III covers training-adapted procedures for the detection of selected bacterial pathogens in food. The selected pathogens are the Gram-positive coccus, *Staphylococcus aureus*, the Gram-positive short rod, *Listeria monocytogenes*, and the Gram-negative rods, *Salmonella enterica* and Shiga toxin-producing *Escherichia coli*. Basic techniques and practical experiences gained in Parts I and II are essential for proper execution of experiments in this part of the book. Considering the pathogenicity of bacteria handled during completion of these exercises, it is important that students practice proper safety measures to avoid harming themselves or other people using the laboratory. The safety guidelines in the introductory chapter of this book should be reviewed and understood before performing these exercises.

Foodborne Disease Hazards

Foodborne diseases are illnesses that result from ingesting foods containing health hazards. Hazards associated with food consumption include physical (e.g., sharp objects), chemical (e.g., heavy metal ions), or microbiological agents. Only microbial etiological agents of foodborne diseases (i.e., pathogens) are addressed in this part of the book. Microbial hazards can result in foodborne intoxication or infection, based on the mode of action of the etiological agents (Table III.1). Foodborne intoxication results from ingestion of toxins produced by microorganisms in the food. Ingestion of living cells is not required to cause microbial intoxication; instead, presence of the microbial toxin in ingested food causes the disease. *Staphylococcus aureus* is an example of a pathogen that causes foodborne intoxication. If ingestion of living cells is required to cause the foodborne disease, it is described as a foodborne infec-

Analytical Food Microbiology: A Laboratory Manual, Second Edition. Ahmed E. Yousef, Joy G. Waite-Cusic, and Jennifer J. Perry.
© 2022 John Wiley & Sons, Inc. Published 2022 by John Wiley & Sons, Inc.

TABLE III.1 Modes of transmission of microbial foodborne diseases.

Characteristic	Disease Type		
	Intoxication	Non-Invasive Infection	Invasive Infection
Hazard	Toxin (of microbial origin) in food	Live pathogen in food	Live pathogen in food
Pathogen type	Non-enteric toxigenic organism	Enteric toxigenic organism	Enteric invasive organism
Events leading to disease	• Toxin is produced in food • Ingestion of toxin-containing food • Toxin survives gastrointestinal conditions • Toxin reaches and reacts with intestinal target	• Pathogen is present in food • Ingestion of pathogen at level ≥ infectious dose • Pathogen survives gastrointestinal conditions and grows in the intestine • Toxin is produced in the intestine • Toxin reaches and interacts with intestinal target	• Pathogen is present in food • Ingestion of pathogen at level ≥ infectious dose • Pathogen survives gastrointestinal conditions and may even grow in the intestine • Pathogen crosses intestinal barrier • Pathogen may reach and interact with target organ(s)
Pathogen examples	• *Clostridium botulinum* • *Staphylococcus aureus*	• *Clostridium perfringens* • *Vibrio cholerae*	• Enteroinvasive *Escherichia coli* • *Listeria monocytogenes*

tion. Foodborne infections include diseases caused by invasive or non-invasive pathogens. Non-invasive infection (or toxico-infection) is caused by microorganisms that thrive in the intestine (i.e., enteric) and produce toxins; their close proximity to the intestinal wall enhances their ability to cause the disease. These microorganisms do not need to invade the intestinal wall or body tissues to cause the disease. An example of foodborne pathogens that cause non-invasive infections is *Clostridium perfringens*. Invasive infections, however, result from pathogens that invade the intestinal wall, and may also reach other body tissues. *Listeria monocytogenes* is an example of this type of pathogen. Note that some foodborne pathogens can use more than one mode of disease transmission. *Clostridium botulinum*, for example, causes conventional botulism, which is an intoxication, but it can also cause infant botulism, which is a non-invasive infection (toxico-infection).

Contamination of a food with a pathogen may or may not lead to a foodborne disease. Presence of a small population of a non-enteric toxigenic microorganism may not be sufficient to cause health hazards if the microorganism is unable to grow and produce its toxin in food. Therefore, for intoxication to occur, presence, growth, and toxin production in food are prerequisites for disease transmission. For infectious pathogens, the mere presence and survival of the pathogen in the food may be sufficient to cause a disease when this food is consumed. In this case, the pathogen needs to be present at a threshold concentration, called infectious dose. The infectious dose varies with the pathogen, the carrier food, and the susceptibility of the individual consuming the food. Consumers who are most susceptible to foodborne diseases are the very old, the very young (e.g., infants), the pregnant, and immuno-compromised individuals.

Detection vs. Enumeration of Pathogens in Food

When a pathogen is present in food, only a small population (often < 10^3 CFU/ ml or CFU/g) is expected and that population may not homogeneously distributed in the food lot. Attempts to enumerate such a small population would reveal little or no information considering the high detection level of enumeration methods and the lack of specificity of most enumeration techniques. Therefore, most analyses for foodborne pathogens are designed to determine their presence or absence in food. These analyses, therefore, are called detection or presence/ absence methods. An enrichment in larger quantities of food (25–375 g) is typically used to improve analyst's ability to detect these pathogens. Pathogens that cause infections (e.g., *L. monocytogenes* and Shiga toxigenic *E. coli*) should not be present in ready-to-eat food at any detectable levels. Presence of these pathogens in raw food before processing is also not desirable due to potential cross-contamination of the finished product.

Presence of a small population of intoxication-causing pathogens in raw food before processing may be tolerated provided that no toxin has already been produced and that there is a processing step that eliminates or prevents the growth of the pathogen in the finished product. A small number of *Clostridium botulinum* spores, for example, may be present on raw beans, but the properly retorted canned product should not contain any of the pathogen's viable spores. In this case, testing for *C. botulinum* in the raw product may not be justifiable; however, the retorted products are subjected to sterility testing. If the final product is believed to be involved in a botulism outbreak, testing of the retorted cans for botulinum neurotoxin may be warranted. Manufacturers of minimally processed food, particularly meat products, may test the processed product for toxigenic microorganisms such as *S. aureus* or *Bacillus cereus*. A high population of these pathogens is alarming, thus determination of the count of these pathogens in finished food is carried out occasionally.

PATHOGEN DETECTION METHODS

Any detection method starts with sampling the food to be analyzed and ends with a report indicating if the pathogen is present or absent. If present, the analyst ideally provides a confirmed isolate of the detected pathogen. Such an isolate can be stored frozen for future referencing, pathogenicity testing, whole genome sequencing, or other purposes.

There are two approaches in pathogen detection: culture-based and culture-independent. In culture-based methods, the pathogen in the food sample is subjected to an enrichment procedure, screening and selection on selective and differential growth media, isolation, and identification by a combination of culture and biochemical techniques. Methods that are totally culture-based are labor and resource intensive, yet some of these methods are the gold standard in pathogen detection. Culture-independent methods make use of rapid techniques such as polymerase chain reaction (PCR), immunoassay, or biosensor technology. An example of a totally culture-independent method includes subjecting the food sample to DNA extraction and analyzing the extracted DNA metagenomically for pathogen signature sequences. Although culture-independent methods promise to speed up or even automate pathogen detection, they have inherent problems: (i) the analysis does not result in live pathogen isolates – such isolates would be beneficial in future studies as indicated earlier; and (ii) considering that pathogen populations are typically small in food, which may also contain a complex microbiota, the amount of pathogen's DNA extracted is often insufficient to make determination about its presence or absence. Most official pathogen detection methods are hybrids that combine culture-based and culture-independent techniques.

The following discussion is directed to hybrid methods which are the most commonly used by the industry and regulatory agencies. These methods typically involve the following steps: (i) food sampling and sample preparation; (ii) enrichment of the targeted pathogen in the sample; (iii) screening the enrichments for signs indicating the presence of the pathogen; (iv) isolating the pathogen from presumptive positive enrichments; (v) confirming the identity of the isolated pathogen; and (vi) typing the pathogen (i.e., identifying it at the subspecies level) if possible and desired.

Sampling and Sample Preparation

Sampling and sample preparation, discussed in Chapter 2, are applicable to this step of pathogen detection.

Enrichment

Enrichment involves homogenizing the food sample in a suitable liquid enrichment medium and incubating the mixture under conditions appropriate for the targeted pathogen. The purpose of the enrichment is to increase a pathogen's population. This population increase makes it easy to detect the pathogen during the subsequent steps. For example, if PCR is used to screen the enrichments, it is estimated that pathogen population should be 10^3–10^4 CFU/ml, at least, for obtaining high yield and good quality DNA. If immunoassay is used in the screening step, a pathogen population of 10^4 to 10^5 CFU/ml in the enrichment is needed to meet the assay minimum detection limit. The enrichment process also helps in resuscitating pathogen cells that may have been injured during food processing or due to the presence of inhibitory ingredients in the food; this helps these cells endure the selective pressure of media used in subsequent steps.

A multi-step enrichment technique usually includes culturing in a nonselective broth medium (pre-enrichment or primary enrichment), followed by subculturing in a selective broth medium (secondary enrichment). During the pre-enrichment step, both the microorganism of interest and the other food microbiota will increase in number. During the secondary enrichment, the microorganism of interest will grow preferentially while the growth of other microorganisms is suppressed. Theoretically, the enrichment is presumed to allow the detection of one pathogen cell in the analytical food sample, which is usually 25 g.

Sample Screening

Professional food microbiological analysts commonly analyze many more negative than positive samples when they test commercial food for a given pathogen. Therefore, it is beneficial to include a screening step to quickly cull the negative samples. To speed up the screening, targeted pathogen may be concentrated from the enrichment using immunomagnetic separation. This technique not only concentrates pathogen cells, it also decreases food and media contaminants in the test sample.

In culture-based methods, screening can be done using selective or selective-differential media. Alternatively, a PCR or immunoassay test is often carried out to speed the determination of sample status. Information about PCR and immunoassay procedures is provided under the Identification section. Although samples that are found to be pathogen-negative during the screening do not need to be analyzed further, positive samples need to be subjected to the additional detection steps to confirm the test result. In teaching laboratories, students may continue the analysis despite the clues about the absence of the pathogen in the food sample.

Pathogen Isolation

Isolation of a pathogen from food is possible when the enrichment is plated on a suitable selective or selective-differential medium. Selection uses culture media that permits growth of the target bacterium while killing or preventing the growth of most other non-target microbes. Selection may be accomplished via antibiotics, salts, acids, or other suitable selective agents. Distinguishing the target organism from other organisms present in the sample requires the use of differential agents in the medium. These agents include pH indicators that show whether an organism can utilize a carbohydrate in the medium or dyes that change the color of colonies where particular enzymes are produced and catalyze a particular reaction. Many media combine selective and differential properties, providing the ability to select and differentiate simultaneously. Successful completion of this step, on pathogen-containing samples, should result in obtaining isolates that can be selected for further testing and confirmation.

Identification of Pathogen Isolate

Identification normally refers to naming the microorganism to the genus, and preferably the species level. Identification of microorganisms at the subspecies level is known as typing. Ideally, identification and typing are done in one step, but if this is not feasible, the two steps can be done sequentially. Microbiological identification can be completed using biochemical test, immunological assay, or genetic techniques.

Biochemical tests

These tests help identify targeted microorganisms by measuring their ability to (i) utilize specific sugars; (ii) produce certain enzymes; or (iii) produce key metabolites. Biochemical testing is applied on cultures in broth, isolated colonies from agar media plates (i.e., isolates), or isolate cell suspensions. These are mixed with reagents in suitable media, mixtures are incubated, and results are observed and recorded. Note that a biochemical test can be as simple as transferring a portion of a colony to a slide, adding a drop of hydrogen peroxide solution, and watching for gas bubble formation. Formation of gas bubbles, in this case, indicates that the isolate is catalase positive. Results from several biochemical tests can be used to identify the isolate. The more key tests that are completed, the greater the confidence in the identification results. To complete biochemical identification efficiently, companies have developed elaborate biochemical identification kits. One of these kits is the API diagnostic system, which uses a strip of wells containing small portions of dehydrated media (Chapter 12). Different kits are used for different targeted bacteria.

Immunological assays

An immunoassay simply documents an interaction between an antigen and antibody. Many microbial cell components are antigenic (i.e., when injected in an animal, they trigger immune response in the form of antibodies). Different antibodies are produced in response to different antigens. In immunoassays, we take advantage of the specificity between antigens and antibodies for microbial identification purposes. For example, a flagellar protein (called flagellin) from *E. coli* reacts only with a specific antibody, which does not react with any flagellins from *S. enterica*. Therefore, this specific antibody can be used to differentiate between *E. coli* and *S. enterica*. It is even more interesting that different flagellins from the same species interact with different antibodies. Hence, it is possible to use immunoassays not only to differentiate between species, but to also to differentiate different "types" within the same species. In other words, antigen–antibody interaction can be used in species identification as well as typing.

There are many types of immunoassay, the simplest of which is the agglutination test. In this test, an antibody specific to an antigen on the target bacterial cell is used. Mixing the antibody with the target bacterial isolate results in visible clumping of bacterial cells. Considering that the clumping is sometimes hard to detect, a latex agglutination test was developed. In this case, latex particles are coated with antibodies that produce easily visible precipitate when mixed with the bacterial cell suspensions.

Enzyme immunoassay, or more specifically, enzyme-linked immunosorbent assay (ELISA), is a popular immunological method for detection of pathogens. ELISA can be used not only to detect, but also to quantify a targeted antigen. In one of the direct ELISA formats (Fig. III.1), plastic wells are coated with the specific antibody against an antigen of the targeted microorganism. When the analyte containing the targeted antigen is added to the wells, the target will bind to the antibody, creating an immune complex that remains attached to the well's

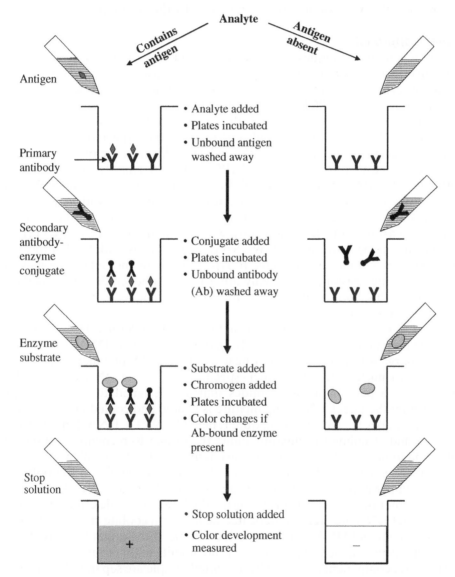

Figure III.1 Enzyme-linked immunosorbent bioassay (ELISA) technique for detection of antigens.

plastic surface. An additional enzyme-labeled (conjugated) secondary antibody is added to bind to the immune complex. Presence of the enzyme in the well, after washing, indicates the presence of the target. Horseradish peroxidase is commonly used to label the secondary antibody. Presence of the enzyme is detected by addition of an appropriate chromogenic substrate. This is typically seen as a color development in the well. Color intensity can be used to estimate relative concentration of the antigen.

Immunological methods also include serotyping (i.e., classifying microorganisms at the sub-species level into serotypes). Serotyping is particularly important in the case of identification of *S. enterica* in food. This species includes hundreds of serotypes that were originally differentiated by this technique.

Genetic approach

Different genetic techniques have been used to identify bacterial species. DNA-DNA hybridization was used in the past as an identification tool, but it is a labor-intensive technique. Currently, PCR-based techniques are commonly used for bacterial identification. PCR is used to amplify signature DNA sequences within the genome of the targeted species. If the PCR procedure results in the amplification of a DNA product (an amplicon) that has the expected characteristics (measured as size or melting point), this indicates the presence of that species in the food sample analyzed. Alternatively, PCR is used to amplify a variable region within the 16S rRNA gene in bacterial chromosome. In this case, the PCR product is purified and sequenced. Sequence results are matched with sequences in genome databases (e.g., NCBI); this leads to the identification of the analyzed isolate at least to genus level. Recently, whole genome sequencing (WGS) has become increasingly affordable. If WGS is carried out, this leads to identification of the species and makes it possible to type the isolate and determine its virulence, antibiotic resistance, and many other traits. Pathogen detection methods reported in this book make use of the PCR technique to reveal signature sequences in the targeted pathogens.

The PCR technique is used to replicate *in vitro* a predefined DNA sequence from the target microorganism (Fig. III.2). In this technique, DNA from the target microorganism is mixed with a heat-resistant DNA polymerase (e.g., Taq polymerase), nucleotides, and primers specific to the target DNA sequence. The primers are a pair of oligonucleotides that are designed to flank a unique sequence on the target DNA. The mixture is heated and cooled in timed cycles using a thermocycler. When the mixture is heated (~94°C), DNA denatures into single strands. Cooling this mixture allows the primer to recognize and specifically bind to the target DNA (i.e., anneal). Temperature of annealing varies with the primer sequence and structure. For optimum functionality of the polymerase, the temperature of the mixture is raised, generally to 68–72°C. At this time, DNA polymerase extends the primer using the free nucleotides, thus two copies of the target DNA are created. Repeated heating and cooling create millions of copies of the target DNA in a short period of time. If the target DNA is absent, the primer cannot anneal (bind) and thus no amplification occurs.

The DNA components of the PCR-amplified sample are separated on agarose gel by electrophoresis. A rectangular slab of agarose gel is prepared with wells

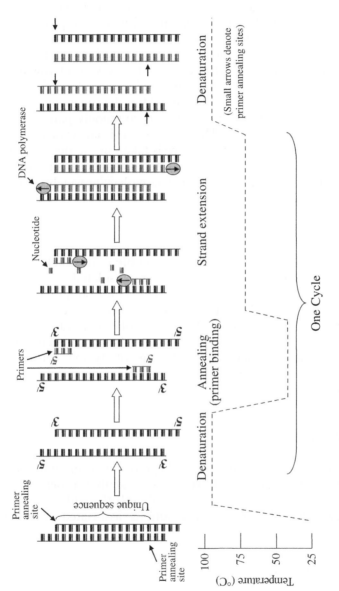

Figure III.2 Polymerase chain reaction (PCR) method for amplifying unique DNA sequence for subsequent detection by gel electrophoresis. The cycles shown are repeated sufficiently (e.g., 35 times) to produce detectable PCR product.

on one end of the rectangle. The reaction samples, containing the PCR products, are dispensed (loaded) in the gel wells, along with a tracking dye. A mass marker (DNA ladder) is also loaded in a separate well. Electricity from a power supply is applied to separate DNA fragments on the basis of mass and charge. Considering that DNA is negatively charged, its bands run from the negatively charged to the positively charged electrode. Large fragments move more slowly than the smaller ones. To make the DNA band visualization possible through ultraviolet (UV) radiation, the gel is stained. Ethidium bromide, a DNA-binding molecule, was traditionally the most commonly used stain. The ethidium bromide-DNA complex fluoresces when exposed to UV. Considering the potential health hazards associated with ethidium bromide use, alternative stains and techniques are currently used more commonly. After the run is complete, the gel is visualized on a UV transilluminator, which is commonly part of a camera-equipped gel documentation system. This allows for a visual evaluation of amplicon (band) presence and size. After a picture is taken of the fluorescent bands, the size of separated bands, in base pairs (bp), is determined relative to the mass marker. The presence of a band with the correct molecular weight indicates the presence of the pathogen in the sample.

Typing

If a food is suspected to be involved in a disease outbreak, it is analyzed to detect the corresponding pathogen. Once the pathogen is isolated and identified, it becomes necessary to determine if it is the same strain responsible for the disease outbreak. Therefore, food and patient isolates need to be compared. This pathogen transmission tracking is possible if the isolates are identified beyond the species level (i.e., typed). Many typing approaches have been applied for decades; these include biotyping, phage typing, ribotyping, serotyping, etc. Currently, WGS is reliably used in pathogen tracking, making it unnecessary to use conventional typing methods.

IMPORTANT CONSIDERATIONS

Analyses for the detection of foodborne pathogens start with food that is most likely safe to handle by the analyst. However, as the analysis progresses, isolates that are potentially pathogenic increase in number. If the pathogen is found, it becomes necessary to take all the precautions associated with handling pathogens.

Controls and Surrogates

In addition to food samples, students also run parallel analysis of positive and negative controls for the targeted pathogen. A positive control is a microorganism that is similar to the targeted pathogen. Because of safety concerns, a non-virulent strain may be used as a positive control provided the strain has similar characteristics and produces most of the typical reactions of the microorganism of interest. On the contrary, a negative control is a microorganism that does not produce typical reactions of the one targeted by the analysis, but it still grows on most media used in the analysis. Analyzing the positive and negative controls allows the analyst to verify the correctness of the procedure and to test the quality of media and reagents.

Method Sensitivity and Specificity

Sensitivity is the proportion of positive test results obtained when the method is applied to samples known to carry the microorganism targeted by the analysis. To illustrate the concept, suppose that 100 food samples have been inoculated with a pathogen and analyzed by a method designed to detect this pathogen. If the method detects the pathogen in 97 of these samples, method sensitivity is 97%. This leaves three samples that tested negative, while in fact these are supposed to be positive. Therefore, the method produces a false-negative rate of 3%. Method sensitivity also determines the minimum detectable concentration of the targeted microorganism in the contaminated sample. A method with high sensitivity detects pathogens present in smaller numbers in the sample, compared to a method with low sensitivity.

Specificity defines the method's ability to distinguish a targeted microorganism in the tested sample from other microorganisms in the sample. For example, if 100 food samples, containing natural microbiota, were free from a pathogen of concern but the analytical method produced positive results in five samples, the method's specificity is 95%. The method is also described as producing a false-positive rate of 5%.

SELECTED REFERENCES

Carpenter, A.B. (2007). Immunoassays for the diagnosis of infectious diseases. In: *Manual of Clinical Microbiology*, 9e (ed. P.R. Murray, E.J. Baron, M.L. Landry, J.H. Jorgensen, and M.A. Pfaller), 257–270. Washington, DC: American Society for Microbiology.

Dwivedi, H.P., Smiley, R.D., and Pincus, D.H. (2015). Rapid methods for the detection and identification of foodborne pathogens. In: *Compendium of Methods for the Microbiological Examination of Foods*, 5e (ed. Y. Salfinger and M.L. Tortorella), 127–152. Washington, DC: American Public Health Association Press.

Wang, Y. and Salazar, J.K. (2016). Culture-independent rapid detection methods for bacterial pathogens and toxins in food matrices. *Comprehensive Reviews in Food Science and Food Safety 15*: 183–205.

Yousef, A.E. (2008). Detection of bacterial pathogens in different matrices: current practices and challenges. In: *Principles of Bacterial Detection – Biosensors, Recognition Receptors and Microsystems* (ed. M. Zourob, S. Elwary, and A. Turner), 31–48. New York, NY: Springer.

CHAPTER 10

Staphylococcus aureus

INTRODUCTION

Staphylococcus spp. are Gram-positive bacteria that belong to the phylum Firmicutes. Members of the genus have coccus cell morphology; these cells are found singly, in pairs, and short chains, or characteristically in irregular grape-like clusters. This cluster appearance reflects the cells' ability to divide along multiple axes, as opposed to streptococci, which divide along only one axis to form chains. *Staphylococcus* spp. are catalase positive, but they are oxidase negative due to the absence of c-type cytochromes. Members of the genus are mesophilic, facultative anaerobic bacteria. Being facultative anaerobes, these bacteria can metabolize carbohydrates both fermentatively (i.e., anaerobically) and oxidatively (i.e., aerobically). These species are also capable of growth in harsh conditions, including in the presence of bile salts and high levels of sodium chloride (10% or higher).

The genus includes more than 40 species; some are coagulase positive, and others are coagulase negative, but both categories include species that cause human diseases. *Staphylococcus* spp. are ubiquitous colonizers of skin and mucous membranes of humans and other warm-blooded animals. *S. aureus* colonizes mainly the nasal cavity, whereas *S. epidermidis* inhabits the skin. *Staphylococcus* spp. also contribute to the microbiota of nectar as well as other habitats. The primary species of interest in clinical and food microbiology is *S. aureus* (Table 10.1).

Staphylococcus aureus is a coagulase-positive and β-hemolytic bacterium that produces a heat-stable nuclease (thermonuclease). The bacterium can utilize mannitol aerobically and ferment it anaerobically to produce acidic end-products. The bacterium produces lecithinase, thus it is capable of hydrolyzing egg yolk. Golden pigmentation of *S. aureus* colonies is often used as one of the identifying features of the species. Ability of the bacterium to survive for an

Analytical Food Microbiology: A Laboratory Manual, Second Edition. Ahmed E. Yousef, Joy G. Waite-Cusic, and Jennifer J. Perry.
© 2022 John Wiley & Sons, Inc. Published 2022 by John Wiley & Sons, Inc.

TABLE 10.1 Characteristics of selected *Staphylococcus* spp.

Property	*S. aureus*	*S. epidermidis*	*S. hyicus*	*S. intermedius*
Coagulase	+	−	W	+
Heat-stable nuclease	+	−	+	+
Yellow pigment	+	−	−	−
Hemolysis	+	V	−	+
Mannitol fermentation anaerobically	+	−	−	−
Enterotoxin production	+	−	+	−

Key: (+) positive; (-) negative; (V) variable; (W) weak reaction.

extended period in a dry state is an important characteristic that concerns food processors. Due to its tolerance to limited water availability, the bacterium may grow in food and media with water activity (a_w) as low as 0.85.

The Disease

S. aureus causes several human illnesses, including a food-transmitted disease, staphylococcal gastroenteritis. The disease is an intoxication that results from the consumption of foods where *S. aureus* grew and produced a toxin known as staphylococcal enterotoxin. The symptoms of this disease appear shortly after ingesting the contaminated food. Disease onset time is 4–6 hours and its severity depends on the susceptibility of the individual, amount of toxin ingested, and other factors. Common symptoms include nausea, vomiting, stomach cramping, and diarrhea. In severe cases, dehydration, muscle cramping, and changes in blood pressure and pulse rate may occur. Affected individuals typically recover within two days but the disease may last longer in severe cases.

The Toxins

Although staphylococcal enterotoxins are commonly associated with *S. aureus*, these toxins may also be produced by other *Staphylococcus* spp. (i.e., *S. hyicus*). The foodborne disease (staphylococcal gastroenteritis), however, is almost always associated with *S. aureus*. It is notable that some strains of *S. aureus* may not produce the toxin. Staphylococcal enterotoxin is a family of heat-stable toxic proteins with some structural variability, but they have similar function and some amino acid sequence homology. Staphylococcal enterotoxins A through E and G through J are all members of this family of toxins. These proteins vary in size from 25 to 29 kDa and exhibit resistance to gastrointestinal proteases (i.e., pepsin and trypsin). The enterotoxin most associated with staphylococcal gastroenteritis is SEA, but SEB, SED, and SEE are also responsible for some of the disease outbreaks. Genes encoding for these toxins may be on the chromosome or on plasmids.

The toxin is typically produced in food in detectable levels only when the population of *S. aureus* exceeds 10^6 CFU/g. If pathogen growth and toxin production occur prior to cooking or processing, the toxin may survive the treatment and pose health risks to consumers. Ingestion of less than 1 μg (200–400 ng) of the toxin may cause staphylococcal gastroenteritis in humans.

Incidence of The Pathogen in Food

It is estimated that 30% of the human population are carriers of *S. aureus*. Therefore, contamination with this pathogen may result from improper handling of food by workers who are pathogen carriers. If the contaminated food is not cooked or processed properly, the pathogen may grow and produce its toxin (staphylococcal enterotoxin). Subsequent cooking or processing of such food may be sufficient to eliminate the pathogen, but there is a greater chance for the toxin to survive the processing and pose risk to consumers. For example, pasteurization of milk (e.g., 71.6°C for 15 s) is sufficient to eliminate a considerable population of *S. aureus,* but the treatment does not affect the toxin. Foods with a history of causing staphylococcal gastroenteritis include meats (particularly sliced meats), dairy products (e.g., raw milk and cheese), bakery products (e.g., cream-filled pastries, cream pies, and chocolate eclairs), puddings, certain types of salad (e.g., egg, tuna, chicken, potato, and macaroni salads), and sandwich fillings. *S. aureus* is one of the causes of mastitis; hence, milk from infected animals may contain the pathogen.

Analyzing Food for *S. aureus* and Its Toxins

Foods are analyzed for this pathogen when the processor needs to determine if a food or a food ingredient is a potential source of the pathogen, or that the processing is effective to eliminate pathogens in the finished product. Interest in analyzing food for this pathogen increases when it is suspected to be the causative agent in a foodborne disease outbreak. If a food is analyzed for *S. aureus*, it is more common to enumerate rather than detect the pathogen. Foods implicated in staphylococcal gastroenteritis often contain large populations of *S. aureus* and thus enumerating the pathogen in food may reveal the associated risk. The presence of a large *S. aureus* population should alert food processors to take corrective measures. For reasons explained previously, a food may not contain a countable *S. aureus* population but may still contain the enterotoxin at hazardous levels. If such a food is implicated in staphylococcal gastroenteritis, it would be meaningful to analyze it for the toxin.

Based on the previous discussion, it is valuable to practice two methods related to this pathogen: (i) detection and enumeration of *S. aureus* – this will be covered in the first exercise of the chapter; and (ii) detection and quantification of the staphylococcal enterotoxin. Unfortunately, quantifying the toxin in food is a specialized task that is outside the scope of this training book. An alternative approach is to determine if *S. aureus* isolated from the food sample can produce enterotoxins. This alternative approach will be followed in the second exercise of this chapter.

SELECTED REFERENCES

Bennett, R.W., Hait, J.M., and Tallent, S.M. (2015). *Staphylococcus aureus* and staphylococcal enterotoxins. In: *Compendium of Methods for the Microbiological Examination of Foods*, 5e (ed. Y. Salfinger and M.L. Tortorella), 509–526. Washington, DC: American Public Health Association Press.

Landgraf, M. and Destro, M.T. (2013). Staphylococcal food poisoning. In: *Foodborne Infections and Intoxications*, 4e (ed. G.J. Morris and M. Potter), 389–400. London, UK: Elsevier.

Tang, J.-N., Shi, X.-M., Shi, C.-L., and Chen, H.-C. (2006). Characterization of a duplex polymerase chain reaction assay for the detection of enterotoxigenic strains of *Staphylococcus aureus*. *Journal of Rapid Methods and Automation in Microbiology* 14: 201–217.

Todar, K. (2021). *Staphylococcus aureus* and staphylococcal diseases. Online textbook of bacteriology. http://textbookofbacteriology.net/staph.html

EXERCISE I: ENUMERATION, ISOLATION, AND IDENTIFICATION OF *Staphylococcus aureus* IN FOOD

PATHOGEN ENUMERATION. PATHOGEN IDENTIFICATION. MOST-PROBABLE NUMBER TECHNIQUE.

INTRODUCTION

This exercise is a culture-based method to enumerate *S. aureus* in food. If present, *S. aureus* typically constitutes a small sub-population of food microbiota, and the pathogen's numbers are most likely too small to be countable by the plating technique. Hence, the most-probable number (MPN) approach, which typically has a lower detection limit than the plating technique, will be used in this exercise.

Media used for enumeration, by the plating or MPN technique, should be selective enough to allow the multiplication of the target organism (i.e., *S. aureus*) while suppressing non-target microbiota. Unfortunately, excessive selectivity of such a medium may discourage the recovery of stressed or injured members of the targeted population. Stress and injury of food microbiota may occur throughout the supply chain, including during the production, sublethal processing, and storage steps. As a compromise, the enumeration media are made to be selective enough to allow a good representation of the targeted population but not excessively selective so that stressed or injured cells are not suppressed. To compensate for the decreased selectivity, the enumeration medium may contain differential factors that allow a phenotypic distinction between the target and non-target microorganisms. In other words, selective-differential media are suitable for counting small sub-populations, such as that of *S. aureus*, in a food containing a complex microbiota. It should be cautioned, however, that the most selective media available commercially will not completely suppress the growth of all non-target microorganisms in food. Therefore, the population detected by these enumeration media should be considered presumptive and their identity may need to be confirmed by additional analyses.

Some of the ingredients that can be used to selectively enumerate *S. aureus* include salt (NaCl) at ~10% level, lithium chloride (LiCl) at 0.5%, or potassium tellurite (K_2TeO_3) at a level up to 0.05%. The bacterium-identifying characteristics include fermentation of mannitol, reduction of tellurite to produce black colonies, and hydrolysis of an egg yolk component due to the production of lecithinase. The production of coagulase and thermonuclease are other important traits that can be used to differentiate *S. aureus* from other microorganisms. Some of these characteristics are used to design selective-differential media suitable for isolation and enumeration of *S. aureus* (e.g., Baird-Parker agar). Some advances have been made in selective-differential media for *Staphylococcus* spp. Examples of these media are "Rabbit plasma fibrinogen agar" and "CHROMagar™ Staph aureus."

OBJECTIVES

1. Enumerate *Staphylococcus aureus* in food using the MPN technique.
2. Practice isolation technique using selective-differential media.
3. Use identification tests to determine whether isolates from food are *S. aureus*.
4. Practice safe procedure for handling pathogens in the laboratory.

MEDIA AND TESTS

Baird-Parker Agar

This is a selective-differential medium specifically designed for the detection and enumeration of *S. aureus*. The bacterium forms black, shiny, convex colonies surrounded by an opaque zone, frequently with an outer clear zone. Other species may produce gray or less shiny black colonies. Lithium chloride and potassium tellurite in this medium select against most non-*Staphylococcus* spp. Differentiation of *Staphylococcus* spp. is brought about by tellurite and egg yolk. *S. aureus* reduces the tellurite salt to elemental tellurium, resulting in the formation of black colonies and producing a clear halo due to hydrolysis of lecithin.

Blood Agar

Blood agar is a rich medium that supports the most fastidious microorganisms. It is made of an agar medium (e.g., blood agar base or Columbia blood agar base), which is autoclaved, and then 5–10% blood (e.g., sheep blood) is added. The reaction of microorganisms on blood agar media typically fits one of these categories:

α-hemolysis: This is caused by the reduction of the red blood color to a greenish discoloration around the colony. This phenotype results from the breakdown of red blood cells and the conversion of blood hemoglobin to methemoglobin by the hemolytic microorganism.

β-hemolysis: This is manifested as a clear zone around the colonies due to the lysis of red blood cells and breakdown of hemoglobin by the hemolytic microorganisms.

γ-hemolysis: In this case, colonies do not change blood color. This indicates no blood hemolysis.

Notes:
- *It is not uncommon that a microorganism causes double hemolysis; an inner β-hemolysis and an outer α-hemolysis zone.*
- *It was reported in published literature that media containing digested plant proteins (e.g., tryptose soy agar) might contain cellobiose, which inhibits hemolysis.*
- *Researchers suggested using washed erythrocytes, instead of raw blood, as the latter may contain antibodies that interfere with hemolysis.*

S. aureus grows well on blood agar and expresses different types of hemolysins. These hemolysins cause clearing of media (color and opacity) around the colonies, a phenotype considered β-hemolysis.

Brain-Heart Infusion Broth Supplemented with Yeast Extract

Brain heart infusion supplemented with yeast extract (BHI-YE) is a rich non-selective medium used for the cultivation of various microorganisms. This medium will be used to propagate the confirmed *S. aureus* isolates in preparation for enterotoxin production in Exercise II.

Tryptic Soy Agar

This is a rich medium that meets the growth requirements for *S. aureus* and many other microorganisms.

Tryptic Soy Broth Supplemented with Salt and Pyruvate (TSB-SP)

The medium is made of tryptic soy broth with additions of 10% sodium chloride and 1% sodium pyruvate. The rich component of the medium provides the bacterium with general metabolic needs. Sodium chloride is added to select for salt-tolerant microorganisms, including staphylococci, present in the food sample. Sodium pyruvate aids in the recovery of cells that may have been stressed or injured during food processing.

Catalase Test

The presence of active catalase enzyme is determined by adding hydrogen peroxide to a specimen of an isolated colony on a glass slide. If effervescence is observed, then the colony is catalase-positive.

Gram Reaction

Gram staining, coupled with light microscopy, is recommended considering it is useful in observing the typical cell clustering of *S. aureus*. The procedure was described in Chapter 4.

Coagulase Test

In the medical field, coagulase-positive staphylococci isolates are presumed to be highly pathogenic. *S. aureus* is coagulase-positive but other *Staphylococcus* spp. that colonize the human body (e.g., *S. epidermidis*) are coagulase-negative. This test relies on the ability of isolates to coagulate citrated rabbit plasma. Coagulase-positive *S. aureus* turn the plasma into a cloudy gel-like (viscous) clot, whereas coagulase-negative isolates do not cause clotting of the plasma.

Coagulase production is an important characteristic of *S. aureus*. The enzyme converts fibrinogen (a soluble blood protein) into the insoluble fibrin; this

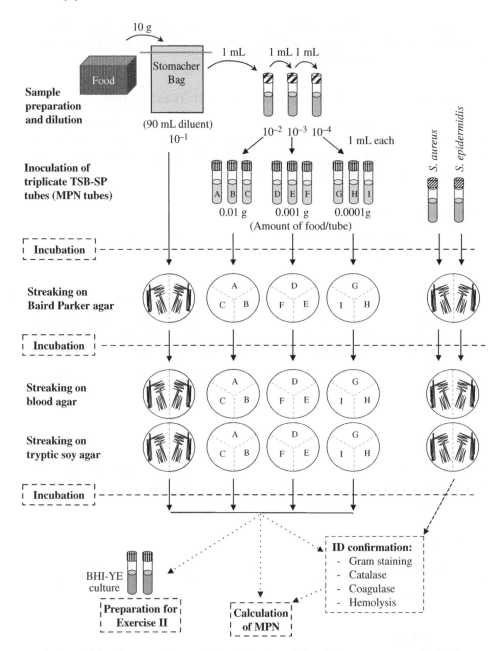

Figure 10.1   Enumeration and identification of *Staphylococcus aureus* in food.

conversion clots the blood or its plasma. However, there are two types of coagulases produced by *S. aureus*: (i) free coagulase, which is produced in liquid cultures; and (ii) cell-bound coagulase (also known as clumping factor) which remains attached to the producing cells. The free coagulase is detected by the "tube coagulase" test, whereas the bound coagulase is detected using the "slide coagulase" test. The latter test will be used in this exercise.

PROCEDURE OVERVIEW

An overview of the procedure used in Exercise I is shown in Figure 10.1. During the first session of this exercise, selected foods will be homogenized, and additional serial dilutions prepared. These dilutions will be used to inoculate multiple tubes containing TSB-SP, which is a selective medium for salt-tolerant bacteria such as staphylococci, as described previously. These inoculated tubes will be incubated and scored for growth (turbidity) or no growth (no turbidity). To ascertain that the turbidity is caused by the growth of *S. aureus*, screening and confirmation steps will be completed. During the second session, the incubated TSB-SP will be streaked onto Baird-Parker agar to determine the presence or absence of *Staphylococcus* sp. in each turbid tube. In the third session, colonies displaying the typical morphology of *S. aureus* colonies will be isolated by streaking on blood agar and TSA. These isolates will be confirmed as *S. aureus* during the fourth session by observing blood hemolysis and testing for catalase and coagulase reactions, and by Gram staining coupled with light microscopy. After these confirmations, it will be possible to determine with confidence the *S. aureus* MPN in food. Additionally, the identified isolates will be used in Exercise II of this chapter.

If the population of *S. aureus* is below the detection limit of the MPN technique just described, it would be beneficial that the analysts determine if the food simply does or does not contain the pathogen. Therefore, the exercise will include the following detection approach (Figure 10.1). The homogenized food sample, prepared in Session 1, will be incubated (to produce an enrichment) and analyzed by plating on Baird-Parker agar, isolation of suspect *S. aureus* colonies, and confirmation of these isolates as described previously.

Organization

Each pair of students will test one food. Possible foods include raw ground meats, fresh (raw) or fermented sausages, tuna salad, potato salad, soft or mold-ripened cheeses, and frozen bread dough. Previously described procedures for food sampling, homogenization, and dilution will be followed in this exercise. Confirmed *S. aureus* isolates will be tested in Exercise II for the presence of enterotoxin genes and expression of these genes to produce enterotoxins.

Personal Safety

S. aureus is an infectious and toxigenic bacterium that should be handled with care. Follow the safety guidelines that were reviewed in Chapter 1 of this book. Use disposable gloves when handling the isolates and the pathogenic cultures. Make sure to sanitize the work area after use.

SESSION 1: SAMPLING, HOMOGENIZATION, DILUTION, AND INOCULATION OF MPN TUBES

During this session, the selected food will be homogenized and 10^{-2}, 10^{-3}, and 10^{-4} dilutions will be made. Dilutions are used as the inoculum for the TSB-SP tubes for MPN analysis. Tubes will be incubated at 35°C for 48 hours and examined during the subsequent laboratory session. See Figure 10.1 for procedure detail.

MATERIALS AND EQUIPMENT

Per Pair of Students

- Food Sample
- One bottle of peptone (90 ml each)
- Three peptone water tubes (9 ml each) for preparing dilutions
- Nine TSB-SP tubes (9 ml each)
- Other common supplies such as micropipettes and sterile pipette tips

Class-Shared

- Scale for weighing food samples (e.g., a top-loading balance with 500 g capacity)
- Tools for sample preparation
- Stomacher and stomacher bags
- Common laboratory equipment (e.g., incubator, set at 35°C)

PROCEDURE

1. **Sample Preparation and Making Dilutions**
   a. Prepare the food for sampling to ensure that a representative sample is analyzed. This varies with the food and may involve cutting, partitioning, grinding, and/or mixing.
   b. Weigh 10 g of food directly into a stomacher bag, held upright in a light holder (e.g., wire basket), then add 90 ml peptone water.
   c. Homogenize the food-diluent mixture for 2 min in a stomacher.
   d. Prepare 10^{-2}, 10^{-3}, and 10^{-4} dilutions using the peptone water diluent tubes. *Note: A lower dilution series would ideally be used to decrease detection limit, but MPN tubes receiving dilutions lower than 10^{-2} are likely to become turbid prior to incubation due to the turbidity from food particles in the 10^{-1} dilution.*
   e. Incubate the homogenized sample, in the stomacher bag, at 35°C for 48 hr; this will be used for direct detection of *S. aureus* if the population is too small to enumerate by the MPN technique.
2. **Inoculating TSB-SP Tubes (MPN Tubes)**
   a. Label three tubes of TSB-SP for each of the 10^{-2}, 10^{-3} and 10^{-4} dilutions. Write on the label the *amount of food (in grams)* dispensed into each tube of the three sets. Additionally, make sure the tubes within each set are numbered as follows: A–C for the lowest, D–F for the middle, and G–I for the highest dilutions (See Table 10.2).

TABLE 10.2 Results of tests needed to determine *Staphylococcus aureus* most-probable number in _____ (name the food).

Test sample	Dilution (sample weight)	Tube No.	Turbidity in MPN tubes	Baird-Parker agar[a]	Blood agar (hemolysis)	Colony morphology on TSA	Gram staining[b]	Catalase[b]	Coagulase[b]	MPN tube score[c] (+/−)
Food MPN tubes	10^{-2} (0.01 g)	A								
		B								
		C								
	10^{-3} (0.001 g)	D								
		E								
		F								
	10^{-4} (00001 g)	G								
		H								
		I								
Enrichment	(10 g)	n/a	n/a							n/a
S. aureus	n/a	n/a								n/a
S. epidermidis	n/a	n/a								n/a

[a] Typical or atypical colony morphology and reactions.

[b] One analysis only, representing each dilution, is completed to make the task manageable.

[c] If a given tube is positive for turbidity, Baird-Parker agar reactions, and hemolysis, and produced typical colony morphology on TSA, Gram staining, catalase, and coagulase, then the tube will be scored "positive." A tube with all tests negative will be scored "negative." Other situations require careful considerations.

b. Dispense 1 ml of the 10^{-2} dilution into each of the appropriately labeled three tubes (i.e., A–C).

c. Repeat the previous step using the 10^{-3} dilution, dispensing 1 ml in each of the tubes labeled D–F.

d. Repeat the previous step using the 10^{-4} dilution, dispensing 1 ml in each of the tubes labeled G–I.

e. Vortex the tubes and incubate at 35°C for 48 hr.

SESSION 2: SCREENING MPN TUBES USING *Staphylococcus* SELECTIVE-DIFFERENTIAL MEDIUM

In addition to *Staphylococcus* spp., growth of other salt-tolerant bacteria in food may cause the turbidity of TSB-SP tubes. Therefore, it is crucial to screen these tubes using a selective-differential medium. This will be accomplished by streaking the contents of incubated TSB-SP tubes onto Baird-Parker agar, incubating the inoculated plates, and examining them during the subsequent session. Additionally, the incubated, homogenized food (the enrichment) will be streaked on a plate of Baird-Parker agar. Control cultures of *S. aureus* and *S. epidermidis* also will be streaked on the same medium.

MATERIALS AND EQUIPMENT

Per Pair of Students

- Incubated TSB-SP tubes
- Incubated homogenized sample (stomacher bag content, i.e., the enrichment)
- *S. aureus* control culture
- *S. epidermidis* control culture
- Uninoculated TSB-SP tube (negative control)
- Five Baird-Parker agar plates (three for the incubated TSB-SP tubes, one for the food enrichment, and one for the two provided control cultures)

Class-Shared

- Common laboratory supplies and equipment (e.g., incubator set at 35°C)

PROCEDURE

1. **Scoring TSB-SP Tubes**
   a. Vortex each of the TSB-SP tubes and observe for turbidity. For comparison, the uninoculated TSB-SP tube is what a negative tube would look like.
   b. If the medium is turbid, score the tube as positive (+). If no growth is apparent, the tube is scored as negative (−); for comparison, observe the uninoculated TSB-SP tube.
   c. Record the results in Table 10.2.
2. **Streaking the Contents of MPN Tubes on Baird-Parker Agar**
   a. Collect three plates of Baird-Parker agar for inoculation from the TSB-SP tubes. Divide each plate area into up to three segments, as needed, using a marker (Note that the lines should be drawn on the base of the agar plate). Include in the label the amount of food in the source tube, and the tube number to be used for streaking.
   Note: Although it is tempting to use only the positive MPN tubes for streaking, it is preferable to use all nine tubes. It is important, however, to

keep track of which plate segment received the contents of which MPN tube (see Figure 10.1).

b. Using the incubated TSB-SP tubes, streak for isolation onto Baird-Parker agar.

c. Invert all plates and incubate at 35°C for 48 hr.

3. **Streaking Incubated Homogenized Food (Enrichment) on Baird-Parker Agar**

Use the incubated homogenized food sample to streak Baird-Parker agar plate as follows:

a. Mix the bag contents by hand.

b. Using sterile inoculation loop, three-phase streak a loop-full sample of bag contents onto a properly marked Baird-Parker agar plate.

c. Invert the plate and incubate at 35°C for 48 hr.

4. **Streaking *Staphylococcus* Cultures on Baird-Parker Agar**

a. Label one half of a Baird-Parker agar plate for *S. aureus* and the other half for *S. epidermidis*.

b. Three-phase streak these *Staphylococcus* cultures onto the Baird-Parker agar plate.

c. Invert the plate and incubate at 35°C for 48 hr.

Note: It is preferable that inoculated plates, prepared by a student group, are stacked inverted and tied together with a masking tape, before incubation at the same conditions.

SESSION 3: INSPECTING AND ISOLATING PRESUMPTIVE *Staphylococcus aureus* COLONIES

In this session, incubated Baird-Parker agar plates with colonies having typical *S. aureus* morphology and reactions will be observed. This information will help in validating the turbidity results observed in the MPN tubes. These typical colonies will be isolated and streaked on a rich differential medium, blood agar, which will help determine the hemolytic activity of the isolates. The same colonies also will be streaked on tryptic soy agar, a rich non-selective medium. Isolated colonies on TSA agar will be subjected to further identification and confirmation during the next session.

MATERIALS

Per Pair of Students

- Incubated Baird-Parker agar plates
- Five blood agar plates: Three for the food sample, one for the food enrichment, and one for the *Staphylococcus* spp. control cultures
- Five tryptic soy agar plates: Three for the food sample, one for the food enrichment, and one for the *Staphylococcus* spp. control cultures
- Other common supplies such as inoculation loops

Class-Shared

- Disposable gloves
- Other common laboratory supplies and equipment

PROCEDURE

1. **Inspecting Colonies on Baird-Parker Agar Plates for Typical *S. aureus* Characteristics**
   a. **Food MPN tube isolates**
      i. Observe the *S. aureus* (positive control) on the Baird-Parker agar plate. Typical *S. aureus* colonies on Baird-Parker agar produce black colonies with an opaque zone, surrounded with a clearing; an example is shown in Figure 10.2.
      ii. Observe the *S. epidermidis* (control) on the Baird-Parker agar plate. Typical *S. epidermidis* colonies on Baird-Parker agar produce grey or black colonies with no halos; *S. epidermidis* does not produce lecithinase, so it should not have a halo around colonies; an example is shown in Figure 10.2.
      iii. Check for the presence of colonies having typical *S. aureus* morphology and reactions on Baird-Parker agar plates prepared from the TSB-SP tubes (food sample). Compare these with the colonies obtained from the provided *S. aureus* culture.

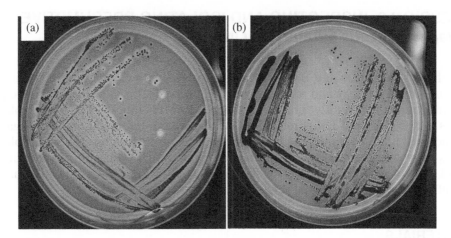

Figure 10.2 Colonies of *Staphylococcus aureus* (a) and *S. epidermidis* (b) on Baird-Parker agar medium.

     iv. Determine which original MPN tubes produced typical colony morphology and reactions on Baird Parker agar.

     v. Report the Baird-Parker agar score in Table 10.2.

b. Food enrichment isolates

     i. Observe the Baird-Parker agar plates streaked from the homogenized food enrichment. Determine whether colonies present match those of typical *S. aureus*.

     ii. Report results in Table 10.2.

2. Isolating Typical *S. aureus* Colonies

a. Food MPN tube isolates

     i. If typical colonies are found on the Baird-Parker agar plates, mark up to nine well-isolated colonies, one colony per plate segment (i.e., MPN tube).

     ii. Using a sterile inoculation loop, transfer the marked colonies to appropriately labeled blood agar plates and TSA plates (Figure 10.1).

     Notes:

       • *The labeling of the Baird-Parker agar plates and the blood agar and TSA plates should match. This facilitates tracking and scoring the original MPN tubes.*

       • *If none of the Baird-Parker agar plates from the food samples produced typical colonies, skip inoculating the corresponding blood agar and TSA plates; use the control cultures for the remainder of the exercise.*

     iii. Invert all inoculated plates and incubate at 35°C for 24 hr.

     Note: If the period between the laboratory sessions is longer than 24 hr, refrigerate the inoculated plates and transfer them to the incubator 24 hr before the subsequent session.

b. Food enrichment isolates

    i. If typical colonies are found on the Baird-Parker agar plate, mark two well-isolated colonies.

    ii. Using a sterile inoculation loop, transfer the marked colonies to appropriately labeled blood agar and TSA plates.

    iii. Invert the inoculated plates and incubate at 35°C for 24 hr.

c. *Staphylococcus* spp. control cultures

    i. Mark a well-isolated *S. aureus* colony and a well-isolated *S. epidermidis* colony.

    ii. Using a sterile inoculation loop, transfer the marked colonies to appropriately labeled blood agar and TSA plates.

    iii. Invert the inoculated plates and incubate at 35°C for 24 hr.

Note: It is preferable that inoculated plates, prepared by a student group, are stacked inverted and tied together with masking tape before incubation under the same conditions.

SESSION 4: CONFIRMING *Staphylococcus aureus* IDENTITY AND CALCULATING *S. AUREUS* MPN

In this laboratory session, colonies isolated on blood agar plates will be examined for hemolysis. Colonies isolated on TSA plates will be subjected to morphological examination and biochemical confirmatory testing. Cell morphology will be determined by Gram staining and microscopy. Catalase and coagulase tests will be completed during this session. This information will help in confirming that the turbidity observed originally in the MPN tubes was caused by *S. aureus*; thus, MPN of *S. aureus* in the original food can be calculated. The analysis also will result in *S. aureus* food isolates that will be sub-cultured and used in Exercise II.

MATERIALS AND EQUIPMENT

Per Pair of Students

- Incubated blood agar and TSA plates
- Hydrogen peroxide solution (6%, 5 ml)
- Undiluted citrated rabbit plasma (1 ml)
- Two 9-ml BHI-YE broth tubes (for culturing isolate to be used in Exercise II)
- Other common supplies such as inoculation loops and microscope slides

Class-Shared

- Microscopes
- Gram staining kit
- Other common laboratory supplies and equipment

PROCEDURE

1. **Inspecting Incubated Blood Agar Plates**
    a. Observe the colonies on the blood agar that came from the provided *S. aureus* culture. Typical *S. aureus* colonies on this medium should be large and produce β-hemolysis.
    b. Observe the colonies on the blood agar that came from the provided *S. epidermidis* culture. Typical *S. epidermidis* colonies on this medium should appear non-hemolytic (i.e., γ-hemolysis).
    c. Observe the colonies on the blood agar that came from the food MPN tubes and food enrichment. Compare the appearance of these colonies with that for the *S. aureus* colonies that came from the provided culture.
    d. Record the hemolysis results in Table 10.2.
2. **Inspecting Incubated TSA Plates**
    a. Observe the colonies on the TSA that came from the provided *S. aureus* culture. Typical *S. aureus* colonies on this medium should have golden pigmentation.

b. Observe the colonies on the blood agar that came from the provided *S. epidermidis* culture. Typical *S. epidermidis* colonies on this medium should appear lighter in color, compared to *S. aureus*.

c. Observe the colonies on TSA that came from the food MPN tubes and food enrichment. Compare the appearance of these colonies with that of the *S. aureus* colonies that came from the provided culture.

d. Record the colony morphology results in Table 10.2.

3. Gram staining and Microscopy

a. Mark up to four typical *S. aureus* colonies on the TSA plates that received food isolates: three representing the three dilutions of the original MPN tubes, and one representing the enrichment. If no colonies representing the MPN tubes exist, select two colonies from the enrichment, if available.

b. Mark one *S. aureus* colony and another *S. epidermidis* colony on the TSA plate that received the control cultures.

c. Use the marked colonies to complete Gram staining as described in Chapter 4 of this book.

d. Examine the cell morphology under the microscope's oil-immersion lens. Cells of *S. aureus* should be Gram-positive cocci. Observe any cell clustering in any of the fields.

e. Record the observations for cell morphology in Table 10.2.

4. Catalase Test

a. Mark up to four typical *S. aureus* colonies on the TSA plates that received food isolates: three representing the three dilutions of the original MPN tubes, and one representing the enrichment. If no colonies representing the MPN tubes exist, select two colonies from the enrichment, if available.

b. Mark one *S. aureus* colony and another *S. epidermidis* colony on the TSA plate that received the control cultures.

c. Using a sterile inoculation loop, transfer masses of the six marked colonies to clean, properly labeled microscope slides, two colonies per slide.

d. Add a small drop of the hydrogen peroxide solution to each colony sample.

e. If immediate effervescence is observed, the colony is positive for catalase. If no effervescence is noted, the colony is negative for catalase.

f. Record results for catalase reaction in Table 10.2.

5. Slide Coagulase Test

a. Mark up to four typical *S. aureus* colonies on the TSA plates that received food isolates: three representing the three dilutions of the original MPN tubes, and one representing the enrichment. If no colonies representing the MPN tubes exist, select two colonies from the enrichment, if available.

b. Mark one *S. aureus* colony and another *S. epidermidis* colony on the TSA plate that received the control cultures.

c. Using the inoculation loop, transfer a small drop of water to a clean, properly labeled microscope slide; two drops per slide.

d. Using a sterile inoculation loop, transfer a portion of each marked colony to the assigned drop of water on the appropriately labeled slide.

e. Emulsify the cell mass in the drop of water, with no spreading, to produce a milky drop of suspension.

f. Dip a sterile inoculation loop into the undiluted plasma and transfer the plasma to the first colony suspension on the slide. Repeat with the remaining colony suspensions.

g. Observe any visible clumping within 10 seconds of plasma addition; this indicates a coagulase-positive result. Use the control cultures as a guide: *S. aureus* should produce visible clumping (coagulase-positive), whereas absence of clumping (coagulase-negative) should be seen in *S. epidermidis* suspension.

Note: Proportion of water, cell mass, and plasma are crucial for producing optimum coagulase results. If cell clumping (i.e., in case of presumptive positive results) cannot be visualized clearly, use the "Alternative Slide Coagulase Test" as described below.

h. Record the scores for coagulase test in Table 10.2.

6. **Alternative Slide Coagulase Test** (Developed by MJ Mezydlo, The Ohio State University)

a. Mark up to four typical *S. aureus* colonies on the TSA plates that received food isolates: three representing the three dilutions of the original MPN tubes, and one representing the enrichment. If no colonies representing the MPN tubes exist, select two colonies from the enrichment, if available.

b. Mark one *S. aureus* colony and another *S. epidermidis* colony on the TSA plate that received the control cultures.

c. Using a micropipetter, dispense 10 μl sterile water for each preparation on properly labeled microscope slides.

d. Using a sterile tip of the micropipetter, transfer a portion of a marked colony to the assigned drop of water on the appropriately labeled slide. Repeat using the remaining marked colonies.

e. Using a sterile tip of the micropipetter, emulsify the cell mass in the drop of water, with no spreading, to produce a milky drop of cell suspension.

f. Place the slide on the stage of dissecting microscope, at 20× magnification, and focus on the drop of cell suspension.

g. Using a micropipetter, dispense 10 μl of plasma onto the drop of cell suspension.

h. Using the microscope, observe any visible clumping within 10 seconds of plasma addition; this indicates a coagulase-positive result. Use the control cultures as a guide: *S. aureus* should produce visible clumping (coagulase-positive), whereas absence of clumping (coagulase-negative) should be seen in *S. epidermidis* suspension.

i. Record the scores for coagulase test in Table 10.2.

TABLE 10.3 (Propose a meaningful title)

Food	Group	MPN/g	Log$_{10}$ MPN	Average	Standard deviation
Food 1 (---------------)	G1				
	G2				
	G3				
Food 2 (---------------)	G4				
	G5				

(Provide appropriate footnotes)

7. **Calculating *S. aureus* MPN in Food**
   a. Examine results in Table 10.2 and determine which of the MPN tubes is confirmed as positive.
   b. Develop the three-digit MPN score for the sample analyzed.
   c. Using the MPN table (provided in Chapter 3), find the MPN of *S. aureus* in the sample analyzed. Notice that the standard MPN table represents nine tubes prepared from a food sample diluted from 10^{-1} to 10^{-3} (i.e., 0.1, 0.01, and 0.001 g food/tube). In this exercise, 10^{-2} to 10^{-4} dilutions were used (i.e., 0.01, 0.001, and 0.0001 g food/tube). Correction for this dilution ($\times 10$) needs to be included in reported MPN values.
   d. Report the results in the class data worksheet (Table 10.3) and compare your group result with those from other groups.

8. **Culturing Confirmed Isolates for Use in Exercise II**
   a. Select up to two food isolates that have been identified as *S. aureus* (i.e., based on turbidity in the MPN tubes, typical colony morphology and reaction on Baird-Parker agar, β-hemolysis, catalase-positive and coagulase-positive reactions, and typical cell morphology during Gram staining).
   b. Using a sterile inoculation loop, transfer these two isolates to the provided BHI-YE broth tubes.
   c. Incubate the inoculated tubes at 35°C for 24 hr.

DATA INTERPRETATION

This exercise is aimed to accurately count the *S. aureus* sub-population in foods with complex microbiota, using the MPN technique. The technique involved preparing food dilutions in a selective medium (TBB-SP). Incubated MPN tubes turn turbid if *Staphylococcus* spp. are present in food, but other bacteria can produce this reaction, resulting in false-positive tubes. Therefore, the initial MPN score should be scrutinized by applying several tests to cull any false-positive tubes. The screening tests applied in this exercise included plating on Baird-Parker, blood agar, and TSA, and testing the isolates biochemically for catalase, coagulase, and hemolysin production. Cell morphology was also checked by Gram staining (Table 10.2). MPN tubes that produced test results consistent with *S. aureus* should be counted as positive and thus *S. aureus* MPN in food can be calculated. However, some tubes may produce variable results that require careful consideration (see the footnote of Table 10.2).

In addition to population counting, this exercise allowed for the isolation and identification of food isolates as *S. aureus*. Isolates confirmed as *S. aureus* will be tested in Exercise II for the presence of genes encoding for staphylococcal enterotoxins, and for their ability to express these genes to produce the true hazard – the enterotoxin.

QUESTIONS

1. Suppose you relied on the initial MPN tube score (without the additional testing), what would be the *Staphylococcus aureus* most probable number (MPN) in the food you analyzed? How different is this count compared to the one you obtained after completing the additional testing?

2. Would the analysis completed in this exercise prepare you to determine the risk of transmission of staphylococcal gastroenteritis by the food you analyzed? Explain your answer.

3. The other pathogens covered in this book (e.g., *Salmonella*) have a "zero-tolerance" standard in ready-to-eat foods, i.e., detection of even one cell in 25 g of food is unacceptable. Why isn't the same standard applicable to *Staphylococcus aureus*?

4. This exercise represents a culture-based method to enumerate *Staphylococcus aureus* in food. What would be a culture-independent approach to determine the population of this pathogen in the food? Outline a likely culture-independent method.

EXERCISE 2: EXPRESSION OF ENTEROTOXIN GENES BY *Staphylococcus aureus* FOOD ISOLATES

POLYMERASE CHAIN REACTION. ENZYME-LINKED IMMUNOSORBENT ASSAY. GENE EXPRESSION.

INTRODUCTION

Staphylococcus aureus produces many toxic proteins; among these is a group that causes the foodborne staphylococcal gastroenteritis. This disease is caused by several staphylococcal enterotoxins (SE), particularly SEA, SEB, SED, and SEE, which are the most commonly reported causes of staphylococcal gastroenteritis. These toxins have a remarkable ability to induce emesis and gastroenteritis and they are also noted for their superantigenicity. Concentrations of SE reported to cause the disease are less than 1 μg. The toxins are notable for their resistance to heat, acid, and gastrointestinal proteases including pepsin, trypsin, and rennin. The genes encoding these toxins are typically located on the cell's chromosome but have also been found on mobile elements such as plasmids.

Production of *S. aureus* enterotoxins in food is contingent upon many factors, including: (i) presence of the gene encoding the toxin; (ii) ability of the bacterium to express these genes into toxigenic proteins; and (iii) properties of the food (e.g., pH, water activity, storage temperature) are such that they support the growth and metabolic activities of the bacterium. *S. aureus* isolates from Exercise I will be examined for the presence of staphylococcal enterotoxin genes, and if found, the ability of the cell to express these genes to produce enterotoxin will be determined.

OBJECTIVES

1. Detect staphylococcal enterotoxin genes using polymerase chain reaction (PCR).
2. Detect staphylococcal enterotoxins using enzyme-linked immunosorbent assay (ELISA).
3. Practice molecular techniques used in detecting foodborne disease hazards.

PROCEDURE OVERVIEW

In Exercise I, food was analyzed for presence of *S. aureus* and pathogen isolates should have been collected and saved to be used in this exercise. These isolates will be examined for the presence of staphylococcal enterotoxin genes, and if found, the ability of the cell to express these genes will be analyzed. The PCR will be used to amplify the gene encoding staphylococcal enterotoxin A (SEA), as described in Figure 10.3. A large portion of staphylococcal gastroenteritis cases are caused by this toxin. Using a specific primer pair (Table 10.4), PCR will amplify a portion of this gene, and the PCR product (499 bp) will be separated on an agarose gel using electrophoresis. The gel will be stained and visualized under UV light. Any amplicon bands will be compared to a size standard (DNA ladder) to determine if it is the PCR product from the enterotoxin gene. An

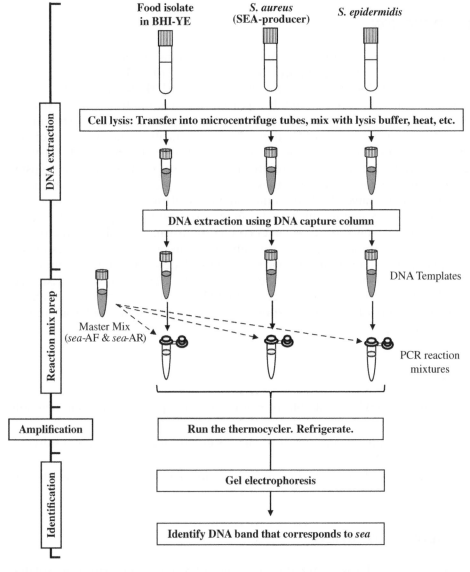

Figure 10.3 Detection of staphylococcal enterotoxin gene (*sea*) in food isolates and *Staphylococcus aureus* control cultures using a PCR-based method (refer to text for procedural details).

TABLE 10.4 Forward and reverse primers for *Staphylococcus aureus* enterotoxin A gene (*sea*) that will be used in this exercise (Designed by A.G. Abdelhamid in Yousef's lab at The Ohio State University).

Gene	Primer	Oligonucleotide	Predicted size (bp)
sea	*sea*-AF	5'- TCC CCT CTG AAC CTT CCC AT -3'	499
	sea-AR	5'- TGA ATT GCA GGG AGC AGC TT -3'	

explanation of the PCR technique is included in the introduction to Part III of this book.

Presence of the gene-encoding SEA does not always lead to the production of the corresponding toxin. Presence of the toxin protein in *S. aureus* culture, however, indicates that the expression of the gene was successfully completed (Figure 10.4). The SEA protein will be detected using an ELISA. A commercial ELISA kit will be used in the detection of not only SEA, but also other members of this family of toxins. For an explanation of ELISA basics, see the introduction to Part III of this book.

This exercise in spread over four or five sessions and its two modules (detection of SEA gene and demonstration of SEA production) can be conducted independently of each other, if desired. The exercise, however, can be completed in less time; some of the sessions dedicated to SEA gene detection can be overlapped with those for toxin detection.

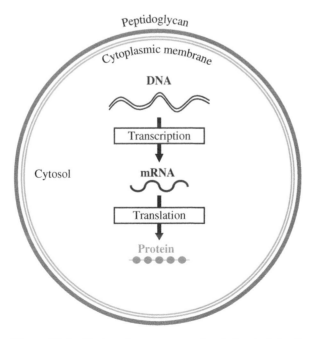

Figure 10.4 Basics of gene expression in bacterial cell.

Organization

Students will work in pairs, but occasionally multiple groups will share reagents. Note that reagents are often aliquoted in a limited number of vials, hence sharing of these aliquoted amounts by two groups or more will save laboratory resources. Each pair of students will test one of the food isolates they confirmed as *S. aureus* in Exercise I of this chapter. Additionally, an instructor-provided *S. aureus* culture will be tested as a positive control.

Personal Safety

S. aureus is an infectious and toxigenic bacterium that should be handled with care. Follow the safety guidelines that were reviewed early in this manual. Use disposable gloves when handling the isolates and the pathogenic cultures. Make sure to sanitize the work area after use.

SESSION 1: DETECTION OF SEA GENE: DNA EXTRACTION AND PREPARING FOR PCR

Staphylococcus aureus isolated from food, in the previous laboratory exercise, will be tested for presence of the gene encoding for staphylococcal enterotoxin A, the most commonly encountered toxin in staphylococcal gastroenteritis outbreaks. Each group of two will analyze one of the food isolates as well as the provided *S. aureus* culture (positive control).

MATERIALS

Per Pair of Students

- Incubated *S. aureus* (in BHI-YE broth) isolated from food
- Instructor-provided *S. aureus* overnight culture in BHI-YE broth for use as "positive control"

  Note: The positive control should be a producer of staphylococcal enterotoxin A (SEA). Examples of SEA-producing strains are S. aureus ATCC 13565 and S. aureus ATCC 13566.
- Instructor-provided *S. epidermidis* overnight culture in BHI broth for use as "negative control"
- One 10-ml BHI-YE broth tube (for preparing culture to be used in ELISA test)
- Seven sterile microcentrifuge tubes (three for centrifugation, three for elution, one for PCR reaction mixture preparation)
- Floating microcentrifuge tube rack. This is not needed if thermostatically controlled heating block (also known as dry bath incubator) is used.
- Glass test tube (for waste)
- Lysis Buffer, 1 ml per group (20 mM Tris-Cl, pH 8.0; 2 mM sodium EDTA; 1.2% Triton X-100; 20 mg/ml lysozyme)
- Ethanol (95%), 1 ml per group
- Qiagen DNeasy Tissue Kit (Qiagen, Inc., Valencia, CA):
  - Spin columns (3 per group)
  - Collection tubes (9 per group)
  - Proteinase K, aliquoted appropriately
  - Buffers (AW1, AW2, and AE), aliquoted appropriately
- Three 200-μl PCR tubes
- Micropipetters and sterile pipette tips
- Vortex
- Other common supplies

Class-Shared

- Ice in a Styrofoam container (ice bath/ice box)
- Water bath, set at 35°C; alternatively, use heating block set at 35°C

- Water bath, set at 55°C; alternatively, use heating block set at 55°C
- Polymerase mix: MyTaq PCR Kit (Bioline, Inc., Taunton, MA), kept on ice. The mix contains DNA polymerase (*Taq*), dNTPs, and buffer.

  Note: Several alternative PCR kits are available commercially. Preparation of PCR reaction mixture may differ slightly depending on the kit used.

- Thermocycler
- Primers (standard desalted) suspended in sterile molecular-grade water (kept on ice). Concentration is 20 µM (Table 10.4)

PROCEDURE

1. DNA Extraction

a. Label three microcentrifuge tubes: one for the *S. aureus* food isolate and the other two for the provided *S. aureus* culture (positive control) and *S. epidermidis* culture (negative control).

b. Transfer 1.5 ml of each culture to the appropriate microcentrifuge tube.

c. Centrifuge for 2 min at 12,000 rpm.

   Note: make sure tubes are balanced in the microcentrifuge before starting.

d. Pipette off supernatant into waste tube, being careful not to disturb the pellet.

e. Re-suspend each pellet in 180 µl of lysis buffer; vortex.

f. Place tubes in the floating microcentrifuge tube rack and incubate in 35°C water bath (or use the heating block) for 30 min.

g. Add 25 µl of proteinase K and 200 µl of buffer AL to each sample tube. Vortex.

h. Place tubes in the water bath (or heating block) set at 55°C and incubate for 30 min.

i. Add 200 µl of ethanol to each sample tube and vortex.

j. Place three DNeasy spin columns into three collection tubes; label the spin columns with sample information.

k. Pipette sample mixture into appropriate spin column.

l. Centrifuge at 8000 rpm for 1 min; discard flow through and collection tubes.

m. Place spin columns into new set of collection tubes.

n. Add 500 µl of Buffer AW1 to each spin column.

o. Centrifuge at 8000 rpm for 1 min; discard flow through and collection tubes.

p. Place spin columns into new set of collection tubes.

q. Add 500 µl of Buffer AW2 to each spin column.

r. Centrifuge at 12,000 rpm for 3 min; discard flow through and collection tubes.

s. Label three new sterile microcentrifuge tubes with sample information.

t. Place the appropriate spin column onto its labeled microcentrifuge tube.

u. Add 50 µl of Buffer AE directly to each DNeasy membrane.

v. Allow to sit for 1 min.

w. Centrifuge at 8000 rpm for 1 min to elute the DNA into the microcentrifuge tube.

Note: Do not discard flow through, it contains sample DNA (i.e., template DNA).

2. PCR Reaction Preparation

A PCR master mixture will be made in a sterile microcentrifuge tube. After this mixture is prepared, aliquots are dispensed into PCR tubes. Isolated DNA from each sample will be added to a PCR tube containing the master mixture. The following details the procedure:

a. Prepare PCR master mixture: In a new sterile microcentrifuge tube, mix 136 µl sterile water, 40 µl 5× *MyTaq* mixture (buffer, dNTPs, and *Taq*), 4 µl of primer *sea*-AF, and 4 µl of primer *sea*-AR.

Note: Polymerase (Taq) *is the most sensitive component of the PCR reaction mixture. Care should be taken to keep reagents and reaction mixtures on ice.*

b. Mix the contents of the microcentrifuge tubes.

c. Carefully label the top of three sterile PCR tubes with a code for sample analyzed (i.e., food isolate, positive control, and negative control).

d. Transfer 46 µl of the PCR reaction mixture to each of the three PCR tubes.

e. Add 4 µl of appropriate template DNA to each PCR tube.

f. Mix the contents of the PCR tubes.

g. Place PCR tubes in the thermocycler. Record the location of your tubes in the thermocycler on the sheet provided by the instructor.

h. Thermocycler will run for 3 minutes at 95°C and for 30 cycles as follows: 95ºC for 15 s, 58ºC for 15 s, 72ºC for 30 s. At the end of the 30 cycles, the thermocycler will be run at 72ºC for 4 min, then tubes will be held at 4°C until removed.

i. PCR products will be refrigerated (or frozen) until next session.

3. Preparing Cultures for ELISA

a. Using the incubated *S. aureus* food isolate, transfer a loopful of the culture to a fresh BHI-YE broth tube.

b. Incubate the inoculated BHI-YE broth tube at 35°C for 24 hr.

SESSION 2: DETECTION OF SEA GENE: GEL ELECTROPHORESIS

PCR products will be separated using agarose gel electrophoresis and visualized by UV light with comparison to a DNA standard. PCR results will be interpreted qualitatively to determine the presence of *sea*, the SEA gene.

MATERIALS AND EQUIPMENT

Per Pair of Students

- PCR products from previous session
- Micropipetters and sterile pipette tips
- Incubated *S. aureus* food isolate, from the previous session
- One 10-mL BHI-YE broth tube (for preparing culture to be used in ELISA test)

Class-Shared

- Agarose gel (1%)
- Running buffer (TAE)
- Loading buffer containing EZ-Vision™ DNA Dye
- DNA ladder (instructor will load the ladder)
- Gel electrophoresis equipment (gel running tray and matching power supply)
- Gel documentation station (UV transilluminator, digital camera, computer, and image-capture software)

PROCEDURE

1. **Gel Electrophoresis**
   a. Label three spots on a piece of Parafilm with sample information.
   b. Place 2 µl of loading buffer near each label.
   c. Mix 10 µl of the appropriate PCR reaction with the loading buffer using the micropipette.
   d. Dispense 10 µl of the mixture, just prepared, into appropriate well of the agarose gel (which is submerged in the running buffer).
   e. Record which sample was put into which well on the provided key sheet.
   f. Run gel electrophoresis for approximately one hour at 100 V.
   g. Gel picture will be taken using ultraviolet light; picture will be viewed during Session 3.

2. **Preparing Cultures for ELISA**
   a. Using the incubated *S. aureus* food isolate, transfer a loopful of the culture to a fresh BHI-YE broth tube.
   b. Incubate the inoculated BHI-YE broth tube at 35°C for 24 hr.

SESSION 3: DETECTION OF SEA GENE: ELECTROPHORESIS RESULTS OBSERVATION

An image of the UV-illuminated agarose gel will be evaluated, and results interpreted.

MATERIALS AND EQUIPMENT

Per Pair of Students

- Photograph of the band on the agarose gel
- Incubated *S. aureus* food isolate, from the previous session
- One 10-ml BHI-YE broth tube (for preparing culture to be used in ELISA test)

PROCEDURE

1. PCR Results

a. Observe the gel photograph and record the results of these observations. Begin by locating lanes containing the DNA ladder in the gel and label with appropriate band sizes (see manufacturer insert).

b. Locate the lane containing the provided *S. aureus* culture (positive control) and *S. epidermidis* culture (negative control). Estimate the size of the DNA amplification band for control culture using the DNA ladder as a reference. There should be no band present in the *S. epidermidis* lane. Lack of band in the positive control lane or presence of band in the negative control lane indicates that the results of the PCR are invalid and should not be considered conclusive with regard to the food isolate.

c. Locate the lane containing the PCR product originating from the food isolate. If a band is present, estimate its size (bp) using the DNA ladder as a reference. Determine the presence or absence of PCR products (DNA bands) for SEA gene.

d. Record the final PCR results for control culture and food isolate in Table 10.5.

2. Preparing Cultures for ELISA

a. Using the incubated *S. aureus* food isolate, transfer a loopful of the culture to a fresh BHI-YE broth tube.

b. Incubate the inoculated BHI-YE broth tube at 35°C for 24 hr.

TABLE 10.5 {Use appropriate title}

Group initials	Sample [food isolate or lab positive]	SEA Gene Detection (present/absent)	SEA Detection (present/absent)	Conclusion

{Include appropriate footnotes}

SESSION 4: DETECTING THE EXPRESSION OF STAPHYLOCOCCAL ENTEROTOXINS USING ELISA

ELISA will be used in this exercise to detect enterotoxin production from confirmed *S. aureus* food isolates. The immunoassay is designed to detect staphylococcal enterotoxins, the actual disease-causing agents in staphylococcal food intoxication. In a professional setting, the immunoassay may be carried out directly on the food product, thus presence of the staphylococcal enterotoxins in the suspect food can be linked to a disease outbreak. For educational purposes, this laboratory exercise includes detection of toxins in pure culture of potentially toxigenic *S. aureus* isolates.

A commercial ELISA test kit (RIDASCREEN® Set A,B,C,D,E, R-Biopharm AG, Darmstadt, Germany) will be used in toxin detection. This kit includes microtiter strips with wells pre-coated with primary antibodies against each of the staphylococcal enterotoxins (A, B, C, D, and E). According to the kit's manufacturer, the wells are coated with specific antibodies for each of the five enterotoxins; wells A–E correspond to antibodies for enterotoxins A–E. Wells F and G are coated with sheep IgG and serve as a negative control. The well H is coated with a mixture of the antibodies for the five enterotoxins and it serves as a positive control. The supernatant from the cell culture will be added to wells A through G. The kit-provided positive control (staphylococcal enterotoxin) will be added to well H only.

Note: The positive control is purified staphylococcal enterotoxin that can cause disease – please use extra caution throughout this exercise.

The kit includes a secondary antibody that binds to any staphylococcal enterotoxin molecules captured by the primary antibody and retained on well walls after washing. If no enterotoxins are present in the food sample, the secondary antibody is washed out during the washing step. To detect the presence of the secondary antibody in the well, it has to be linked to an enzyme that catalyzes a reaction with a colored end product (Figure 10.5). Therefore, the kit includes two conjugates: Conjugate-1 is the secondary antibody conjugated to a biotin molecule and Conjugate-2 is a peroxidase enzyme conjugated to an avidin molecule. Avidin on Conjugate-2 binds to biotin on Conjugate-1, leading to a complex containing both the secondary antibody and the peroxidase. The enzyme portion

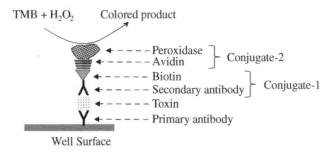

Figure 10.5 Enzyme-linked immunoassay used in detection of staphylococcal enterotoxin.

of Conjugate-2 (e.g., horseradish peroxidase, HRP) catalyzes the reaction between hydrogen peroxide and the chromogenic substrate, 3,3',5,5'-tetramethylbenzidine (TMB). This reaction results in a blue-colored product. Sulfuric acid is added to the mixture to stop the reaction and help in producing a stable end product; this also changes the product color from blue to yellow.

$$H_2O_2 + TMB \xrightarrow{\text{HRP}} \begin{array}{c} \text{Blue} \\ \text{intermediate} \end{array} \xrightarrow[\text{(to stop reaction)}]{\text{Acid}} \begin{array}{c} \text{TMB diimine} \\ \text{(yellow)} \end{array}$$

The absorbance of the colored end products is measured using a spectrophotometer or a microtiter-plate reader at 450 nm (A_{450nm}). The absorption is proportional to the quantity of staphylococcal enterotoxin present in the sample.

Use of the negative control wells allows the user to determine if the wells were adequately washed (i.e., wells F and G have no antibodies and should produce negative results). A positive control well allows the user to confirm that the reagents are functioning properly (i.e., well H contains staphylococcal enterotoxin antibodies, is filled with an enterotoxin solution, and thus should produce a positive result).

Before conducting this exercise, analysts should check the commercial kit insert for complete instructions regarding reagents' preparation and for the latest procedural modifications.

MATERIALS AND EQUIPMENT

Complete instructions for preparing reagents used in this laboratory exercise and amounts of each needed to run the test can be found in the package insert, provided by the commercial kit producer.

Per Pair of Students

- Incubated *S. aureus* food isolate, from the previous session
- Instructor-provided *S. aureus* overnight culture in BHI-YE broth for use as "lab positive control." *This should be the same strain used as a positive control in Session 1.*
- Sterile 2-ml syringe
- One 0.22-μm syringe filter
- One microcentrifuge tube containing "kit-positive"
  Note: This is the staphylococcal enterotoxin – handle with care
- Sterile microcentrifuge tubes
- Wash solution, 40 ml
- One microcentrifuge tube containing Conjugate-1, properly aliquoted volume
- One microcentrifuge tube containing Conjugate-2, properly aliquoted volume
- One microcentrifuge tube containing chromogen/substrate (TMB-A (3,3',5,5' tetramethylbenzidine) and hydrogen peroxide), properly aliquoted volume

- One microcentrifuge tube containing stop solution, properly aliquoted volume
- Two ELISA test strips, 8 wells each
- Micropipettors (P1000, P200) and tips
- Other common supplies

Class-Shared

- Microcentrifuge
- Microtiter plate reader
- Other common laboratory supplies (e.g., disposable gloves) and equipment

PROCEDURE

1. **Sample Preparation for ELISA**
   a. Label two microcentrifuge tubes with identifying information for your group; one will be used for the "food isolate" and the other for "lab positive."
   b. Transfer 1.5 ml of *S. aureus* food isolate and lab positive control to the microcentrifuge tubes.
   c. Centrifuge for 5 minutes at 5000 rpm.
   d. Transfer each of the supernatants (the cell-free liquids) into a 2-ml syringe (Figure 10.6). Filter-sterilize the using 0.2-μm filter. Collect each filtered supernatant into a sterile microcentrifuge tube. The microfiltration procedure can be done in this sequence:

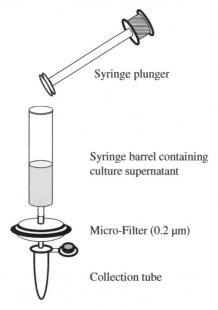

Syringe plunger

Syringe barrel containing culture supernatant

Micro-Filter (0.2 μm)

Collection tube

Figure 10.6   Microfiltration to prepare cell-free culture supernatant.

i. Open the microfilter package, exposing only the side of the filter to be fitted to the syringe barrel.

ii. Remove the syringe plunger, and any tip-guard or needle, and screw the filter onto the syringe barrel.

iii. Remove the remainder of the microfilter package and place the sterile end of the microfilter onto the sterile microcentrifuge tube (on a micro-centrifuge tube rack). Do not contaminate the tip of the microfilter.

iv. Pour the supernatant of culture into the syringe barrel.

v. Attach syringe plunger and push it gently to drive the supernatant through the filter. Filtration may be stopped when at least 900 μl of sterile filtrate is collected in the microcentrifuge tube.

vi. Appropriately dispose of the syringe, filter, and any remaining liquid, taking note of any drips and sanitizing surfaces, changing gloves, etc. as necessary.

2. **ELISA**

Note: If the entire exercise cannot be executed in one laboratory session, it may be paused after step h, provided that it can be completed in the subsequent session. In this case, instructors carry out step i and refrigerate the ELISA strips until students resume the exercise in the subsequent laboratory session. This approach may result in non-specific binding, causing the negative control to produce larger absorbance readings than expected (e.g., A_{450nm} of 0.3). If that is the case, the cut-off absorbance value should be adjusted accordingly.

a. Pipette 100 μl of the food isolate filtrate into wells A-G of the first ELISA strip.

b. Pipette 100 μl of kit-positive into well H.

c. Mix gently by rocking plate back and forth.

d. Repeat steps a–c using the filtrate of the lab positive control, applied to the second ELISA strip.

e. Cover the strips with foil and incubate at 35°C for 1 hour.

f. Empty the wells carefully into the provided waste container, in a way to avoid contaminating any well by the contents of neighboring wells. This can be accomplished by holding the strip horizontally from both ends and rolling it 180° over the waste container. To make sure all well contents are emptied, hold the strip upside down and tap it three times onto a stack of clean paper towels.

g. Wash each well four times with washing buffer solution; use 250 μl per well for each wash. Empty the well contents as described in the previous step. Do not let wells dry out.

h. Pipette 100 μl of Conjugate-1 into each well.

i. Mix gently by rocking plate back and forth.

j. Cover with foil and incubate at 35°C for 1 hour.

Note: This laboratory session may end here. In this case, the instructor or the person in charge needs to store the incubated strips (or plates) under refrigeration until the exercise resumes in the subsequent laboratory session.

k. Empty wells into the provided waste container, as described previously.

l. Wash each well four times with washing buffer solution; use 250 µl per well for each wash. Empty the well contents as described in a previous step. Do not let wells dry out.

m. Pipette 100 µl of Conjugate-2 into each well.

n. Mix gently by rocking plate back and forth.

o. Cover with foil and incubate at 35°C for 30 min.

p. Empty wells into the provided waste container, as described previously.

q. Wash each well four times with washing buffer solution; use 250 µl per well for each wash. Empty the well contents as described in a previous step. Do not let wells dry out.

r. Pipette 100 µl of substrate/chromogen to each well.

s. Mix gently by rocking plate back and forth.

t. Cover the strips with foil and incubate at 35°C for 15 min.

u. *Do not empty the wells*. Pipette 100 µl of stop solution into each well.

v. Mix gently by rocking plate back and forth.

3. **Reading Strips**

a. Read the well strip in a microtiter plate reader at 450 nm using an air blank. Read results within 30 min of addition of stop solution.

b. Record the results of the ELISA test in Table 10.6 and copy SEA detection result into Table 10.5.

4. **Interpreting Data**

a. Calculate the cutoff value. According to the kit's manufacturer, ELISA plate readings are valid if the following conditions are met:

   i. The negative controls must have an absorbance of less than 0.200.

   ii. The positive control absorbance value must be at least 1.000.

   iii. Values for positive and negative controls that fall outside the manufacturer's predetermined limits indicate errors in protocol (such as

TABLE 10.6 {add a descriptive title and footnotes for this data, including food sample used, media, recovery method and what data were obtained}

Well	Absorbance (450 nm)	Valid?	Result
A			
B			
C			
D			
E			
F – negative control			
G – negative control			
H – positive control			

under-washing of wells or nonspecific binding) or problems with the reagents.

The cutoff value is calculated by adding 0.150 to the average absorbance value of the negative controls. Using this calculated cutoff value results in fewer false positive results.

b. Use of the cutoff value:

   i. A sample with an absorbance value greater than or equal to the cutoff value is considered to be a positive result.

   ii. A sample with an absorbance reading below the cutoff value is considered to be a negative result.

QUESTIONS

1. Were the results of the PCR and the ELISA complementary? Comment on the class findings and explain the possible causes for any discrepancy.

2. What are the predominant enterotoxins detected in the analyzed food isolates? Compare these data with commonly detected enterotoxins in two similar foods, using published literature, and cite appropriate references.

3. List and discuss essential factors that affect enterotoxin production, by *Staphylococcus aureus*, in the food analyzed by your group.

4. Considering the MPN results in Exercise I, and the analyses conducted in Exercise II, do you predict that the sample analyzed by your group would contain detectable levels of staphylococcal enterotoxin? Why or why not?

CHAPTER 11

Listeria monocytogenes

SAMPLING PROCESSING ENVIRONMENT. CULTURE-BASED DETECTION METHOD. PCR-BASED DETECTION METHOD

INTRODUCTION

Listeria spp. are non-spore forming, Gram-positive, rod-shaped bacteria that belong to the phylum *Firmicutes*. *Listeria* spp. traditionally include *L. monocytogenes*, *L. innocua*, *L. ivanovii*, *L. seeligeri*, *L. welshimeri*, and *L. grayi*. In 2010, another species, *L. marthii*, was added. Recently, additional species have been included under the genus *Listeria*, but these will likely be reassigned to genera other than *Listeria* in the future. Descriptions in this chapter will be limited to the traditional *Listeria* spp.

Members of the genus *Listeria* are facultative anaerobes. They grow over a relatively wide range of temperatures including refrigeration (e.g., 4°C) and they are motile by flagella. Considering their biochemical profile, *Listeria* spp. are catalase-positive, oxidase-negative, methyl red and Voges-Proskauer positive, and nitrite reductase-negative. These species are capable of fermenting D-glucose, D-fructose, D-mannose, maltose, cellobiose, and lactose, but not inositol. Fermentation of sugars results in production of acids such as lactic and acetic, with no gas production. Additionally, *Listeria* spp. hydrolyze esculin and sodium hippurate.

Species of the genus *Listeria* can be differentiated based on carbohydrate fermentation and blood hemolysis (Table 11.1). Fermentation of D-xylose, L-rhamnose, D-ribose, mannitol, sucrose, and others, and hemolysis on blood agar are useful in speciating *Listeria* isolates. Among the members of this genus, *L. monocytogenes* is the only species classically recognized as a human pathogen, although *L. ivanovii* has recently been implicated in cases of human gastroenteritis.

Analytical Food Microbiology: A Laboratory Manual, Second Edition. Ahmed E. Yousef,
Joy G. Waite-Cusic, and Jennifer J. Perry.
© 2022 John Wiley & Sons, Inc. Published 2022 by John Wiley & Sons, Inc.

TABLE 11.1 Differentiation of *Listeria* spp. by biochemical testing and blood hemolysis.

Species	Fermentation with Acid Production[a]					Hemolysis Pattern	
	D-Xylose	L-Rhamnose	D-Ribose	D-Mannitol	Sucrose	β-Hemolysis[b]	CAMP test[c]
L. monocytogenes	–	+	–	–	+	+	+
L. innocua	–	v	–	–	+	–	–
L. seeligeri	+	–	–	–	+	+	+
L. welshimeri	+	v	–	–	+	–	–
L. ivanovii	+	–	+	–	+	+	–
L. grayi	–	–	+	+	–	–	–
L. marthii	–	–	–	–	–	–	–

[a] v=variable.
[b] When tested in agar media with sheep blood
[c] When tested against *Staphylococcus aureus*

Characteristics

L. monocytogenes is an intracellular pathogen that also can be found ubiquitously outside its hosts. It is a short rod bacterium that is motile by a few flagella when cultured at 20–25°C. Motility is weak or lacking if the culture is incubated at 37°C. The characteristic tumbling motility can be observed by microscopic examination of a hanging drop preparation. Overnight incubation of *L. monocytogenes* on nutrient-rich agar media produces colonies with 0.2–0.8 mm diameter. The colonies are smooth with entire margins and are typically translucent and bluish gray in color. When these colonies are examined using a suitable light source, with an incidence light angle of 45° and a viewing angle of 135°, a blue-green iridescence may be seen.

    L. monocytogenes is a facultative anaerobe that adapts to a wide range of oxygen availability. The bacterium can generate ATP through oxidative phosphorylation (i.e., aerobic respiration) or substrate-level phosphorylation (i.e., fermentation), but no anaerobic respiration has been reported for this bacterium. Considering that a larger amount of ATP is produced via oxidative phosphorylation, the bacterium grows well under aerobic and microaerophilic conditions, but its growth rate decreases appreciably under anaerobic conditions. When grown on blood agar, *L. monocytogenes* shows β-hemolysis, although it is usually a narrow zone. The bacterium ferments glucose into lactate, acetate, and other products such as acetoin, thus it is positive for the Voges Proskauer test. Growth temperatures range from –1.5 to 45°C with the optimum at 30–37°C. The bacterium grows at refrigeration temperature, albeit at a slow rate; for example, its generation time (GT) in skim milk at 4°C is 30–40 hours, whereas at 35°C, it is ~40 minutes. This ability to grow under refrigeration is used as a differential factor in a step called "cold enrichment," which may be applied prior to isolation of the pathogen from certain foods in the absence of high levels of psychrotrophic organisms. *Listeria* spp. grow at relatively wide ranges of acidity (pH 4 to 9) and salt concentrations (up to 10%).

    Isolates that are confirmed as *L. monocytogenes* should be typed, i.e., categorized at the sub-species level. Serological and genetic techniques (e.g., multilocus sequence typing) are often used in typing this pathogen. Differentiation of cell surface antigens allows researchers to classify *L. monocytogenes* into serotypes. Thirteen serotypes have been recognized, but most cases of listeriosis are associated with serotypes 1/2a, 1/2b, and 4b.

The Disease

Consumption of foods contaminated with *L. monocytogenes* may lead to an invasive infection and the resulting disease is known as listeriosis. It is generally assumed that the infective dose is > 10^2 listeriae; however, smaller doses may have been associated with the disease. The incubation period of listeriosis is one to several weeks. Children less than four years old, pregnant people, the elderly, and immune-compromised individuals are the most susceptible to the disease. In pregnant individuals, listeriosis causes miscarriages, stillbirth, or premature birth. In near-term cases, neonatal listeriosis may be acquired by the child during delivery. Listeriosis manifestation in susceptible non-pregnant adults includes bacteremia (symptoms resulting from the presence of the pathogen in the blood, e.g., fever, malaise, and fatigue), meningitis (inflammation of the membranes of the brain or the spinal cord), and meningoencephalitis (inflammation of the brain and its membranes). Symptoms associated with meningitis and meningoencephalitis include fever, malaise, seizures, and altered mental status. Hospitalization rate among infected individuals is estimated to be as high as 90% and mortality rate is 20–30%.

Listeria monocytogenes also has been associated with a non-invasive disease called febrile gastroenteritis. The most common symptoms of this gastrointestinal listeriosis are fever and diarrhea and the incubation period is 20–27 hours. It appears that the infective dose of gastrointestinal listeriosis is much higher ($> 10^5$ CFU) than that of the invasive disease.

Due to the severity of the illness and the high mortality rate, US regulatory agencies have implemented a "zero tolerance" policy for *L. monocytogenes* in ready-to-eat (RTE) foods. When implemented, this policy means that *L. monocytogenes* must be undetectable in any 25 g sample of a RTE food.

Foods Implicated

Listeria monocytogenes is ubiquitous in the environment. Because of this, it has been isolated from many raw foods such as milk, salad vegetables, meat, poultry, and seafood. RTE foods that have been exposed to contaminated processing environments before packaging may become vehicles for pathogen transmission. Frankfurters, deli meats, soft cheeses, and similar RTE foods have been implicated in listeriosis outbreaks. This risk increases if a product supports the growth of the pathogen.

Environmental sampling for *Listeria* spp. is part of food safety plans for facilities that produce RTE foods that are exposed to the environment after processing steps lethal to the bacteria and prior to packaging. These environmental monitoring programs are intended as verification activities for sanitation programs in high hygiene areas. Quality assurance personnel commonly sample various surfaces in the processing environment to determine the presence or absence of *Listeria* spp. Drains are an important collection and harborage site for *Listeria* spp. and are often emphasized in environmental monitoring programs.

Detection

When *L. monocytogenes* is present in food, it is usually found at levels too low for enumeration by plating techniques. The infectious dose of *L. monocytogenes* is low and many foods support growth of *L. monocytogenes*; therefore, mere presence of the pathogen in food constitutes a health risk to consumers. Consequently, foods are commonly analyzed for the presence, rather than the population count of *L. monocytogenes*, and this analysis is described as "detection."

Since a relatively large analytical sample (e.g., entire unit or ≥ 25 grams) of food is analyzed, the theoretical minimum detection limit of the analysis is one *Listeria* per unit or sample weight. For environmental samples, the detection limit will be based on the unit of area that was swabbed (e.g., 100 cm²) or another relevant description of the physical location (e.g., a single drain). Photographs of the sample location and a brief narrative of the sampling procedure used are particularly helpful to guide follow-up sampling needs.

Food or environmental samples may be analyzed for the presence of *Listeria* spp. only or specifically for *L. monocytogenes*. There are many considerations that determine the preference of detecting *Listeria* at the genus or the species level. If the food is implicated in a listeriosis outbreak, it should be analyzed for the presence of *L. monocytogenes* and confirmed isolates should be subtyped. From a regulatory perspective, enforcement action is pursued when *L. monocytogenes* is confirmed in a food or on a direct food contact surface. For routine analyses, however, processors may prefer to analyze food, ingredients, or processing environments using more rapid *Listeria* spp. detection methods. In this case, results of the analysis can guide their environmental monitoring program and facilitate sanitation schedules or other mitigation steps.

PROCEDURE OVERVIEW

Organization

Two laboratory exercises, to detect *Listeria* in food and environmental samples, are presented in this chapter. The first exercise is a culture-based method that is sufficient to determine if the food contains *Listeria* spp. or not. The second exercise is a PCR-based method designed to screen food samples as follows: (a) if the result is negative, the analyst can conclude the food does not contain *L. monocytogenes*, and (b) if the result is positive, further confirmatory analyses are warranted to help the analyst conclude whether the food contains *L. monocytogenes*. The two exercises share common sampling and enrichment steps and can easily be executed simultaneously; however, presenting them separately minimizes procedural confusion. If desired, one student (of the group of two) can oversee the culture-based method while the other executes the PCR-based method. Summaries of the two laboratory exercises are shown in Table 11.2 and Figures 11.1 and 11.2.

Personal Safety

L. monocytogenes is a highly pathogenic microorganism that should be handled with care. Analysts should follow the safety guidelines that were reviewed in Chapter 1. Use disposable gloves and wear protective goggles when handling potentially contaminated foods, isolates, and pathogenic cultures. Make sure the work area is sanitized after use. Pregnant and immunocompromised individuals should not be allowed into the laboratory or other areas where analysis for *L. monocytogenes* is in progress.

TABLE 11.2 Detection of *Listeria monocytogenes* in food and environmental samples.

	Culture-based method		PCR-based method	
Session	Step	Description	Step	Description
1	Selective enrichment	Culturing in half-Fraser broth	Selective enrichment	Culturing in half-Fraser broth
2	Isolation	Plating enrichments on: • MOX[a] agar • BLA[b]	Identification (PCR-based)	• DNA extraction • Amplification of DNA by PCR[c]
3	Identification	Plating isolates on: • TSA-YE[d] • TSA-blood	Identification (continued)	Gel electrophoresis
4	Identification (continued)	• Colony morphology • Cell morphology • Catalase reaction • Hemolysis	Identification (continued)	Results interpretation *Note: Additional confirmation tests are needed if PCR results are positive.*

[a]Modified Oxford
[b]Brilliance listeria agar
[c]Polymerase chain reaction
[d]Tryptic soy agar + yeast extract

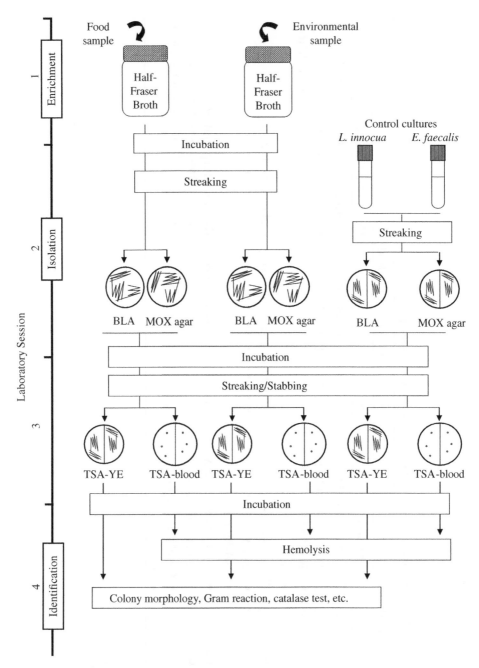

Figure 11.1   Detection of *Listeria* spp. in food and environmental sample using the culture-based method (refer to text for procedural details).

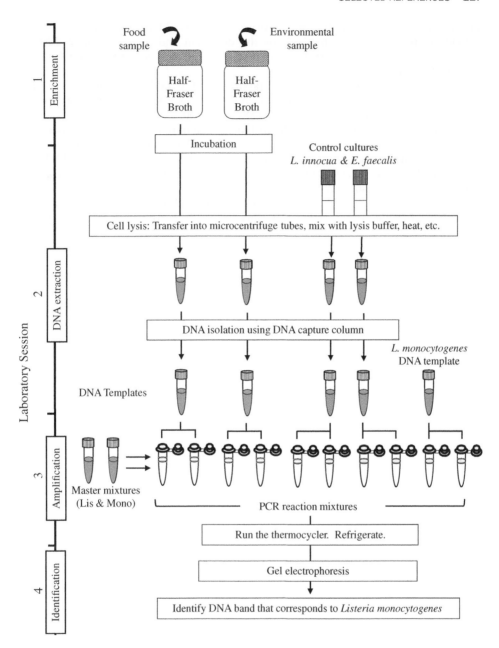

Figure 11.2 Detection of *Listeria monocytogenes* in food and environmental samples using a PCR-based method (refer to text for procedural details).

SELECTED REFERENCES

Hitchins, A.D., Jinneman, K., and Chen, Y. (2017). Detection of *Listeria monocytogenes* in foods and environmental samples, and enumeration of *Listeria monocytogenes* in foods. *Bacteriological Analytical Manual*. Washington, DC: Food and Drug Administration.

International Organization for Standardization (ISO). (2017). Microbiology of the food chain – Horizontal method for the detection and enumeration of *Listeria monocytogenes* and of *Listeria* spp. – Part 1: Detection method. ISO 11290–1. Geneva, Switzerland: ISO.

Kohler, S., Leimeister-Wachter, M., Chakraborty, T., et al. (1990). The gene coding for protein p60 of *Listeria monocytogenes* and its use as a specific probe for *Listeria monocytogenes*. *Infection and Immunity* 58: 1943–1950.

Moberg, L. and Kornacki, J.L. (2015). Microbiological monitoring of the food processing environment. In: *Compendium of Methods for the Microbiological Examination of Foods*, 5e (ed. Y. Salfinger and M.L. Tortorella), 27–43. Washington, DC: APHA Press.

Ooi, S.T. and Lorber, B. (2005). Gastroenteritis due to *Listeria monocytogenes*. *Clinical Infectious Diseases* 40: 1327–1332.

Orsi, R.H. and Wiedmann, M. (2016). Characteristics and distribution of *Listeria* spp., including *Listeria* species newly described since 2009. *Applied Microbiology and Biotechnology* 100: 5273–5287.

EXERCISE I: DETECTION OF *Listeria* SPECIES USING CULTURE-BASED METHOD

INTRODUCTION

For detecting *Listeria* spp. in food and environmental samples, several cultured-based methods have been used by different agencies and institutions. The culture-based method implemented in this exercise involves three main steps: enrichment, isolation, and identification. The following is a description of these steps.

Enrichment

For enrichment, the food or environmental sample is mixed with a rich selective-differential broth; half-Fraser broth is commonly used. The mixture is incubated at 30°C to allow selective proliferation of *Listeria* spp. Darkening of the medium color suggests the presence of *Listeria* spp. Incubation is performed at a mild temperature to encourage the resuscitation of injured cells. To accommodate a less frequent laboratory classroom schedule, sample-broth mixture is refrigerated after homogenization and then incubated at 30°C for 24 hours before the start of the subsequent laboratory session.

Note: Some methods call for a 24-hour incubation in half-Fraser followed by a transfer to Fraser broth with additional incubation at 30°C for 24 hours.

Cold enrichment was a common technique decades ago. It relied on the ability of *L. monocytogenes* to grow at refrigeration temperature, whereas the growth of many other contaminants in the sample is arrested. Although cold enrichment gives good recovery of small *Listeria* populations, the method is time-consuming as it takes weeks of refrigeration to produce satisfactory recovery of the pathogen, making it not ideal for routine analyses and rapid decision-making in food processing facilities or for outbreak investigations.

Isolation

Following enrichment in half-Fraser broth, samples are streaked for isolation onto selective-differential media. Many media are available which vary in performance depending on the background microbiota in the food or environmental samples. Selective-differential agar media include Oxford agar (OX), modified Oxford (MOX) agar, or polymyxin acriflavine lithium chloride ceftazidime aesculin mannitol (PALCAM) agar. MOX agar, which is used in this exercise, contains salts (i.e., LiCl and NaCl) and antimicrobial agents (e.g., moxalactam and colistin) that are tolerated by *Listeria* spp., but not by many other competing microorganisms. Selection of *Listeria* spp. on these media is facilitated by the inclusion of esculin and a ferric salt (ferric ammonium citrate) as differential agents. All *Listeria* spp. hydrolyze esculin into glucose and esculetin (6,7-dihydroxycoumarin); the latter reacts with ferric ions producing a black product. Colonies exhibiting gray coloration with blackening of media are considered presumptive *Listeria* spp.

Chromogenic agar media are selective-differential media suitable for isolation and differentiation of *Listeria* spp. and/or *L. monocytogenes*. Depending on the desired outcome of the testing procedure (identification of *Listeria* spp. vs. *L. monocytogenes*), the analyst can use chromogenic media that produce identical coloration for all *Listeria* spp. (e.g., CHROMagar Listeria) or those that produce unique coloration for *L. monocytogenes* compared to other *Listeria* spp. (e.g., Brilliance *Listeria* agar [BLA] and CHROMagar Identification Listeria). Differentiation on these media is a combination of the detection of phospholipase C activity, β-glucosidase activity, and xylose fermentation; however, the biochemical basis and reagents needed for these reactions are typically proprietary. Whenever possible, it is suggested that identification employ at least two selective-differential media.

Identification

Colonies showing typical *Listeria* reactions on the isolation media (e.g., MOX agar and BLA) can be confirmed as *Listeria* spp. using a number of morphological and biochemical tests. To help in identification of *Listeria* spp., isolated colonies are examined morphologically on tryptic soy agar + yeast extract (TSA-YE) and tryptic soy agar + sheep blood (TSA-blood), and cells are subjected to catalase, oxidase, and Gram staining tests. Motility, particularly the tumbling motility, is a useful morphological identification test.

Cells of *Listeria* spp. appear as short Gram-positive rods, are catalase-positive, and oxidase-negative. *L. monocytogenes* colonies produce a characteristic faint blue-gray color on TSA-YE and β-hemolysis on TSA-blood. Hemolysis caused by *L. monocytogenes* can be further enhanced if streaked on TSA-blood in the proximity of *Staphylococcus aureus*, i.e., Christie–Atkins–Munch-Petersen (CAMP) test. Motility of *L. monocytogenes* is temperature dependent; therefore, motility medium is inoculated by stabbing and incubated at two temperatures (25°C and 37°C) to determine relative motility.

OBJECTIVES

1. Learn and practice a culture-based method for detection of *Listeria* spp. in food or environmental samples.
2. Apply safe procedures for handling pathogens in the laboratory.

MEDIA

Brilliance *Listeria* Agar (BLA)

BLA is a selective-differential medium containing a chromogenic substrate for the isolation and presumptive identification of *Listeria* spp. in food or environmental samples. Lithium chloride, along with several antimicrobials (nalidixic acid, ceftazidime, polymyxin B, and amphotericin), provides the selection for *Listeria* spp. The chromogenic substrate contained in BLA is

5-bromo-4-chloro-3-indoyl-β-D-glucopyranoside (X-glucoside). This compound serves as a substrate for β-glucosidase; colonies with this enzyme activity will have a blue-green color indicative of *Listeria* spp. This medium contains a secondary differential characteristic based on the detection of phopholipase C activity. L-α-phophatidylinositol is a substrate for phopholipase C. Isolates producing phospholipase C will develop an opaque halo. *L. monocytogenes* and *L. ivanovii* are known to possess phospholipase C and should produce a characteristic halo on this medium.

Half-Fraser Broth

In this exercise, half-Fraser broth is used in the enrichment step. Selectivity of this medium is due to the combined presence of acriflavin, nalidixic acid, and lithium chloride. The medium is prepared to contain half the concentration of acriflavin and nalidixic acid found in Fraser broth. Lithium chloride inhibits the growth of enterococci while permitting the growth of *Listeria* spp., which are salt tolerant. Esculin and ferric ammonium citrate act together as differential agents. *Listeria* spp. produce an enzyme that hydrolyzes the esculin. The resulting hydrolysis product reacts with the iron (ferric) ions of the ferric ammonium citrate. This reaction produces a characteristic black precipitate. This reaction is responsible for the darkening of the broth. Thus, media with presumptive *Listeria* are dark, whereas negative enrichments may remain yellowish. Note that the choice of food sample may influence color development. Mushrooms, for example, which demonstrate significant polyphenol oxidase activity upon homogenization, will darken the broth regardless of the presence of *Listeria* cells.

Modified Oxford (MOX) Agar

MOX agar is a selective-differential medium used for the isolation of *Listeria* spp. Selectivity of this medium is due to the presence of lithium chloride, colistin sulfate, and moxalactam. *Listeria* spp. can grow in the presence of these antimicrobials. Similar to Fraser broth, the medium contains esculin and ferric ammonium citrate that produce a characteristic black precipitate by the action of *Listeria* spp. Lithium chloride inhibits the growth of Gram-negative organisms while permitting the growth of *Listeria* spp., which are salt tolerant.

Motility Medium

This semi-solid medium is used for the visual determination of motility under various incubation conditions. The medium is composed of tryptone, peptone, and a low percentage of agar (3.5 g/l), and is prepared in glass test tubes. The medium is inoculated by stabbing with a needle down the center of the tube. For each isolate, two tubes are inoculated, with one being incubated at 25°C and the other at 37°C. *L. monocytogenes* isolates will display a characteristic umbrella shape in the motility medium with incubation at 25°C, but will not display motility at 37°C.

Tryptic Soy Agar with Blood (TSA-Blood)

This medium is used for cultivating fastidious microorganisms and for differentiating their ability to cause blood hemolysis. The medium is made of TSA with the addition of 5% sheep blood. To produce optimum results, *L. monocytogenes* is stabbed into TSA-blood and plates are incubated at 35°C for 48 hours, preferably under increased CO_2 atmosphere. Under these conditions, *L. monocytogenes*, *L. ivanovii*, and *L. seeligeri* produce distinct, but often narrow zones of β-hemolysis around the stabs. *L. innocua* shows no zone of hemolysis (γ-hemolysis). Possible types of hemolysis on blood agar media, in general, are:

- **α-hemolysis:** Reduction of the red blood color to a greenish discoloration around the colony. This results from the breakdown of red blood cells and conversion of blood hemoglobin to methemoglobin by the hemolytic microorganism.
- **β-hemolysis:** Production of a clear zone around the colonies due to the lysis of red blood cells and breakdown of hemoglobin by the hemolytic microorganisms.
- **γ-hemolysis:** This indicates no hemolysis and no change in the color of blood around the colonies.

Tryptic Soy Agar with Yeast Extract (TSA-YE)

This medium is made of TSA with addition of 2% yeast extract. TSA-YE is a relatively rich, nonselective medium. It is used in this experiment to provide a clear background that allows examination of colony morphology since the medium does not interfere with the appearance of the characteristic blue-gray tint of *Listeria* colonies (Henry oblique transmitted illumination test). Colonies growing on this medium are also suitable for testing the catalase and oxidase reactions and serving as the inoculum source for motility and biochemical tests.

TESTS

Catalase Test

The presence of active catalase enzyme is determined by adding hydrogen peroxide to part of a colony on a glass slide. If effervescence is observed, then the colony is catalase-positive.

Christie-Atkins-Munch-Peterson (CAMP) Test

The CAMP test is a possible confirmatory test, particularly when hemolysis around the stabs in TSA-blood medium is questionable. Traditionally, this test has been used to identify group B β-hemolytic streptococci. To run this test, *Staphylococcus aureus* is streaked onto blood agar. *L. monocytogenes* streaked perpendicular to the *S. aureus* streak will produce an enhanced zone of hemolysis where streaks are closest to each other. *L. monocytogenes* should produce a synergistic region of β-hemolysis resembling the appearance of a matchstick. β-hemolytic streptococci streaked perpendicular to *S. aureus* would produce an arrowhead shape.

Gram Reaction

Gram staining procedure has been discussed in Chapter 4.

Oxidase Test

Oxidase test procedure has been discussed in Chapter 8.

PROCEDURE OVERVIEW

A simplified culture-based method has been assembled for use in this laboratory exercise (Figure 11.1). Environmental swabs need to be collected prior to the first laboratory session. Information on sampling food processing environments can be found in Chapter 2. Environmental swabs may be collected by students; however, students or instructors need to obtain approval before sampling selected facilities. Supplies and instruction on sample collection will need to be coordinated prior to the start of the first laboratory session. In this exercise, a swab sampling procedure for examination of surfaces will be followed. Per group of two, a single swab sample will be collected, with a focus on drains and high-traffic areas. If permitted by a facility supervisor, students will take the supplies to a food processing or handling facility to be sampled. Appropriate facilities include factory, restaurant, retail, food preparation, or pilot plant. Within the facility, sites suitable for sampling include chillers, coolers, freezers, food or ingredient storage or holding tanks, packaging machines, meat slicers in deli shops, floors, walls, and drains.

Possible foods for analysis include raw meat or seafood, raw vegetables, frozen vegetables, and deli sliced products. Artisanal cheese, mold-ripened cheese, and several minimally processed dairy products are also suitable for sampling. Appropriate sampling and sample preparation technique should be followed as described in previous chapters.

The first laboratory session will consist of setting up the primary enrichment (in half-Fraser broth) of the environmental swab as well as a food sample. In the second session, students will be streaking the enrichments for isolation of typical colonies on selective-differential (MOX agar) and chromogenic (BLA) media. The third session will be dedicated to the transfer of presumptive *Listeria* isolates to TSA-YE and TSA-blood. The fourth session will be reserved for the identification *Listeria* spp. by Gram stain, catalase, and oxidase reactions. The fourth session could also include initiating the motility test, if supplies for this test are provided; however, results of the test will need to be viewed in the subsequent session.

Organization
Students will work in groups of two. Each group will be analyzing a food sample and an environmental sample. Members of the group will work independently on the assigned samples but will work cooperatively on a set of control cultures (Figure 11.1).

SESSION 1: SAMPLE PREPARATION AND ENRICHMENT

Food or environmental samples will be enriched in half-Fraser broth. If students are collecting environmental samples, supplies and instructions should be provided and samples collected in advance of this session. Information on sampling food processing environments can be found in Chapter 2 of this book and in Moberg and Kornacki, 2015.

MATERIALS AND EQUIPMENT

Per Group of Two

- Food sample and environmental swab
- 225 ml (food) and 45 ml (environmental swab) half-Fraser broth in bottles

Class-Shared

- Scale for weighing food samples (e.g., a top-loading balance with 500 g capacity)
- Tools (e.g., sterile knives, cutting boards) for sample preparation
- Stomacher and stomacher bags
- Common laboratory items such as disposable gloves, incubators, and tubs for holding enrichments during incubation

PROCEDURE

1. **Enrichment Preparation (Food Sample)**
   a. Prepare the food for sampling to ensure that a representative sample is analyzed. This varies with the food and may involve cutting, partitioning, grinding, or mixing.
   b. Weigh 25 g of the food sample into a stomacher bag. Add the 225 ml of half-Fraser broth to the bag contents. Stomach for 2 min.
   c. Place the bag in the holder, designated by the laboratory instructor, for holding the enrichment bags. This container should support the bag and keep it upright during incubation and handling. Alternatively, transfer the bag contents into a sterile jar and close with the lid.
   Note: if a jar is used, it should be autoclavable so that its contents can be heat-sterilized after use.
   d. Incubate enrichment bags or jars at 30°C for 48 hours.
2. **Enrichment Preparation (Environmental Samples)**
   a. Vortex the environmental swab recovery liquid.
   b. Transfer the entire contents (~5 ml liquid plus the swab) of the tube to a stomacher bag.
   c. Add 45 ml of half-Fraser broth to the stomacher bag.
   d. Place the bag in the appropriate incubation containers designated by the laboratory instructor. This container should support the bag and keep it upright during incubation and handling.
   e. Incubate enrichment bags at 30°C for 48 hours.

SESSION 2: ISOLATION

Enrichments in half-Fraser broth will be streaked for isolation on selective-differential and chromogenic media.

MATERIALS AND EQUIPMENT

Per Group of Two

- Food and environmental samples enriched in half-Fraser broth
- Control cultures
    - *Enterococcus* control: Overnight culture of *E. faecalis*
    - *Listeria* spp. control: Overnight culture of *L. innocua*

    Notes: L. monocytogenes *may be used if level-II biosafety facility is available for running the tests. Alternatively, use* L. innocua *as* L. monocytogenes *surrogate;* L. innocua *will produce some but not all typical* L. monocytogenes *reactions in this culture-based detection technique (see Table 11.1).*
- Three MOX agar plates (Two for samples, one for the controls)
- Three BLA plates (Two for samples, one for the controls)

Class-Shared

- Common laboratory supplies and equipment (e.g., incubator set at 35°C)

PROCEDURE

1. Label the MOX agar and BLA plates: each set of three includes one plate for the food, one plate for the environmental sample, and one for the control cultures. For the control plate, draw a line that divides the plate into halves, one-half for each control culture.
2. Three-phase streak each enrichment onto one MOX agar plate and one BLA plate.
3. Three-phase streak the control cultures (*Listeria* sp. and *E. faecalis*) onto the split MOX agar and the BLA plates.
4. Incubate the plates at 35°C for 24 to 48 h.

SESSION 3: IDENTIFICATION

This session includes evaluation of presumptive *Listeria* spp. from the selective-differential and chromogenic media. Presumptive *Listeria* spp. isolates will be transferred to non-selective media for further testing.

MATERIALS AND EQUIPMENT

Per Group of Two

- Incubated MOX agar and BLA plates
- Three TSA-YE plates (two for the samples and one for the controls)
- Three TSA-blood plates, thickly poured (two for the samples and one for the controls)
- Overnight culture of *E. faecalis* (if no colonies of the bacterium grew on the incubated MOX agar and BLA plates)

Class-Shared

- Common laboratory supplies and equipment (disposable gloves, incubator set at 35°C, etc.)

PROCEDURE

1. **Examination of Incubated Agar Plates**

   Examine the colonies on MOX agar and BLA plates streaked in the previous laboratory session. Colonies with typical reactions on these media are considered presumptive *Listeria* spp. isolates.

   a. **MOX agar and BLA plates of control cultures**

   i. Examine the appearance of the colonies of the control cultures on the MOX agar control plate. Colonies of *Listeria* spp. are differentiated from most non-*Listeria* by their dark gray color surrounded by a black zone (halo) due to the hydrolysis of esculin. *E. faecalis* is typically inhibited by the selective ingredients in MOX agar; however, some strains may grow and hydrolyze esculin. Colony morphology (shape and size) will be the primary distinguishing features between enterococci and *Listeria* spp.

   ii. Examine the appearance of the colonies of the control cultures on the BLA control plate. Colonies of *L. monocytogenes* appear blue-green (due to β-glucosidase activity) and are surrounded by an opaque halo (due to phospholipase C activity). *L. innocua* does not possess phospholipase activity, thus the colonies will be blue-green in color but lack the opaque halo. *E. faecalis* should not produce detectable colonies on this medium.

   iii. Record your observations in Table 11.3.

TABLE 11.3 {add a descriptive title for this data, including food sample used, source of environmental sample, media, recovery method, appropriate footnotes, and what data were obtained}

Source	Half-Fraser broth	MOX	BLA	TSA-YE	Gram stain	Catalase test	Hemolysis
		Presence or absence of presumptive colonies and colony descriptions					
Food sample							
Environmental sample							
Listeria innocua							
Enterococcus faecalis							

b. MOX agar and BLA plates from food and environmental samples

  i. Examine the MOX agar and BLA plates from food and environmental sample enrichments for colonies that have characteristic *Listeria* spp. appearance. If such colonies are present, *mark up to two isolated colonies per sample*; ideally, one originates from the MOX agar plate and the other from the BLA plate.

  ii. Record your observations in Table 11.3.

  Note: If only one of the two samples has presumptive Listeria *spp. colonies, both members of the group will work on isolates from that sample throughout the rest of the exercise. If neither of the two samples produced typical* Listeria *colonies, both members of the group will work with colonies from the* Listeria *spp. control plate.*

2. Subculturing for Identification

The selected presumptive *Listeria* spp. colonies on the plates of MOX agar and BLA will be streaked on the rich medium TSA-YE, and stabbed into the TSA-blood.

a. Isolates from food sample

  i. Label a TSA-YE plate and draw a line that divides the plate into two equal halves.

  ii. Using an inoculation loop, pick a portion of the marked presumptive *Listeria* colony from the incubated food-MOX agar plate. Streak this isolate (two-phase streaking), onto one-half of the TSA-YE plate.

  iii. Repeat the previous step with the second marked colony on the food-BLA plate and streak on the other half of the TSA-YE plate.

  iv. Repeat steps 1 through 3 on TSA-blood plates, but instead of streaking, use a needle to stab the isolate repeatedly into the agar.

  Note: When possible, portions of the same marked colony are used to inoculate both TSA-YE and TSA-blood plates. If this can't be achieved, multiple colonies of typical Listeria *spp. should have been marked on the plates of MOX agar and BLA.*

  v. Incubate the inoculated TSA-YE and TSA-blood plates at 35°C for 24 hours.

b. Isolates from environmental sample

  i. Label a TSA-YE plate and draw a line that divides the plate into two equal halves.

  ii. Using an inoculation loop, pick a portion of the marked presumptive *Listeria* colony from the incubated environmental-MOX agar plate. Streak this isolate (two-phase streaking), onto one-half of the TSA-YE plate.

  iii. Repeat the previous step with the second marked colony on the environmental-BLA plate and streak on the other half of the TSA-YE plate.

  iv. Repeat steps 1 through 3 on TSA-blood plates, but instead of streaking, use a needle to stab the isolate repeatedly into the agar.

Note: When possible, portions of the same marked colony are used to inoculate both TSA-YE and TSA-blood plates. If this can't be achieved, multiple colonies of typical Listeria *spp. should have been marked on the plates of MOX agar and BLA.*

v. Incubate the inoculated TSA-YE and TSA-blood plates at 35°C for 24 hours.

c. **Controls**

i. Label a TSA-YE plate and draw a line that divides the plate into two equal sections.

ii. Using an inoculation loop, pick a colony from the incubated *Listeria* spp. control culture (MOX agar or BLA plate) and two-phase streak onto the appropriate area of the TSA-YE plate.

iii. Repeat the previous step using the *Enterococcus* control culture. If no colonies grew on MOX agar or BLA plate, use the provided overnight-incubated *E. faecalis* culture.

iv. Repeat all previous steps for controls using a TSA-blood plate, but instead of streaking, use a needle to stab the isolate repeatedly into the agar.

v. Incubate the inoculated TSA-YE and TSA-blood plates at 35°C for 24 hours.

SESSION 4: MORPHOLOGICAL AND BIOCHEMICAL CHARACTERIZATION

Morphological and limited biochemical testing will be completed during this session to confirm the identity of isolates as *Listeria* spp.

MATERIALS AND EQUIPMENT

Per Group of Two

- Incubated TSA-YE plates
- Incubated TSA-blood plates
- Hydrogen peroxide, 6% solution
- Oxidase reagents. Alternatively, oxidase strips obtained commercially may be used. In this case, the test is completed using manufacturer's instructions.
- Gram staining reagents

Class-Shared

- CAMP test demonstration plates with the following cultures (prepared by instructors)
  - *Staphylococcus aureus*
  - *Listeria monocytogenes*
  - *Enterococcus faecalis*
  - *Streptococcus agalactiae*
- A light source for colony morphology examination
- Other common laboratory supplies (e.g., microscope slides) and equipment (e.g., microscopes)

PROCEDURE

1. **Colony Morphology**
   a. Examine the appearance of the colonies on the *Listeria* spp. control TSA-YE plate, using an obliquely-positioned light source. Typical colonies are 0.2–1.5 mm in diameter and appear faint blue-gray when seen using a 45° (oblique angle) light setup. To see that subtle coloration, tilt the plate while holding it over the light source.
   b. Examine the food and environmental sample plates and mark isolated colonies that have the characteristic *Listeria* appearance, as just described for the *Listeria* spp. control culture.
   c. Record your observations in Table 11.3.

2. **Cell Morphology**
   a. If presumptive *Listeria* colonies are observed on the TSA-YE plates containing sample isolates, perform a Gram stain on up to two of these colonies per sample.

b. Examine stained smears under the microscope and determine isolates' Gram reaction and cell morphology.

c. If no presumptive *Listeria* were cultured on TSA-YE plates, then Gram stain the control cultures.

d. Record your observations in Table 11.3.

3. Catalase Test

a. Perform a catalase test on a colony from the positive control, a colony from the negative control, and up to two colonies for each analyzed sample.

b. Transfer a portion of the colony to a microscope slide and add a single drop of 6% hydrogen peroxide solution. Instant effervescence due to the reduction of oxygen bubbles indicates the presence of the catalase enzyme. No bubbles formed indicates a negative reaction.

c. Record your observations in Table 11.3.

4. Oxidase Test

a. Perform the oxidase test on a colony from each control culture. Using a sterile toothpick (not a metal loop), transfer a portion of the colony to supplied filter paper; multiple colonies can be tested on one piece of filter paper.

b. Using provided Pasteur pipette, add one drop of a freshly prepared oxidase reagent onto the smear.

c. Wait 20–30 seconds for a color change. *Listeria* spp. are oxidase-negative, so there should be no color change within 30 seconds (color changes after this time are common, but do not indicate a positive result). *E. faecalis* should also be oxidase negative.

d. Repeat the test using up to two colonies for each analyzed sample.

e. Record oxidase results in Table 11.3.

5. TSA-Blood plates (Blood Hemolysis)

a. Examine the TSA-blood plates receiving stabs from the controls and the food and environmental sample colonies.

b. Observe any blood clearing around the stabs. Use the light source to observe the typical *L. monocytogenes* narrow zone of hemolysis, if present.

c. Record hemolysis results in Table 11.3.

6. Evaluation of Results

Evaluate the results of the food and environmental sample testing as recorded in Table 11.3. Determine if the samples are positive or negative for the presence of *Listeria* spp., or whether the result is inconclusive. Record the results of this analysis in Table 11.4. Keep in mind that isolates with similarities to *Listeria* spp. or *L. monocytogenes*, as determined in this exercise, would be subjected to additional tests (e.g., motility, biochemical, and serological tests) to confirm their identities.

7. Demonstration: CAMP Test

a. Examine the demonstration CAMP test on the TSA-blood plate.

TABLE 11.4 {add a descriptive title for this data, including food sample used, source of environmental sample tested, media, recovery method and what data were obtained}

Source	Culture method	PCR method	Remarks
Food sample:			
Environmental sample:			

b. Look for characteristic zones of β-hemolysis at the intersection of the primary *S. aureus* streak with the cross-streaks of *L. monocytogenes, E. faecalis,* and *S. agalactiae. E. faecalis* should display no synergy with *S. aureus. S. agalactiae* will show a large arrowhead shape of hemolysis due to synergy with *S. aureus. L. monocytogenes* should display a characteristic matchstick shape of hemolysis with *S. aureus.*

EXERCISE II: DETECTION OF *Listeria monocytogenes* USING PCR-BASED METHOD

INTRODUCTION

Polymerase chain reaction (PCR) is a molecular technique that is often used for rapid detection of pathogens in food. PCR-based methods offer a more rapid and easier to interpret alternative to conventional culture-based methods, such as the one described in Exercise I of this chapter. The PCR technique is used to amplify a unique DNA sequence in the target organism's genome; the amplified sequence serves as a signature for the pathogen being detected. See the Part III introduction for detailed discussion about PCR technique.

The method used in this exercise is based on the premise that *L. monocytogenes* produces an invasion-associated protein called p60. This is a large protein, consisting of 484 amino acid residues, and is encoded by the gene *iap*. This gene also has homologous sequences in most *Listeria* spp. Therefore, the part of the gene that is shared among *Listeria* spp. can be used to detect most members of the genus *Listeria*, whereas a sequence within the gene that is found only in *L. monocytogenes* can be used specifically to detect *L. monocytogenes*. Based on this information, two primer sets (a total of four primers) will be used in this exercise. The primer set Lis1A and Lis1B is specific for all *Listeria* spp. The second primer set, MonoA and MonoB, is specific for *L. monocytogenes* only.

Although the PCR-based method can be run independently as a stand-alone detection method, it is preferable to run it simultaneously with the culture-based approach described in Exercise I. An overview of the PCR-based method is shown in Figure 11.2 and Table 11.2. Although this method is less time consuming (compared to conventional methods) in laboratories where these analyses are run routinely (e.g., over the course of a single day), this time saving is less apparent in most teaching laboratories.

OBJECTIVES

1. Learn and practice molecular pathogen detection methods using polymerase chain reaction.
2. Apply safe procedures for pathogen handling in the laboratory.

PROCEDURE OVERVIEW

Both the culture- and PCR-based methods include similar sample preparation and enrichment steps. In the current method, the PCR technique is applied to the enriched samples. If *L. monocytogenes*-specific PCR product is not detected, the result is considered conclusively negative. However, if the specific PCR product for the bacterium is detected, additional culture and biochemical tests need to be completed before confirming that the sample is positive for *L. monocytogenes*.

Note that for regulatory purposes it is necessary to isolate live bacterial cells, not just obtain information regarding DNA sequences.

This PCR technique includes the following steps: (i) isolation of genomic DNA from bacterial cells in the enriched sample; (ii) use of the isolated DNA as a template for hybridization with primers designed for the detection of *L. monocytogenes*; (iii) amplification of the DNA sequence intervening the primer binding sites using PCR; (iv) separation of the amplified DNA as bands on agarose gel; and (v) detection of the bands with the correct sizes as a recognition of the presence of *L. monocytogenes* in the analyzed sample. An overview of the procedure is shown in Figure 11.2.

DNA Extraction

Extraction and purification of bacterial genomic DNA present in the sample enrichment will be accomplished using a commercial kit (Qiagen DNeasy Tissue Kit), as recommended by the kit's manufacturer. Briefly, a portion of incubated half-Fraser enrichment broth is transferred to a sterile microcentrifuge tube and centrifuged to pelletize the cells. The supernatant is discarded, and the cell pellet is washed using 0.85% NaCl. The cells are pelleted again, suspended in lysis buffer, which contains lysozyme, and incubated under conditions suitable for the action of this enzyme (i.e., 37°C for 30 min). Proteinase K and an additional buffer solution are added and the mixture is incubated at 56°C for 30 min, a condition suitable for the action of the enzyme. Ethanol is then added to the mixture and the sample is applied to a separation column and centrifuged to allow the DNA to bind to the column. Several washing steps are performed, and the purified DNA is eluted from the column using a final elution buffer.

DNA Amplification (PCR Technique)

A portion of the purified DNA is transferred to a PCR tube, which contains the PCR reaction mixture. The reaction mixture includes reaction buffer containing magnesium ions, *Taq* polymerase, nucleotides (dNTPs), and the specific primers. The tubes are transferred to a thermocycler, which has been preprogrammed to heat and cool in cycles. After 30 cycles of melting the DNA (95°C), annealing the primers (58°C), and elongation (72°C), the reaction mixture is cooled and refrigerated or frozen until used in a subsequent laboratory session.

DNA Detection (Gel Electrophoresis)

Each agarose gel is cast to contain 12 wells; this should be enough for DNA-mass ladder (2 wells), and PCR reactions of food, environmental samples, and controls from one group (10 wells). The gel is placed in the electrophoresis unit, with the wells oriented toward the negative electrode. The agarose gel should be completely immersed in the running buffer before sample loading. The PCR reactions are mixed with a loading (tracking) dye, which contains a compound that will fluoresce when it intercalates into double-stranded DNA and is exposed to UV light. The PCR product and loading dye mixtures are dispensed in the individual agarose wells. Electrophoresis is performed by connecting the electrophoresis unit to electricity and adjusting and maintaining the voltage at the desired setting. Keeping track of reaction mixtures during preparation, PCR, and gel electrophoresis is crucial for eliminating sample mix-up.

Gel Photographing

The loading dye used in this exercise contains a fluorescent component that binds to DNA, eliminating the need to stain the finished gel with ethidium bromide. When the gel is exposed to UV light, the DNA bands will be visible, and a picture is taken using a digital camera.

Results Explanation

The size of the band(s) resulting from the samples will be determined by comparing their position relative to the bands of the mass ladder. A 1.5-kb band is indicative of the presence of *Listeria* spp. A 0.4-kb band is indicative of the presence of *L. monocytogenes*.

Data Interpretation

If no *Listeria* spp. are detected in the analyzed sample by the PCR-based procedure, the result may be considered conclusive, particularly if the streaking on MOX agar and BLA media (during the culture-based procedure) also produced negative results. If DNA bands typical to *Listeria* spp. are found on the agarose gel, the result may be used to confirm similar positive results from the culture-based method.

Organization

The exercise requires four laboratory sessions to complete (Figure 11.2). During the first session, students will work in groups of two on preparing enrichments for a food sample and an environmental sample. This organization is similar to that described in Session 1, Exercise I.

In sessions 2–4, students also will work in groups of two, one sample per student. The two students will cooperate to prepare the PCR master mixtures. They also will cooperate to analyze the two control cultures as well as the *L. monocytogenes* DNA provided by the instructor. Pooling individual students into groups makes it possible to use PCR reagents efficiently.

SESSION 1: SAMPLE PREPARATION AND ENRICHMENT

During this session, food will be sampled, and a selective enrichment will be prepared. Similarly, environmental samples collected prior to the exercise will be enriched. For sampling and sample enrichment, follow the procedure described in Exercise I. Students will be working in pairs during this session.

SESSION 2: DNA EXTRACTION AND PCR

MATERIALS AND EQUIPMENT

Per Group of Two
- Food sample enriched in half-Fraser broth
- Environmental sample enriched in half-Fraser broth
- Controls
  - *Enterococcus* control: Overnight culture of *E. faecalis*
  - *Listeria* spp. control: Overnight culture of *L. innocua*
  - *L. monocytogenes* control: DNA isolated from *L. monocytogenes* and prepared by instructors ahead of the session
- 10 sterile microcentrifuge tubes (4 for centrifugation, 4 for elution, 2 for PCR reaction mixtures preparation)
- Floating microcentrifuge tube rack. This is not needed if thermostatically controlled heating block (also known as dry bath incubator) is used.
- Glass test tube (for waste collection)
- Sodium chloride solution (0.85 %) – 10 ml per group
- Lysis Buffer – 1 ml per group (20 mM Tris-Cl, pH 8.0, 2 mM sodium EDTA, 1.2% Triton X-100, 20 mg/ml lysozyme)
- Ethanol – 1 ml per group (95–100%)
- Deionized autoclaved water – 1 ml per group
- Qiagen DNeasy Tissue Kit:
  - Spin columns (4 per group)
  - Collection tubes (12 per group)
  - Proteinase K and various buffers, aliquoted appropriately
- Vortexer
- Ten 200-µl PCR tubes
- Micropipetters
- Sterile pipette tips

Class-Shared
- Ice in containers for holding reagents (ice bath)
- Water bath, set at 37°C; alternatively, use dry bath incubator at 37°C
- Water bath, set at 56°C; alternatively, use dry bath incubator set at 56°C
- Polymerase mix: MyTaq PCR Kit (Bioline, Inc., Taunton, MA), kept on ice. The mix contains polymerase, dNTPs and buffer. Each group will require 132 µl

  Note: Several alternative PCR kits are available commercially. It may be necessary to adjust reaction mixture volumes and primer concentrations depending on the kit utilized.
- Primers (standard desalted) suspended in sterile molecular-grade water (kept on ice). Primer concentration is 5 µM and 6 µl is needed per group.
  - Lis1A: 5'-ATG AAT ATG AAA GCA AC-3'
  - Lis1B: 5'- TTA TAC GCG ACC GAA GCC AAC – 3'

    ○ MonoA: 5' – CAA ACT GCT AAC ACA GCT ACT – 3'
    ○ MonoB: 5' – GCA CTT GAA TTG CTG TTA TTG -3'
- Thermocycler

PROCEDURE

As indicated previously, students will work in groups of two to analyze the assigned food and environmental samples. The two students will cooperate to complete the following: (i) preparing the PCR master mixtures; (ii) analyzing the two control cultures; and (iii) analyzing the *L. monocytogenes* DNA provided by the instructor.

1. **Sample Preparation for DNA Extraction**
   a. Label four microcentrifuge tubes: two for the food and environmental sample enrichments, one for *L. innocua* culture (*Listeria* spp. control), and one for *E. faecalis* culture (*Enterococcus* control).
   b. Transfer 1.5 ml of each of the four samples described to the appropriate microcentrifuge tube.
      Note: When pipetting from the food enrichment, avoid transferring food particles or lipids. The cleaner the portion pipetted, the great the chance that DNA extraction is completed successfully.
   c. Centrifuge for 2 min at 12,000 rpm.
      Note: Make sure tubes are balanced in the microcentrifuge before starting.
   d. Pipette off supernatant into waste glass test tube being careful not to disturb the pellet.
   e. Add 1.5 ml 0.85% NaCl to wash each pellet. Vortex until the cells in the pellet have been resuspended.
   f. Centrifuge for 2 min at 12,000 rpm.
   g. Pipette off supernatant into waste glass test tube.
   h. Re-suspend each pellet in 180 µl of lysis buffer.
   i. Place tubes in floating microcentrifuge tube rack and incubate in 37ºC water bath (rack and water bath may be replaced with a heating block) for 30 min.
   j. Add 25 µl of proteinase K and 200 µl of Buffer AL to each sample tube. Vortex.
   k. Place tubes in floating microcentrifuge tube rack and incubate in 56ºC water bath (rack and water bath may be replaced with a heating block) for 30 min.
   l. Add 200 µl of ethanol to each sample tube and vortex.
   m. Place four DNeasy spin columns into four collection tubes.
   n. Label the spin columns with sample information.
   o. Vortex and transfer the treated sample mixtures into appropriate spin columns.
   p. Centrifuge at 8000 rpm for 1 min.

q. Discard flow through and collection tubes.

r. Place spin columns into new set of collection tubes.

s. Add 500 µl of Buffer AW1 to each spin column.

t. Centrifuge at 8000 rpm for 1 min.

u. Discard flow through and collection tubes.

v. Place spin columns into new set of collection tubes.

w. Add 500 µl of Buffer AW2 to each spin column.

x. Centrifuge at 12,000 rpm for 3 min.

y. Discard flow through and collection tubes.

z. Label four new sterile microcentrifuge tubes with sample information.

aa. Place the appropriate spin column onto its labeled microcentrifuge tube.

bb. Add 50 µl of Buffer AE into each spin column.

cc. Allow to sit for 1 min.

dd. Centrifuge at 8000 rpm for 1 min to elute the DNA into the microcentrifuge tube; this will serve as a DNA template during preparing PCR reaction mixture.

Note: Do not discard flow through, it contains sample DNA.

2. **PCR Reaction Mixture Preparation**

Summary: Two PCR master mixtures, containing primers, will be prepared in two sterile microcentrifuge tubes as follows: (i) master mixture containing the Lis primers (for *Listeria* spp.), and (ii) mixture containing the Mono primers (for *L. monocytogenes*). After the master mixtures are prepared, aliquots are dispensed into PCR tubes. DNA templates of the five isolates (two for food and environmental samples, two for control cultures, and one for *L. monocytogenes*) will be added to finalize the reaction mixtures.

Procedure detail is as follows:

a. Prepare a PCR master mixture that includes *Listeria* spp. primers (Master Mix Lis): In a sterile microcentrifuge tube, mix 210 µl sterile water, 66 µl 5× *MyTaq* mixture (buffer, dNTPs and *Taq*), 6 µl of primer Lis1A, and 6 µl of primer Lis1B.

b. Prepare a PCR master mixture that includes *L. monocytogenes* primers (Master Mix Mono): In a sterile microcentrifuge tube, mix 210 µl sterile water, 66 µl 5× *MyTaq* mixture (buffer, dNTPs and *Taq*), 6 µl of primer MonoA, and 6 µl of primer MonoB.

Note: Polymerase (Taq) *is the most sensitive component of the PCR reaction. Care should be taken to keep reagents and reaction mixtures on ice.*

c. Vortex the microcentrifuge tubes to mix their contents.

d. Carefully label 10 sterile PCR tubes with **codes** for primer used (Lis or Mono) and sample analyzed (Food sample, environmental sample, *Listeria* spp. control, *Enterococcus* control, or *L. monocytogenes* control).

Note: The PCR tube is too small to take full label, hence, a code representing a mixture is written on the cap using a permanent marker (e.g., Sharpie pen).

e. Transfer 48 μl of each PCR master mixture to the corresponding PCR tube.

f. Add 2 μl of DNA templates to the appropriately labeled PCR tubes (there should be two tubes per template, one for Lis and one for Mono).

g. Mix the contents of PCR tubes.

h. Place PCR tubes in thermocycler. Record the location of your tubes in the thermocycler on the sheet provided.

i. Thermocycler will run for 3 min at 95°C and for 30 cycles under the following conditions: 95°C for 15 s, 58°C for 15 s, 72°C for 30 s. At the end of the 30 cycles, the thermocycler will be run at 72°C for 4 min, then tubes will be held at 4°C until removed.

j. PCR products will be refrigerated (or frozen) until subsequent session.

SESSION 3: GEL ELECTROPHORESIS

PCR products will be separated using agarose gel electrophoresis and visualized by UV light with comparison to a DNA standard. PCR results will be interpreted qualitatively to determine isolate identity as *Listeria* spp., *L. monocytogenes,* or non-*Listeria* spp.

MATERIALS AND EQUIPMENT

Per Group of Two

- PCR products from previous session
- Micropipetters and sterile pipette tips

Class-Shared

- Agarose gel (1%)
- Running buffer (TAE)
- Loading buffer containing EZ-Vision™ DNA Dye
- DNA ladder (instructor will load the ladder)
- Gel electrophoresis equipment (gel running tray and matching power supply)
- Gel documentation station (UV transilluminator, digital camera, computer, and image-capture software)

Gel Electrophoresis

1. Use Parafilm pieces to prepare the mixtures to be loaded into the gel. Label **10** spots on the Parafilm with the codes used on PCR tubes in the last session (i.e., two primer sets for each of the food and environmental samples, the two control cultures, and the *L. monocytogenes* control).
2. Place 2 µl of loading buffer near each label.
3. Mix 10 µl of the appropriate PCR reaction with the loading buffer using the micropipette.
4. Dispense 10 µl of the mixture prepared in the previous step into appropriate well of the agarose gel (which is submerged in the running buffer).
5. Record which sample was put into which well on the sheet provided.
6. Run gel electrophoresis for approximately 1.5 h at 100 V.
7. Gel picture will be taken using a gel documentation station; picture will be viewed during Session 4.

SESSION 4: ELECTROPHORESIS RESULTS

An image of the UV-illuminated agarose gel will be evaluated and interpreted.

MATERIALS AND EQUIPMENT

Photograph of the agarose gel

PROCEDURE

1. Observe the gel photograph and record the results of these observations. Begin by locating lanes containing the DNA ladder in the gel and label with appropriate band sizes (see manufacturer insert for DNA ladder details).
2. Locate the lanes containing the controls. Estimate the size of the DNA amplification bands for each control culture using the DNA ladder as a reference. Control cultures should lead to the following observations:
   a. *E. faecalis*:
      i. *Listeria* spp. PCR: no band
      ii. *L. monocytogenes* PCR: no band
   b. *L. innocua*:
      i. *Listeria* spp. PCR: 1.5 kb band
      ii. *L. monocytogenes* PCR: no band
   c. *L. monocytogenes*:
      i. *Listeria* spp. PCR: 1.5 kb band
      ii. *L. monocytogenes* PCR: 0.4 kb band
   Note: If an alternative banding pattern is present in lanes containing the controls, then the PCR results may be invalid and cannot be used to interpret the results from the food/environmental isolates. Likely causes for unexpected results from the control cultures include incomplete cell lysis, poor formulation/mixing of master mix, or cross-contamination.
3. Locate the lanes containing the PCR product originating food and environmental isolates. Determine the presence or absence of PCR products (DNA bands) for the *Listeria* spp. primers and the *L. monocytogenes* primers. If bands are present, estimate the size of the band (bp) using the DNA ladder as a reference.
4. Record the final PCR results for food and environmental isolates and controls in Table 11.4. Interpret the results from the food and environmental isolates as *Listeria* spp., *L. monocytogenes*, or non-*Listeria* spp.
5. Keep in mind that positive results, as determined in this exercise, only indicate that *L. monocytogenes* is presumptively present in the food or environmental sample. Additional tests (e.g., isolation followed by motility, biochemical, and serological tests) would be needed to confirm if the isolates are *L. monocytogenes*.

QUESTIONS

1. Why would an analyst choose to identify *Listeria* spp. instead of *L. monocytogenes*? In what instance would identification to the species level be necessary?

2. Considering the results from the entire class, are any of the samples likely to have contained *Listeria* spp. or *Listeria monocytogenes*?

3. In Exercise I, two selective-differential media were used to isolate potential listeriae from the enrichment. Why are two media recommended for use in this step instead of just one? Within your group, did you observe different results between the two media?

4. Answer the following questions with reference to Exercise II:
   a. Three control cultures were used for PCR. What is the purpose of each of these three controls?
   b. Draw a schematic showing the ideal banding pattern of these controls on an agarose gel.
   c. Compare the pattern you drew in the previous question with the results your obtained. Explain any discrepancies.

CHAPTER 12

Salmonella enterica

CULTURE-BASED DETECTION. SELECTIVE ENRICHMENT. BIOCHEMICAL IDENTIFICATION. IMMUNOASSAY

INTRODUCTION

The genus *Salmonella* belongs to the family *Enterobacteriaceae*. Like other members of this family, *Salmonella* spp. are Gram-negative, rod-shaped bacteria. Members of this genus are motile by peritrichous flagella, tolerant to bile salts, indole-negative, catalase-positive, oxidase-negative, and capable of utilizing citrate as their only source of carbon and energy. Salmonellae (plural of *Salmonella*) are facultative anaerobes with both respiratory and fermentative metabolic pathways. They ferment glucose to produce acid and gas (i.e., methyl red positive), decarboxylate lysine, and ornithine, generally produce hydrogen sulfide on certain selective media such as triple sugar iron agar, and do not hydrolyze urea. It is important to note that most members of the genus do not ferment lactose or sucrose.

Classification and Nomenclature

Classification and nomenclature of *Salmonella* spp. have been subjected to many changes in recent decades and are still evolving. Isolates of *Salmonella* vary considerably in the type of antigens they carry. Different isolates carry different somatic (O), capsular (Vi), and flagellar (H) antigens. Isolates that differ in antigenic composition (combinations of O, Vi, and H) are considered to be of different serovars (or serotypes). This antigenic variability accounts for more than 2,500 *Salmonella* serovars. Each serovar is given an antigenic formula that

Analytical Food Microbiology: A Laboratory Manual, Second Edition. Ahmed E. Yousef, Joy G. Waite-Cusic, and Jennifer J. Perry.
© 2022 John Wiley & Sons, Inc. Published 2022 by John Wiley & Sons, Inc.

includes the type of O, Vi (if present), and H antigens found on the cell surface. Despite the large number and the diversity of *Salmonella* serovars, the genus includes only two species: *S. enterica* and *S. bongori*. *Salmonella enterica* is divided into the subspecies that may be named or indicated by Roman numeral: *enterica* (I), *salamae* (II), *arizonae* (IIIa), *diarizonae* (IIIb), *houtenae* (IV), and *indica* (VI). The subspecies *enterica* is the most predominant, and more than 1,500 serovars are members of this subspecies. Members of *S. enterica* subsp. *enterica* are associated with humans and warm-blooded animals, whereas the other subspecies commonly inhabit the intestines of cold-blooded animals.

Historically, members of the genus *Salmonella* were given names based on clinical considerations and/or perceived host specificity (e.g., *Salmonella* Abortusequi associated with septic abortions in horses). Additionally, geographical location of the first isolation was included in naming many isolates (e.g., *Salmonella* London, *Salmonella* Panama). The newest reclassification of salmonellae preserved these familiar names as designations of serovars for *S. enterica* subsp. *enterica*, whereas serovars of the other subspecies are designated by antigenic formulae. For example, there is a serovar called *S. enterica* subsp. *enterica* serovar Dublin. Such a lengthy name is written in full only on the first occurrence in scientific writing, but it can be reduced to the genus and serovar designations in subsequent incidences. Therefore, the serovar just described can be shortened to *Salmonella* Dublin. It should be cautioned that the serovar has capital initial and the word is not italicized.

The following is a brief description of the antigens found in *Salmonella* spp. and other *Enterobacteriaceae*.

Somatic (O) antigens

O antigens are the carbohydrates associated with the lipopolysaccharides (LPS) of the bacterial cell outer membrane. Somatic antigens are heat-stable and resistant to alcohol and dilute acids. The O antigen is composed of several subunits, and each is made of four to six sugars.

Capsular (Vi) antigens

The Vi is a capsular polysaccharide antigen produced by a few serovars of *S. enterica* subsp. *enterica*. This antigen is essentially an equivalent to the K antigen in *Escherichia coli*. *Salmonella* capsular antigens are heat labile and key to the identification of *Salmonella* Typhi, *Salmonella* Paratyphi and *Salmonella* Dublin. The presence of the capsular antigen can mask the O antigens and make it difficult to determine the serovars of these strains. The capsular antigen can be removed by treating the cells with boiling water to expose the O antigens making them accessible to O antisera.

Flagellar (H) antigens

The H antigens are made of the heat-labile flagellin proteins FljB and FliC. These two proteins not only differ in antigenic characteristics, but variations within each protein cause additional antigenic variations. A minority of *S. enterica* serovars produce one type of flagellar protein (monophasic), whereas the majority are capable of expressing the two flagellar proteins (diphasic). A diphasic cell doesn't produce the two proteins simultaneously, but alternates its expression of

FljB and FliC, in a process known as flagellar phase variation. It is believed this phase variation creates heterogeneity that helps the bacterium evade host immune systems. Interestingly, a culture of a single diphasic serovar may contain both cells expressing Phase 1 flagellin antigens and others expressing Phase 2 flagellin antigens.

The antigenic formulae for salmonellae are presented as the O antigens, followed by the Vi antigen in brackets (if present), separated from the first-phase H antigen(s) by a colon, which is separated from the second-phase H antigen(s) by a colon. For example, the antigenic formula of *S. enterica* subsp. *enterica* serovar Typhimurium is 1,4,12:i:1,2. Therefore, *Salmonella* Typhimurium produces O antigens 1, 4, and 12. The first-phase H antigen is i and the second-phase H antigens are 1 and 2.

In addition to the serotyping just described, *Salmonella* isolates may be grouped into biotypes, based on differences of isolates in the sugar fermentation pattern. Similarly, sensitivity to bacteriophages can serve as a basis for grouping isolates into phagetypes or phagovars. Isolates may be differentiated based on sensitivity to bacteriocins or resistance to antibiotics. Genetic typing could be useful in investigating foodborne disease outbreaks, by establishing the relationship between *Salmonella* isolates obtained from food and diseased individuals.

Diseases

Although more than 2,500 *Salmonella* serovars have been identified, less than 100 account for most human infections. Serovars known to cause human diseases are usually transmitted by food or water, and their associated diseases are known collectively as salmonellosis. Some serovars are adapted to a particular host, whereas others live in multiple hosts. Host-adapted serovars are responsible for the most severe diseases such as the human enteric infection typhoid fever, caused by *Salmonella* Typhi, and abortion in ewes, caused by *Salmonella* Abortusovis. These host-adapted serovars often establish an asymptomatic carrier state in a small number of individuals that continue to infect susceptible individuals over time.

Most *Salmonella* serovars cause nontyphoidal salmonellosis. Nontyphoidal *Salmonella* serovars, particularly Typhimurium, Enteritidis, and Newport, are leading causes of bacterial diarrhea in the US and other countries. Foodborne infectious serovars may have small infective doses, with $< 10^3$ CFU sufficient to cause clinical symptoms. The disease incubation period is 12–72 hours and the symptoms (commonly diarrhea, fever, and abdominal cramps) last for 4–7 days. In rare cases, the infection can spread from the intestines to the bloodstream and other body sites; this may lead to severe health complications and death.

Foods Implicated in Salmonellosis Outbreaks

Since the intestine is the natural habitat of *Salmonella*, raw foods of animal origin occasionally harbor this pathogen. *Salmonella* is found in poultry products including chicken, turkey, and eggs. Shellfish, milk, and fresh produce have also been implicated in salmonellosis outbreaks. Recently, salmonellosis outbreaks have been increasingly linked to a number of low-moisture products (e.g., peanut butter and spices). Water has been a vehicle for transmission of salmonellosis. Workers who do not observe proper personal hygiene, especially those working in food harvesting, processing, and service, are potential sources of food contamination with *Salmonella*. In the United States, Enteritidis, Typhimurium, Heidelberg, Javiana, and Newport are among the most common disease-causing serovars.

Most *Salmonella* serovars grow only in the mesophilic temperature range; however, some are presumed capable of growing under refrigeration or thermophilic conditions. Foods with pH < 4.5 normally do not support the growth of *Salmonella*, but some serovars grow at pH 4.0 (e.g., *Salmonella* Infantis) and many serovars may be able to survive in low-pH environments. Pasteurization temperatures readily inactivate most *Salmonella* serovars. A particular strain, *Salmonella* Senftenberg 775W, is known to have exceptional resistance to heat. The recent appearance of isolates with multi-drug resistance (e.g., *Salmonella* Typhimurium DT 104) is a potential threat to the safety of consumers and raises great health concerns worldwide. It is important to note that *Salmonella* may survive for extended periods of time in less-than-ideal conditions, particularly in low-moisture environments; these persistent cells are still capable of causing disease.

Detection of *Salmonella* in Food

Detection of this pathogen, as described in this chapter, aims to determine if the food contains *S. enterica*, but official methods for outbreak investigations are designed to characterize isolates at the subspecies level (serovar, strain). Although some *Salmonella* serovars are rarely associated with human illness, all serovars are considered potential pathogens and thus should not be present in ready-to-eat foods. It is uncommon to find *Salmonella* at countable levels in food. Therefore, food is analyzed for presence or absence of *Salmonella* and methods designed for detection, rather than enumeration, are used.

Conventional methods for detection of *Salmonella* in food are culture-based. Newer "hybrid" methods make use of cultural, biochemical, serological, and genomic characteristics of the microorganism. Typical *Salmonella* isolates have the following culture/biochemical characteristics: (i) production of acid from glucose, but not from lactose or sucrose in triple sugar iron (TSI) agar medium; (ii) decarboxylation of lysine to cadaverine (alkaline product) in lysine iron agar (LIA) medium; (iii) generation of hydrogen sulfide (H_2S) in TSI and LIA; (iv) absence of sucrose or lactose fermentation in xylose lysine desoxycholate (XLD), Hektoen enteric (HE), and similar agar media; and (v) absence of urea hydrolysis. Although the inability of most salmonellae to ferment lactose or sucrose is an important defining biochemical characteristic, a small percentage of *Salmonella* isolates are lactose- or sucrose-positive. Some serovars, or strains within a serovar, lack the ability to produce H_2S on the media just described. Many analysts make a good use of these characteristics in detection of *Salmonella* in food and other matrices.

Similar to other pathogens, *Salmonella* is detected in food by implementing the following steps: (i) pre-enrichment and/or selective enrichment; (ii) isolation; (iii) identification; and (iv) confirmation of identity and typing. An additional step may be included before the isolation, to screen the enrichments into *Salmonella*-negative and presumptive *Salmonella*-positive categories. Enrichment and isolation steps are primarily culture techniques, whereas the identification relies on biochemical, serological, or genetic testing, or combinations of these techniques.

Enrichment

When pathogens such as *Salmonella* are present in food, they are likely to be present at low numbers and in an injured state. Pathogen cells may be injured during food processing (e.g., acidification and dehydration), storage (e.g., freezing), or other processes. Enrichment steps are normally designed to resuscitate these injured cells and increase their populations to detectable levels.

In the pre-enrichment step, the food sample is mixed with a suitable non-selective medium and the mixture is incubated. This results in a modest increase in *Salmonella* population and often causes appreciable increases in numbers of competing microbiota in the sample. After the pre-enrichment, it is presumed that salmonellae are healthy enough to endure a subsequent enrichment step using media containing selective agents. Therefore, a volume of the pre-enriched sample is transferred into the selective enrichment medium. This selective enrichment step allows growth of *Salmonella* and suppresses competing microbiota.

Media used in pre-enrichments vary with the type of food analyzed. Buffered peptone water may be used for pre-enrichment. This solution allows for a gentle recovery of injured cells without causing an overgrowth of background microbiota. Lactose broth is also used as a pre-enrichment medium, in spite of the fact that most *Salmonella* serovars do not use lactose as a carbon source. This medium, however, contains proteins that *Salmonella* uses for growth and the medium is suitable for the slow recovery of injured cells in the food sample. Rich foods such as dry milk may be enriched in *Salmonella* by mixing the sample with distilled water and incubating the mixture. Conversely, nutritionally poor foods (e.g., spices) are pre-enriched in *Salmonella* by adding a rich microbiological medium such as tryptic soy broth.

Selective enrichment includes sub-culturing the pre-enrichment into tubes containing selective broth. Tetrathionate broth and Rappaport-Vassiliadis broth are used for selective enrichment of *Salmonella*. Composition and selective characteristics of these media are reviewed later in this chapter. It is advisable to use two or more selective enrichment media to improve the recovery of *Salmonella* from the sample.

Screening

In laboratories where large number of samples are analyzed routinely, a screening step is applied after the enrichment. Since most commercial food samples are expected to be *Salmonella*-free, a screening step culls the negative samples so that resources are directed to samples that are potentially *Salmonella*-positive. It is plausible to use a rapid technique (e.g., immunoassay or PCR) as a screening tool to save time and effort. Prior to screening, a post-enrichment step may be required to condition the isolates for the rapid technique.

In this exercise, the screening will be done using a rapid immunoassay. However, this test will be run simultaneously with the isolation step and the decision to continue with sample analysis will depend on the results of both tests. Based on the results of the screening and the isolation steps, the analyst will have the following options:

1. If the immunoassay shows the sample is presumptively positive for *Salmonella*, and the isolation step produces colonies with typical *Salmonella* characteristics, the analyst should continue analyzing the sample, i.e., completing the identification.
2. If the immunoassay shows the sample is negative for *Salmonella* and the isolation technique produces no colonies with typical *Salmonella* characteristics, the analyst should not continue analyzing the sample to the end.
3. If the results of the screening and isolation do not agree, it will be up to the analysts to continue or terminate the analysis. It is obvious, however, that the analysis cannot proceed unless the isolation step produces colonies to work with.

Isolation

Isolation includes streaking enrichments onto selective-differential agar media and recognizing presumptive *Salmonella* colonies on the incubated plates. These isolated colonies are sub-cultured for subsequent identification. Media for *Salmonella* isolation should contain selective agents such as bile or desoxycholate salts, brilliant green, or bismuth sulfite. These agents inhibit Gram-positive and non-enteric bacteria. Differentiation of *Salmonella* and non-*Salmonella* bacteria is accomplished by inclusion of suitable carbohydrate-pH indicator combinations. Lactose, sucrose, and salicin are not typically fermented by *Salmonella*, and thus production of acid from these carbohydrates indicates non-*Salmonella* isolates. If lysine is present in the medium, *Salmonella* decarboxylates this amino acid, producing alkaline products that change the color of the pH indicator in the agar surrounding the colony. A common differential system in these media depends on the ability of *Salmonella* to release H_2S from sulfur-containing substrates (e.g., sodium thiosulfate) using its desulfhydrase enzyme. In addition to the enzyme substrate, the isolation media are formulated to contain water-soluble ferrous or ferric salt, which reacts with the released H_2S, producing a black precipitate in and around the typical *Salmonella* colonies.

The principles just described have been implemented in formulating several commonly used *Salmonella* isolation media including XLD agar and bismuth sulfite (BS) agar. BS agar has high selectivity for the bacterium and allows detection of small levels of H_2S generated by isolated colonies. It is common to streak a *Salmonella*-enriched sample onto two or more of these selective-differential agar media, resulting in improved recovery of pathogens from the food and increased sensitivity of the method. Description of these media is discussed later in this chapter.

Biochemical identification

Identification of *Salmonella* on the basis of biochemical characteristics has been done reliably for decades. This approach will be used in the exercise described in this chapter. Note that the more biochemical tests are completed, the greater will be the confidence in the detection results. The following are some of the important biochemical tests for *Salmonella* (and associated reactions *by most serovars*): indole (–), methyl red (+), Voges Proskauer (–), citrate utilization (+), H_2S production (+), urea hydrolysis (–), phenyl alanine deaminase (–), lysine decarboxylase (+), arginine dihydrolase (+), ornithine decarboxylase (+), gelatin hydrolysis (–), esculin hydrolysis (–), oxidase (–), oxidation of nitrate to nitrite (+), lactose fermentation (–), sucrose fermentation (–), and xylose fermentation (+). Considering the large number of biochemical reactions needed to make an identification, companies have produced kits that that allow for running many tests simultaneously. One of these kits (BioMerieux API-20E) will be used in this exercise.

Confirmation and typing

Isolates that produce typical *Salmonella* reactions by biochemical testing should be confirmed by other techniques such as serological analysis. If the isolate reacts with antibodies developed against *Salmonella* somatic or flagellar antigens, this confirms it as *Salmonella*. Isolates may be examined by polyvalent flagellar and somatic antibodies; *Salmonella* isolates will bind with antibodies, producing agglutination upon testing. For serotyping (i.e., identification of an isolate at the serovar level), a more elaborate scheme of serological analysis is needed and is only performed by specialized laboratories.

OBJECTIVES

1. Detection of *Salmonella* in food using a variety of techniques.
2. Using immunoassay as a sample screening tool.
3. Identification of *Salmonella* isolates using a biochemical approach.
4. Practice safe handling of pathogens in the laboratory.

MEDIA

Buffered Peptone Water (BPW)

Buffered peptone water is a non-selective medium used in the pre-enrichment step.

Rappaport-Vassiliadis (RV) + Soya Broth

This medium is used for the selective enrichment of *Salmonella*. The selective agents are malachite green and magnesium chloride, which are assumed to synergistically inhibit Gram-negative enterics other than *Salmonella*. Magnesium chloride creates a hypertonic environment that *Salmonella* tolerates, compared to other bacteria. Additional selection is due to the high incubation temperature (41–43°C) and the relatively low pH (5.2).

Tetrathionate (TT) Broth

Like RV broth, this medium also is used for the selective enrichment of *Salmonella*. Iodine in the medium reacts with sodium thiosulfate to produce tetrathionate:

$$2S_2O_3^{2-} + I_2 \rightarrow S_4O_6^{2-} + 2I^-$$

Tetrathionate is toxic to many bacteria. Salmonellae are selected for because they possess an enzyme, tetrathionate reductase, which detoxifies the tetrathionate. Because *Salmonella* can detoxify the compound, they will multiply more rapidly in this medium than will other species. Bile salts in the medium also provide selection against non-enteric bacteria.

Bismuth Sulfite (BS) Agar

This is a selective-differential medium used for the isolation of potential *Salmonella* colonies. Brilliant green in the medium selects against Gram-positive bacteria. Bismuth sulfite has both selective and differential effects. The bismuth ion selects against coliforms and Gram-positive bacteria. The sulfite ion allows differentiation of organisms producing the enzyme desulfhydrase. Production of desulfhydrase by *Salmonella* results in formation of hydrogen sulfide (H_2S) from the sulfite. The formed H_2S reacts with the ferrous ion of ferrous sulfate ($FeSO_4$) to form a black precipitate of ferrous sulfide (FeS).

$$2H_2S + Fe^{2+} \rightarrow FeS(black\ precipitate) + 2H^+$$

Salmonella colonies on BS agar appear black to green in color with or without a dark halo in the surrounding agar. The colonies often appear with metallic sheen.

Xylose Lysine Desoxycholate (XLD) Agar

XLD agar is a selective-differential medium used for the isolation of potential *Salmonella* colonies. Desoxycholate inhibits non-enteric microorganisms. This medium uses ferric ammonium citrate and sodium thiosulfate as differential agents. *Salmonella* can use thiosulfate as a terminal electron acceptor (anaerobic respiration), reducing the compound into hydrogen sulfide by the help of its membrane-bound thiosulfate reductase. Released hydrogen sulfide reacts with the ferric ion to form a black precipitate. A pH indicator (phenol red) is used to detect the fermentation of sugars (xylose, lactose, and sucrose) in the medium. Fermentation, and the resulting acid product from these carbohydrates, turns this medium yellow. Of these sugars, *Salmonella* ferments only xylose. L-lysine is added to the medium to differentiate salmonellae from other xylose fermenters. Salmonellae have the ability to decarboxylate lysine. Because *Salmonella* ferments a limited amount of sugar (only the xylose), the decarboxylation reaction results in sufficient basic products to balance and overwhelm the acidic products of xylose fermentation. This alkaline reversion turns the colonies and medium around them from yellow to red. Therefore, the combination of sugars, amino acid, and pH indicator allows differentiation of organisms by metabolic activity. *Salmonella* colonies on XLD agar, therefore, may appear red with black centers but are commonly fully black.

PROCEDURE OVERVIEW

In this laboratory exercise, presence of *Salmonella* in food will be tested using a method that depends on culture, biochemical, and serological techniques. The method includes five steps: (i) pre-enrichment; (ii) selective enrichment; (iii) screening; (iv) isolation on selective-differential agar media; and (v) identification of *Salmonella* (if present) using a biochemical approach. These steps have been explained in detail in previous sections of this chapter. A schematic representation of the procedure is displayed in Figure 12.1.

> *Notes:*
> * *Due to the scarcity of* Salmonella-*positive samples in commercial foods, some of the provided samples can be inoculated with a non-virulent* Salmonella *strain. Instructors should keep track of which food is inoculated and which are not. An example of such strain is listed in Session 2 of this exercise.*
> * *Occasionally, a required incubation period is shorter that the interval between the laboratory session available to the analysts. In this situation, the following recommendation is suggested: When media are inoculated in a given session, they are refrigerated during the early part of the interval, and then transferred to incubators for the required time, just prior to the subsequent session.*

Organization

Each pair of students will analyze one food sample. The analysis is completed in six laboratory sessions (Figure 12.1).

Personal Safety

All *Salmonella* serovars are considered pathogenic to humans and thus should be handled with care. Follow the safety guidelines that are reviewed at the beginning of this manual. **Use disposable gloves** when handling the isolates and the pathogenic cultures. Make sure the work area is sanitized after use.

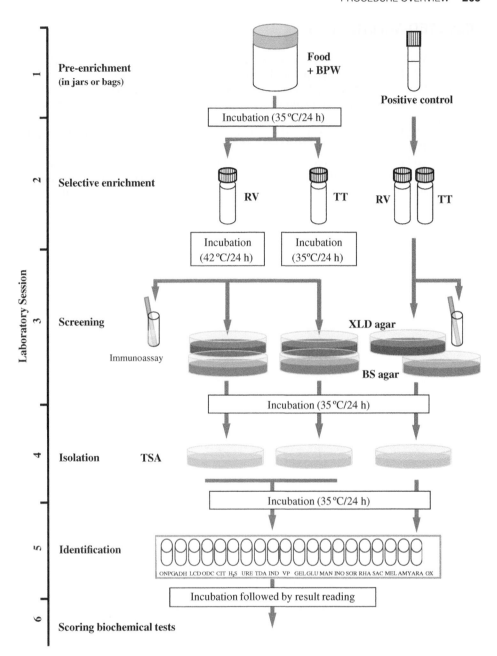

Figure 12.1 Method for detection of *Salmonella* in food.

SELECTED REFERENCES

Andrews, W.H., Wang, H., Jacobson. A., et al. (2020). *Salmonella. Bacteriological Analytical Manual.* Washington, DC: Food and Drug Administration.

Cox, N.A., Frye, J.G., McMahon, W., et al. (2015). Salmonella. In: *Compendium of Methods for the Microbiological Examination of* Foods, 5e (ed. Y. Salfinger and M.L. Tortorella), 445–475. Washington, DC: American Public Health Association Press.

Grimont, P.A.D. and Weill, F.-X. (2007). Antigenic formulae of the *Salmonella* serovars. Institut Pasteur, Paris, France.

International Organization for Standardization (ISO). (2017). Microbiology for the food chain – Horizontal method for the detection, enumeration and serotyping of *Salmonella* – Part 1: Detection of Salmonella spp. ISO 6579-1 ISO, Geneva, Switzerland.

SESSION 1: PRE-ENRICHMENT

During this session, food samples are prepared for analysis, and the pre-enrichment is prepared and incubated.

MATERIALS AND EQUIPMENT

Per Pair of Students

- Food sample
- 225 ml Buffered peptone water
- Stomacher bags
- Stomacher bag holder
- Container for holding enrichment bags (ideally autoclavable)

Class-Shared

- Sample Preparation Utensils (as needed)
  - Alcohol squeeze bottle for sanitizing utensils
  - Plastic container for holding and mixing chopped sample
  - Cutting boards
  - Knife
  - Tongs
- Scale for weighing food samples (e.g., a top-loading balance with 500 g capacity).
- Common laboratory equipment (sample homogenizers, balances, incubators, etc.)

PROCEDURE

Pre-enrichment

1. Label a stomacher bag with all essential information.
2. Prepare the food for sampling to ensure that a representative sample is analyzed. This varies with the food and may involve cutting, partitioning, grinding, or mixing.
3. Place the stomacher bag in the holder and weigh a 25 g food sample directly into the bag.
4. Add the 225 ml buffered peptone water to bag contents.
5. Stomach the bag contents for 2 min.
6. Place the enrichment-containing bag in the tub or beaker designated by the laboratory instructors. This container should support the bag and keep it upright during incubation and handling.
7. Incubate the bags at 35°C for 24 hours. The incubated mixture in these bags will be referred to as "food-BPW enrichment."

SESSION 2: SELECTIVE ENRICHMENT

During this session, pre-enriched samples are sub-cultured to prepare the selective enrichments.

MATERIALS AND EQUIPMENT

Per Pair of Students

- Food-BPW enrichment (Session 1)
- Positive control: Overnight culture of a suitable *Salmonella enterica* serovar. A non-virulent strain (e.g., *Salmonella* Typhimurium ATCC BAA-2828) is preferably used as a positive control.
- Two 10-ml RV broth tubes
- Two 10-ml TT broth tubes
- Sterile 1-ml pipette and pipette controllers
- Other common laboratory supplies (e.g., disposable gloves, micro pipettor set and pipette tips)

Class-Shared

- Common laboratory equipment (e.g., incubators)

PROCEDURE

1. **Food Sample**
   a. Label one RV and one TT broth tube for the food sample.
   b. Mix the food-BPW enrichment by hand-agitating the stomacher bag.
   c. Using 1-ml sterile pipette, transfer 0.1 ml food-BPW enrichment into the RV broth tube, then transfer 1 ml food-BPW enrichment into the TT broth tube.
   d. Incubate the inoculated RV tube at 42°C for 24 hours.
   e. Incubate the inoculated TT tube at 35ºC for 24 hours.
   f. The contents of the incubated tubes will be referred to as "Food-RV" and "Food-TT," respectively.

2. **Positive Control**
   a. For positive control, label one RV broth tube and one TT broth tube.
   b. Vortex the *Salmonella* positive control culture tube and transfer 0.1 ml into the RV broth tube and 1 ml into the TT broth tube.
   c. Incubate the inoculated RV tube at 42°C for 24 hours.
   d. Incubate the inoculated TT tube at 35ºC for 24 hours.
   e. The contents of the incubated tubes will be referred to as "Control-RV" and "Control-TT," respectively.

Note: Enrichments should be incubated for full 24 hours for reliable recovery of Salmonella.

SESSION 3: SAMPLE SCREENING AND *Salmonella* ISOLATION

To isolate *Salmonella* in the food sample, the incubated selective enrichments will be streaked on the selective media: XLD and BS agars. Additionally, a rapid commercial immunoassay kit (Reveal® 2.0 *Salmonella* test strips) will be used to screen the enrichment for the presumptive presence of *Salmonella*. A decision on whether to proceed with food sample analysis will be determined based on the results from both the screening and the isolation tests.

MATERIALS AND EQUIPMENT

Per Pair of Students

- Incubated Control-RV and Control-TT enrichment tubes (Session 2)
- Incubated Food-RV and Food-TT enrichment tubes (Session 2)
- Three XLD agar plates
- Three BS agar plates
- Two Reveal® 2.0 *Salmonella* test strips
- Two 2-ml flat-bottom microcentrifuge tubes (to be used for the immunoassay)
- Other common supplies such as disposable gloves, micro pipettor set and pipette tips, inoculation loops, etc.

Class-Shared

- Common laboratory supplies and equipment (e.g., colony counters and incubators)

PROCEDURE

1. **Preparing for *Salmonella* Isolation**
   a. Observe incubated tubes for signs of growth. Report any observations in Table 12.1.
   b. Label two XLD agar plates for the food sample; one plate will receive inoculum from the Food-RV enrichment and the other from the Food-TT enrichment. These plates will be referred to as "Food-RV-XLD" and "Food-TT-XLD," respectively.

TABLE 12.1 Description of observed enrichment media and result of immunoassay test.

Isolate source	Enrichment medium		Immunoassay
	RV broth	TT broth	
Food sample enrichment			
Positive control			

{This table requires footnotes}

   c. Vortex the Food-RV enrichment tube and streak its contents (3-phase streaking) onto the appropriate XLD plate.

   d. Vortex the Food-TT enrichment tube and streak its contents (3-phase streaking) onto the appropriate XLD plate.

   e. Repeat steps b–d using two BS agar plates that will receive the Food-RV and Food-TT enrichments.

   f. Label one XLD plate for the positive control. Draw a line that divides the plate into two halves; one receives the Control-RV enrichment and other receives the Control-TT enrichment. Label the two halves appropriately.

   g. Vortex the Control-RV culture tube and streak its contents (3-phase streaking) onto the appropriate half of the XLD plate.

   h. Vortex the Control-TT culture tube and streak its contents (3-phase streaking) onto the other half of the XLD plate.

   i. Repeat steps f–h using one BS agar plate that will receive the Control-RV and Control-TT cultures.

   j. Incubate all streaked plates at 35°C for 24 hours.

2. **Immunoassay Screening**

   a. Vortex the incubated Food-TT and transfer an aliquot (500 µl) to a flat-bottom microcentrifuge tube.

   b. Insert one test strip into the tube and incubate at room temperature for 15 min. If the enrichment is positive for *Salmonella*, strip should display two red lines (one may be faint) at the end of incubation.

Note: if the test strip absorbs all the liquid in the tube, more enrichment should be added. The bottom edge of the test strip should always remain submerged in liquid.

   c. Repeat with positive control TT broth.

   d. Report results in Table 12.1.

SESSION 4: ISOLATION

The incubated XLD and BS agar plates will be inspected for colonies having typical *Salmonella* appearance. If such colonies are found, they will be transferred to TSA plates, in preparation for identification during the subsequent laboratory session. If the food sample did not produce any typical isolates, the positive control isolates only will be used in the identification.

MATERIALS AND EQUIPMENT

Per Pair of Students

- Incubated XLD and BS plates (Session 3)
- Three TSA plates
- Other common supplies such as disposable gloves, inoculation loops, etc.

Class-Shared

- Common laboratory equipment (e.g., colony counter, incubators)

PROCEDURE

1. **Examination of Incubated Selective Agar Plates**
   a. Observe the XLD plate inoculated with the control cultures. Typical *Salmonella* colonies are red, surrounded by red-pink agar. These colonies commonly have a black center but may appear all black (Table 12.2).
   b. Observe the BS plate inoculated with the control cultures. Typical *Salmonella* colonies are black to green with or without dark halo and metallic sheen (Table 12.2).
   c. Compare the appearance of colonies originating from the control-RV and control-TT enrichments. Note any differences in appearance or prevalence.
   d. Record these observations in Table 12.3.
   e. Repeat steps a–c with the incubated XLD and BS agar plates that received the food enrichment and observe the presence of any typical *Salmonella* colonies.
   f. Record these observations in Table 12.3.
2. **Preparing Isolates for Identification**
   a. Label three TSA plates: two for food sample isolates and the third for the positive control.
   b. Label two well-isolated colonies, having typical *Salmonella* morphology, obtained from any of the XLD or BS agar plates that received the food sample.
   c. Using the inoculation loop, carefully transfer the two colonies (by streaking) to the two TSA plates; these will be labeled as "Food-TSA."

TABLE 12.2 Characteristics of *Salmonella* on isolation and biochemical identification media.

Medium (original color)	Reaction	*Salmonella* Interpretation	Other microorganisms
Xylose lysine desoxycholate agar (Red)	Red with black centers	• Limited carbohydrate fermentation • Lysine decarboxylation • Production of H_2S	• Yellow (carbohydrate fermentation) • Red with no black center (no H_2S production)
Bismuth sulfite agar (Light gray-green)	Colonies black to green with or without dark halo and metallic sheen	• *Salmonella* survives the strong selectivity of the medium • Production of H_2S	• Most are inhibited • *E. coli* may produce brown to green colonies

TABLE 12.3 (*Add a descriptive title, including food and how results were obtained*)

		Appearance	
Plating Medium	Enrichment Medium	Food Sample	Positive Control
XLD	RV		
	TT		
BS	RV		
	TT		

{This table requires at least five footnotes}

    d. Label two well-isolated colonies, having typical *Salmonella* morphology, obtained from the XLD or BS agar plate that received the positive control and streak on the appropriately marked TSA plate; one isolate per half a plate. This plate will be labeled as "Control-TSA."

    e. Incubate the TSA plates at 35°C for 24 hours.

SESSION 5: IDENTIFICATION

During this session, presumptive *Salmonella* isolates will be subjected to identification using a panel of biochemical tests. This will be accomplished using the API-20E test kit. The kit is made of a strip of 20 wells, each containing dry media or reagents needed to complete a test (Table 12.4). The kit also includes a plastic incubation chamber and the needed reagents. A suspension of presumptive *Salmonella* isolates will be prepared, dispensed in the wells, and the strips will be incubated. Analysis results will be tallied during the subsequent laboratory session.

Notes:
- *Although* Salmonella *is oxidase negative, the test is important for its identification. Oxidase test is not represented on the API-20E strip; therefore, the test will be completed separately, but the result will be included in the strip score.*
- *Companies periodically change test kits or associated procedures. Make sure to check the pamphlet that came with the kit and to follow manufacturer's instructions if they different from what is reported in this exercise.*

MATERIALS AND EQUIPMENT

Per Pair of Students

- Incubated TSA plates (Session 4); plates incubated for more than 24 hours prior to the test should not be used.
- Two API-20E strips with their incubation chambers: one for a food sample isolate (if any presumptive *Salmonella* colonies were successfully isolated on the selective agar media) and the second for an isolate from the positive control.
- Oxidase test supplies:
  o Three oxidase test strips
  o Three empty Petri plates
  o Sterile toothpicks
  o A 24-h culture of oxidase-positive *Pseudomonas* sp., streaked on a suitable agar medium
- Three tubes of sterile 0.85% saline solution (5 ml each); two for preparing cell suspension and one for adjusting cell turbidity
- A test tube of deionized water for use in humidifying the API incubation chamber
- Sterile Pasteur pipette and bulb (or sterile disposable plastic Pasteur pipette)
- A test tube of sterile mineral oil
- Other common supplies such as disposable gloves, inoculation loops, etc.

Class-Shared

- McFarland standards
- Colony counter, incubator set at 35°C, etc.

TABLE 12.4 Biochemical tests included in the API-20E test strip, test interpretation, and reaction of most *Salmonella* serovars.

Test Abbreviation	Reaction	Substrate Used	Positive Reaction	Negative Reaction	*Salmonella* Reaction[a]
ONGP	β-galactosidase	ortho-nitrophenyl-β-galactoside	Yellow	Colorless	Negative
ADH	Arginine dihydrolase	Arginine	Red	Yellow	Positive
LDC	Lysine decarboxylase	Lysine	Red	Yellow	Positive
ODC	Ornithine decarboxylase	Ornithine	Red	Yellow	Positive
CIT	Citrate utilization	Citrate	Blue	Pale green	Positive
H2S	H$_2$S production	Sodium thiosulfate	Black	Colorless/Gray	Positive
URE	Urease	Urea	Red	Yellow	Negative
TDA	Tryptophan deaminase	Tryptophan	Brown-red	Pale brown	Negative
IND	Indole production	Tryptophan	Red (2 min)	Yellow	Negative
VP	Acetoin production	Sodium pyruvate	Red (10 min)	Colorless	Negative
GEL	Gelatinase	Charcoal gelatin	Charcoal black, diffuse	No diffusion of black pigment	Negative
GLU	Glucose utilization	Glucose	Yellow	Blue/Green	Positive
MAN	Mannitol utilization	Mannitol	Yellow	Blue/Green	Negative
INO	Inositol utilization	Inositol	Yellow	Blue/Green	Negative
SOR	Sorbitol utilization	Sorbitol	Yellow	Blue/Green	Negative
RHA	Rhamnose utilization	Rhamnose	Yellow	Blue/Green	Negative
SAC	Sucrose utilization	Sucrose	Yellow	Blue/Green	Positive
MEL	Melibiose utilization	Melibiose	Yellow	Blue/Green	Negative
AMI	α-methyl-D-glucoside utilization	α-methyl-D-Glucoside	Yellow	Blue/Green	Positive
ARA	Arabinose utilization	Arabinose	Yellow	Blue/Green	Negative
OX[b]	Cytochrome c oxidase	Tetramethyl-p-phenylenediamine	Violet	Colorless	Negative

[a] Most serovars
[b] Oxidase is not included in the test strip but is should be completed as an auxiliary test and results included in the strip score.

273

PROCEDURE

1. Preparing Inocula

a. Check the incubated TSA plates to make sure the streaking produced colonies with uniform morphology. Lack of uniformity may indicate lack of culture purity.

b. Label one saline solution tube with "Food" and the other with "Control."

c. From the incubated Food-TSA plates, select and label one colony. If the food did not produce presumptive *Salmonella* isolates, use only the plate from the positive control.

d. Using the inoculation loop, transfer a large portion of the selected colony to the properly-label saline solution tube. Vortex the resulting suspension.

e. Adjust cell density to McFarland standard number 0.5 (Table 12.5) by adding additional cells or additional saline solution.

Note: There may be a need to use more than one colony to produce the right cell concentration.

f. Repeat steps d and e with the colony from the Control-TSA plate.

2. Inoculating API Strips

a. Use two strips: one for the food isolate and the other for the positive control. If the food did not produce presumptive *Salmonella* isolates, use only one strip for the positive control.

b. Label the strip incubation chamber with appropriate sample information.

c. Remove the API-20E test strip from its package and place slightly tilted toward you.

d. Using the sterile Pasteur pipette, transfer the appropriate amount of cell suspension to each well (cupule). To make sure no air bubbles are formed during dispensing, point the tip of the pipette to the side edge of the well opening and fill slowly. Most wells need to be filled only to the bottom edge of the opening. Other wells (CIT, VP, and GEL), marked with boxes around the names, will be filled completely (to the top of the well). For certain wells (with names underlined), the remaining space (at the well opening) will be filled with mineral oil; these are ADH, LDC, ODC, H_2S, and URE. Use a new sterile Pasteur pipette to add oil to these wells.

e. Fill the cavities, at the bottom of the incubation chamber, with deionized water.

f. Place the strip in its incubation chamber and cover with the plastic lid.

TABLE 12.5 McFarland standards and approximate cell density.

McFarland Standard No.	0.5	1	2	3	4
Approximate CFU/mL[a]	1.5×10^8	3.0×10^8	6.0×10^8	9.0×10^8	1.2×10^9

[a] colony forming units per milliliter

g. Incubate the API strip at 35°C for 24 hours. Alternatively, inoculated strips may be incubated at 20–22°C for 48 hours.

3. **Oxidase Test**

a. Place three oxidase test strips in three empty Petri plates labeled for (a) food isolate, (b) positive (*Salmonella*) control, and (c) oxidase-positive *Pseudomonas* sp. culture.

b. Transfer a small drop of sterile water onto a section of each strip.

c. Using sterile toothpicks, transfer portions of test colonies and smear them onto the wet sections of the oxidase test strips. These colonies represent the food's presumptive *Salmonella* isolates, the positive (*Salmonella*) control, and the oxidase-positive *Pseudomonas* sp. culture.

d. The color of the smeared area of the test strip should change to blue/purple within 30 seconds if the bacterium is oxidase positive.

e. Record oxidase result in the chart shown in Figure 12.2.

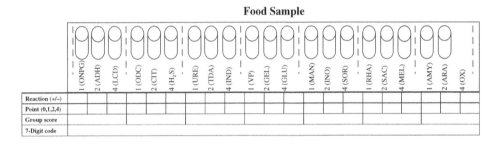

Figure 12.2 Scoring the incubated API-20E strip for food isolate and positive control.

SESSION 6: SCORING API STRIPS

In this exercise, the biochemical identification results will be collected and a conclusion about detection of *Salmonella* in food will be made.

MATERIALS AND EQUIPMENT

Per Pair of Students

- Incubated API-20E test strips (Session 5)
- Kit reagents (Reagents used are listed in the procedure section)

Class-Shared

- Disposable gloves
- API catalogue or access to BioMerieux software for confirming biochemical results

PROCEDURE

For scoring API strips, these steps will be followed:

1. Observe the appearance of each well (cupule). See Table 12.4 for the expected *Salmonella* reactions. Most of the wells can be read immediately, whereas others (TDA, VP, or IND) need reagents added before results are obtained.
   a. TDA (tryptophan deaminase test): Add one drop of TDA reagent (FeCl$_3$). A reddish-brown color indicates a positive reaction.
   b. IND (indole test): Add one drop of James (or Kovac's) reagent. The entire cupule turning pink indicates a positive indole test.
   c. VP (Voges–Proskauer test): Add two reagents, one drop of VP1 (KOH, 40%) and one drop of VP2 (α-naphthol). Wait at least 10 minutes for reaction to occur. A pink or red coloration indicates a positive reaction. Slight pink coloration should be interpreted as a negative reaction.
2. Record the results in Figure 12.2.
3. Determine the numerical code for your isolate.
   a. The strip is divided into groups of three. If the first test in a group is positive, give it a 1. If the second test in a group is positive, give it a 2. If the third test in a group is positive, give it a 4. If any of the tests are negative, give them a 0. Add up the score for each group. You will end up with a seven-digit code.
   b. Using the API catalogue (or accessing BioMerieux software, online), determine the probable identity of the isolates based on the seven-digit code obtained in the previous step.
   c. Report identification result.

DATA INTERPRETATION

As indicated earlier, a decision on the potential presence of *Salmonella* in the food analyzed can be made after results of the screening and isolation steps are examined. These results will determine if the analyst stops or proceeds with the analysis (See the Screening section for the decision options). The biochemical identification determines whether the isolate obtained from the selective-differential media (XLD and BS, in this example) is *Salmonella* or not. Note that in a professional laboratory setting, at least two colonies from each selective-differential medium would be further characterized. If at least one isolate is confirmed as *Salmonella*, then the food is considered *Salmonella*-positive. If none of the isolates are confirmed, then the food is considered negative for *Salmonella*.

It is not uncommon for colonies presenting *Salmonella* typical morphology on XLD and BS agar to be identified as other species. The species that display comparable morphological characteristics are *Citrobacter freundii*, various *Proteus* species, and some Pseudomonads. These species are effectively differentiated using the biochemical identification technique used in this exercise. Certain food matrices (e.g., raw poultry) that are more likely to contain these non-*Salmonella* species require the evaluation of multiple isolates to be confident of the analysis results. This situation can be supported by a robust screening step: a positive screening result would indicate that additional isolates should be tested, whereas a negative screening result would rule out the need for exhaustive testing.

QUESTIONS

1. What was the result of the screening step for the food sample analyzed by your group? Was this result in agreement with that obtained from the isolation?
   a. Which result was easier to interpret and why?
   b. Which of the two techniques is more sensitive and why?
2. Explain the biochemical reactions that occur when typical *Salmonella* colonies grow on XLD and BS. Are there *Salmonella* that produce atypical reactions on these media?
3. Compile the class data and answer the following:
   a. Which non-*Salmonella* species were identified by biochemical identification using the API-20E strips?
   b. Which biochemical tests in the API-20E allowed these species to be differentiated from *Salmonella*?
   c. Did these isolates have typical appearances on XLD or BS?

CHAPTER 13

SHIGA TOXIN-PRODUCING *Escherichia coli*

RAPID MOLECULAR SCREENING. MULTIPLEX PCR. PCR-ASSISTED ISOLATION OF STEC STRAINS

INTRODUCTION

Escherichia coli is a natural inhabitant of the intestinal tract of humans and other warm-blooded animals. Additionally, *E. coli* thrives in various environments due to its great adaptability. Strains of *E. coli* are identified based mainly on the O-antigens and H-antigens on their surface. An O-antigen is determined by the structure of the immunogenic portion of cell's outer membrane lipopolysaccharides (also called endotoxin), whereas the H-antigen is determined by the structure of the protein in cell's flagella (i.e., flagellin). There are more than 180 *E. coli* O-antigens and more than 50 flagellar antigens known. *E. coli* strains are classified into serogroups (e.g., O157), based on the O-antigen present and each serogroup is further divided into serotypes (e.g., O157:H7) depending on the H-antigens that members of the serogroup carry. In addition to O- and H-antigens, some *E. coli* strains carry capsular (K) antigens, which are made of acidic polysaccharides (e.g., *E. coli* K12).

Diseases

Although many *E. coli* strains are harmless, some cause foodborne diseases. The disease-causing strains may be grouped based on their virulence factors into several virotypes such as enterotoxigenic, enteropathogenic, and Shiga toxin-producing *E. coli*. Shiga toxin-producing *E. coli* (STEC) strains are associated with hemorrhagic colitis, a gastroenteritis that causes bloody diarrhea. Production of Shiga toxins by these strains also may lead to hemolytic uremic syndrome

Analytical Food Microbiology: A Laboratory Manual, Second Edition. Ahmed E. Yousef,
Joy G. Waite-Cusic, and Jennifer J. Perry.
© 2022 John Wiley & Sons, Inc. Published 2022 by John Wiley & Sons, Inc.

(HUS) or other serious complications in infected individuals. STEC serogroups include *E. coli* O26, O45, O103, O104, O111, O121, O145, and O157. The diversity among these serogroups makes it difficult to use a common culture technique to differentiate between members of this virotype and other *E. coli* strains. Hence, a genetic approach, targeting virulence factors associated with STEC infection (i.e., Shiga toxin-encoding genes), is the most logical approach.

The present exercise is designed to detect STEC by screening the microbial populations in food enrichments for genes encoding Shiga toxins (stx_1 and/or stx_2) and intimin (another *E. coli* virulence factor encoded by the gene *eae*). Multiplex polymerase chain reaction (multiplex PCR) allows simultaneous detection of these virulence factors. The absence of genes encoding these virulence factors is taken as evidence for the absence of STEC in food. However, if these genes are detected, additional evidence is needed to prove the presence of the pathogen in food. In this case, a procedure to isolate the pathogen from the food sample, or the enrichment, should be followed. Once isolated, the identity and characteristics of the bacterium should be confirmed by cultural, biochemical, as well as genetic techniques. If needed, isolate serogroup or serotype may be determined by additional serological tests. A schematic illustrating this approach is shown in Figure 13.1.

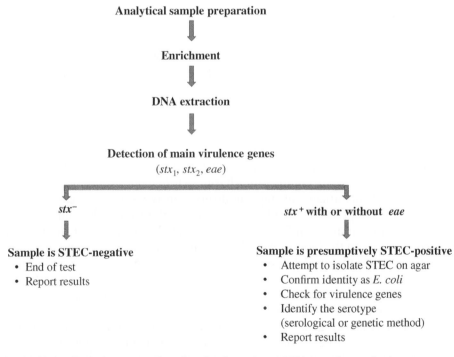

Figure 13.1 Complete procedure for the detection of Shiga toxin-producing *Escherichia coli* (STEC) in food. Note that presence of *eae*, along with *stx*, increases the virulence of STEC.

Pathogen Characteristics

Escherichia coli is a Gram-negative, flagellated, rod-shaped bacterium that belongs to the family *Enterobacteriaceae*. The bacterium is indole-positive, catalase-positive, oxidase-negative, and cannot utilize citrate as a sole source of carbon and energy. *E. coli* can grow in the presence or absence of oxygen, and thus is considered a facultative anaerobe. Aerobic respiration occurs with oxygen serving as the final electron acceptor for the respiratory electron transport process. Under anaerobic conditions, *E. coli* metabolizes carbohydrates using anaerobic respiration or fermentation pathways. Anaerobic respiration can occur in the absence of oxygen and presence of alternative electron acceptors (e.g., nitrate or fumarate) in the medium. Fermentation involves metabolism of glucose using the mixed-acid pathway with acetic, lactic, and succinic acids produced in significant amounts. Formic acid is also produced but the compound is degraded into equimolar amounts of CO_2 and H_2 gasses. The outcome of fermentation is simply acid and gas production. This multiplicity of available metabolic pathways (i.e., aerobic respiration, anaerobic respiration, and fermentation) makes *E. coli* adaptable to the intestinal tract, as well as other environments.

    E. coli can grow in the range of 21–45°C, but grows optimally at 35–42°C. The optimum pH for growth is 6.0–8.0; however, growth can occur at pH values as low as 4.3 and as high as 9.0. Survival has been demonstrated outside this range, at pH as low as 3.5, particularly for the serogroup *E. coli* O157.

OBJECTIVES

1. Rapid screening for STEC in food using multiplex PCR.
2. Isolation and identification of STEC from presumptively positive (based on PCR results) samples.

DETECTION OF STEC IN FOOD

STEC is a pathogenic group that includes several serotypes. Food containing STEC is also likely to contain commensal (harmless) *E. coli* strains. Detection of STEC starts with an enrichment step, but diverse serotypes of *E. coli* can be found in the enriched sample. It is difficult to design a growth medium that differentiates between Shiga toxin and non-Shiga toxin producers because there is no obvious relationship between the biochemical characteristics of *E. coli* types and their ability to produce Shiga toxin. Additionally, the relationship between *E. coli* serotype and Shiga toxin production seems coincidental; currently, there is no scientific evidence to link these two traits. Analysts have tried to overcome this problem by screening for genes encoding virulence factors in the enrichment population, using PCR. Lack of virulence genes in the enrichment population helps the analyst to quickly conclude the absence of STEC in food. On the contrary, presence of virulence genes requires lengthy analyses to isolate the source bacterium and to confirm its identity and its potential virulence. An outline of this approach is shown in Figure 13.1.

Enrichment

Since pathogenic *E. coli* is usually found in small numbers in food, an enrichment step is necessary to increase the population of the pathogen to a level detectable by culture or molecular techniques. The large population of diverse microorganisms in some raw foods (e.g., meat and fresh produce) makes detecting pathogenic *E. coli* in these products a difficult task. The masking effect of background microbiota can be minimized by adding selective agents to the enrichment mixture. Unfortunately, strong selectivity of enrichment broth may also negatively affect the recovery of stressed or injured *E. coli* cells. Therefore, suitability of different enrichment broths for detecting this pathogen in a particular food should be assessed carefully.

Detection of Virulence Genes

STEC strains carry genes that encode Shiga toxins (*stx₁* and/or *stx₂*), and oftentimes intimin (*eae*) and enterohemolysin (*hly*). Shiga toxins are primarily responsible for the symptoms associated with STEC illnesses: hemorrhagic colitis, HUS, and thrombotic thrombocytopenic purpura. Intimin is a non-fimbrial adhesin expressed on the outer membrane of STEC cells, aiding in formation of attachment and effacement lesions in the intestinal epithelium. Therefore, the strains containing the combination of both virulence genes (*stx* and *eae*) are often associated with the most severe forms of STEC-induced disease. Enterohemolysin is also believed to contribute to the disease. Conventional or real-time PCR can be used to detect these virulence genes in the bacterial population of the enrichment. Multiplex PCR is most useful considering that it allows the detection of multiple genes (e.g., *stx₁*, *stx₂*, and *eae*) simultaneously.

In this exercise, three pair of primers, targeting the *stx₁*, *stx₂*, and *eae* genes, will be prepared and provided to students. Each pair includes a forward (F) and a reverse (R) primer.

- *stx₁*F (10 µM) - 5' – ACA CTG GAT GAT CTC AGT GG – 3'
- *stx₁*R (10 µM) - 5' – CTG AAT CCC CCT CCA TTA TG - 3'
- *stx₂*F (20 µM) - 5' – CCA TGA CAA CGG ACA GCA GTT – 3'
- *stx₂*R (20 µM) - 5' – CCT GTC AAC TGA GCA GCA CTT TG - 3'
- *eae*F (10 µM) - 5' – GTG GCG AAT ACT GGC GAG ACT - 3'
- *eae*R (10 µM) - 5' – CCC CAT TCT TTT TCA CCG TCG – 3'

Amplification products of *stx₁*- and *stx₂*- specific primers will be 614 and 779 base pairs, respectively. The amplified product using the *eae*-specific primers will result in a PCR product that is 890 base pairs. These PCR products will be stained with a florescent dye, separated on an agarose gel by electrophoresis, and viewed under UV light.

Isolation and Identification

Sample enrichments that are presumptively positive for STEC, based on PCR results, can be used to isolate the pathogen. In this case, the enrichment needs to be streaked onto agar media selective for *E. coli*, e.g., eosin methylene blue

(EMB) agar. Typical *E. coli* colonies on the agar medium are then isolated and tested for characteristic *E. coli* biochemical reactions and presence of STEC virulence genes.

MEDIA

E. coli Broth with Antibiotics (EC Broth+VCC)

EC broth is a buffered lactose medium that may be used to differentiate and enumerate coliforms. This medium contains lactose, which is fermentable by coliforms. It also contains bile salts, which inhibit Gram-positive bacteria. This medium can be modified by adding antibiotics, which further suppress the growth of Gram-positive and some Gram-negative bacteria. The antibiotic-supplemented medium (EC broth+VCC) contains vancomycin (8 mg/l), cefixime (0.05 mg/l), and cefsulodin (10 mg/l). In this exercise, EC broth is mixed with the food sample and the mixture is incubated for 5 h at 35°C. This is followed by addition of the VCC antibiotic supplement and incubation at 42°C for additional 36 h.

Eosin Methylene Blue (EMB) Agar

This is a selective-differential medium. It is used for the detection and differentiation of coliforms. At acidic pH values, eosin and methylene blue combine to form a precipitate, allowing differentiation of colonies that can ferment lactose from those that cannot. These two dyes also act as selective agents against Gram-positive organisms. Lactose-fermenting colonies have a dark coloration. Non-lactose fermenting colonies will appear colorless to pale straw colored. Sucrose-fermenting colonies, such as those of *Enterobacter* spp., appear pink in color. *E. coli*, which vigorously ferments lactose, produce green, metallic sheen at the primary streak.

Tryptic Soy Agar + Yeast Extract (TSA-YE)

This is a general-purpose rich medium. The addition of yeast extract to TSA allows the growth of fastidious organisms and aids in recovery of injured cells. TSA-YE is used in this exercise for maintenance of *E. coli* isolates that will be confirmed using biochemical methods. Any general purpose non-selective medium would be suitable for this purpose.

PROCEDURE OVERVIEW

This exercise includes a two-stage enrichment. The first stage involves mixing the analytical samples with *E. coli* broth (EC-broth), a mildly selective medium, and incubating the mixture for 5 hours at 35°C. After this short pre-enrichment, antibiotics will be added to the mixture and incubation continued for an additional 36 hours at a higher temperature (42°C). Several antibiotics can be used individually or in combination for adding selectivity to the enrichment medium. After the incubation, the bacterial population in the enrichment is screened for presumptive presence of STEC using a multiplex PCR procedure, targeting known virulence genes. Multiplex PCR involves using multiple pairs of primers, in the same reaction, to amplify several sequences of interest, producing multiple PCR products. In this exercise, the multiplex PCR will be used to detect the presence of STEC virulence genes: stx_1, stx_2, and *eae*. The PCR products will be separated by gel electrophoresis and viewed using ultraviolet light. Negative PCR results may be considered conclusive, and results reported as such (absence of STEC in the analytical sample). Presence of DNA bands corresponding to one or both Shiga toxins indicates a presumptive positive result for STEC, i.e., positive on the condition that the results are confirmed. Confirmation of the results requires isolating the pathogen and verifying its identity. For isolation, a portion of the enrichment will be plated on EMB agar and a number of colonies (up to 24) showing typical *E. coli* morphology will be isolated. These colonies will be tested for the presence of the virulence genes. Positive isolates will also be tested biochemically to confirm their identity as *E. coli*. This exercise will be completed in eight laboratory sessions as shown in Figure 13.2. Detection of Shiga toxin-producing *Escherichia coli* in food samples, using a genetic-based procedure, is depicted in Figure 13.3.

Organization

With the exception of Sessions 1 and 4, students will work in groups of four. Larger groups make the best use of the limited molecular reagents. One pair of these four will prepare and analyze a retail food sample and the other pair will receive and analyze a similar sample that has been inoculated by the instructors with a non-pathogenic STEC surrogate (see description of control cultures below) before the laboratory session. Inoculation of a subset of samples ensures recovery of bacterial isolates that produce most of the typical reactions of the target pathogen. All members of the group of four will be responsible for results from both inoculated and uninoculated samples. Ground meat from different sources, including local, small butcher shops, may be tested. Fresh produce such as bean sprouts and leafy greens are also suitable foods for *E. coli* testing.

Control Strains

Strains representing three groups of *E. coli* will be tested. Traditionally, these are described as "controls."

- **Negative control:** This is a nonpathogenic *E. coli* that does not carry STEC virulence genes. An example of this control culture is *E. coli* K12.
- **Surrogate control:** This is a mutant of a STEC that carries similar characteristics and produces most of the biochemical reactions of the parent strain, but it does not express Shiga toxins. An example of STEC surrogates is *E. coli* OSU 315. Similarly, *E. coli* O157:H7 ATCC 43888 does not produce Stx1 and Stx2, and is, therefore, less hazardous to laboratory personnel. Such a STEC surrogate can be used to inoculate food samples, serving as what can be described as a nonpathogenic STEC-surrogate control strain, or abbreviated as "surrogate control."

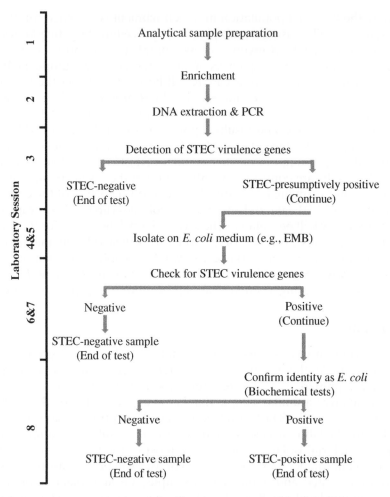

Figure 13.2 Detection of Shiga toxin-producing *Escherichia coli* (STEC) in food, with decision tree, as executed in this exercise (see text for procedural details).

- **Positive control:** A typical STEC is *E. coli* O157:H7 EDL933. This is a highly pathogenic strain, and it is recommended that its live cells should not be used in teaching laboratory exercises, unless a dedicated biosafety level 2 (BSL-II) facility is available for running the analysis. Alternatively, a culture of this strain is handled by well-trained instructors in a BSL-II facility outside of the class setting and its DNA template is prepared for use during the STEC laboratory exercise.

Personal Safety

STEC isolates are highly pathogenic microorganisms that should be handled with care. Follow the safety guidelines that were reviewed in Chapter 1. Use disposable gloves when handling plates and cultures. Make sure the work area is sanitized carefully after each use.

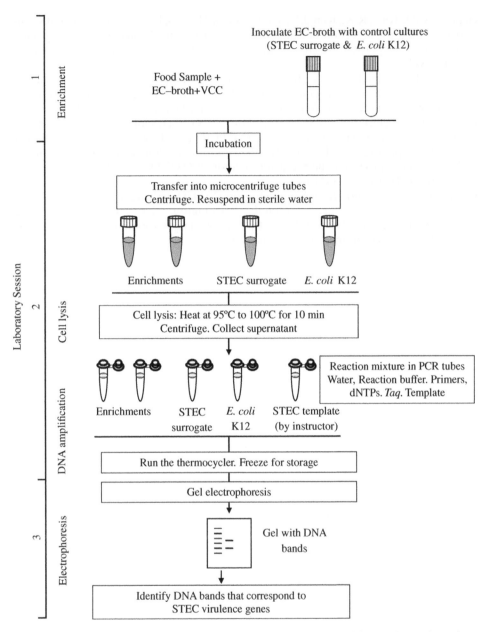

Figure 13.3 Detection of Shiga toxin-producing *Escherichia coli* in food samples using a genetic-based procedure (see text for procedural details).

SELECTED REFERENCES

Fratamico, P.M., Deng, M.Y., Strobaugh, T.P., and Palumbo, S.A. (1997). Construction and characterization of *Escherichia coli* O157:H7 strains expressing firefly luciferase and green fluorescent protein and their use in survival studies. *Journal of Food Protection* 60: 1167–1173.

Gannon, V.P.J., King, R.K., Kim, J.Y., and Golsteyn Thomas, E.J. (1992). Rapid and sensitive method for detection of Shiga-like toxin-producing *Escherichia coli* in ground beef using the polymerase chain reaction. *Applied Environmental Microbiology* 58: 3809–3815.

Gannon, V.P.J., Rashed, M., King, R.K., and Golsteyn Thomas, E.J. (1993). Detection and characterization of the *eae* gene of Shiga-like toxin-producing *Escherichia coli* using polymerase chain reaction. *Journal of Clinical Microbiology* 31: 1268–1274.

International Organization for Standardization. (2019). Microbiology of food and animal feed—Real-time polymerase chain reaction (PCR)-based method for the detection of food-borne pathogens—Horizontal method for the detection of Shiga toxin-producing *Escherichia coli* (STEC) and the determination of O157, O111, O26, O103 and O145 serogroups. ISO/TS 13136:2012.

United States Food and Drug Administration. 2020. Diarrheagenic *Escherichia coli*. *Bacteriological Analytical Manual*. Washington, DC: Food and Drug Administration. Available online: https://www.fda.gov/food/laboratory-methods-food/bacteriological-analytical-manual-bam

SESSION 1: ENRICHMENT

During this session, food will be prepared so that a representative analytical sample is collected. The sample will be mixed with EC broth and incubated for enriching the *E. coli* subpopulation.

MATERIALS AND EQUIPMENT

Per Group of Two

- A food to be analyzed
- 225 ml EC broth
- Antibiotic supplement solution (vancomycin, cefixime, and cefsoludin; VCC).
- Two-compartment filter stomacher bag

Class-Shared

- Scale for weighing food samples (e.g., a top-loading balance with 500 g capacity).
- Tools for sample preparation
- Other common laboratory supplies and equipment (e.g., incubator, set at 35°C)

PROCEDURE

1. Prepare and mix the food to ensure that a representative sample is analyzed. This includes mixing the package contents in a sterile container before taking the 25 g analytical sample (if appropriate).
2. Weigh 25 g of the food into the larger compartment of the stomacher bag; the bag is held in the upright position on the scale using a suitable bag holder. Add 225 ml EC to the bag contents. Homogenize the sample, in a stomacher, for 2 min.
3. Place the bag in a container designated by the laboratory instructor. This container should support the bag and keep it upright during incubation and handling. Alternatively, use a stomacher bag holder placed in an autoclavable tub.
4. Incubate the food-medium mixture at 35°C for 5 hours.
5. After 5 hours of incubation, add 1.25 ml VCC supplement to bag contents and incubate the bag at 42°C for additional 36 hours.

SESSION 2: RAPID SCREENING BY MULTIPLEX PCR

During this session, bacterial genomic DNA will be extracted from food enrichment and control cultures. Multiplex PCR will be executed using extracted DNA, as a template, and the previously described primers for STEC virulence genes. The targeted sequences will be amplified using a thermocycler.

MATERIALS AND EQUIPMENT

Per Group of Four

- Two incubated food-EC broth mixtures (enrichments)
- *E. coli* OSU 315 (or other STEC surrogate control) in EC broth
- *E. coli* K12 (a negative control) in EC broth
- Extracted DNA from *E. coli* O157:H7 EDL933 (or other STEC control)
- Sterile microcentrifuge tubes of different capacities, as needed
- One glass test tube for collecting liquid biohazardous waste (i.e., culture supernatants)
- Microcentrifuge tube rack
- Sterile serological pipette and pipette dispenser
- Pipettors and sterile tips
- 2 ml sterile phosphate buffered saline (PBS) solution
- 1 ml autoclaved deionized water
- Five sterile PCR tubes
- Ice bath (e.g., ice-containing Styrofoam box)

Class-Shared

- Primers (kept on ice): The concentrations indicated below result in final primer concentrations of 0.1–0.2 μM in PCR master mix when recommended volumes are used (6 μl of each primer are needed per a group of four).
  - stx_1F (10 μM)
  - stx_1R (10 μM)
  - stx_2F (20 μM)
  - stx_2R (20 μM)
  - *eae*F (10 μM)
  - *eae*R (10 μM)
- MyTaq preparation: This is a DNA polymerase, mixed with a reaction buffer, Mg^{++}, and dNTPs. The 5× preparation has 5 units/μl polymerase, 5 mM dNTPs, and 15 mM $MgCl_2$. (Note that 60 μl are needed per group of four).
- Heating block maintained at ~100°C (alternatively, a boiling water bath, equipped with a floating rack for the microcentrifuge tubes, may be used).
- Thermocycler
- Other common laboratory supplies and equipment (vortex mixers, microcentrifuges, etc.)

PROCEDURE

1. Storing Enrichment for Subsequent STEC Isolation

a. Label two sterile microcentrifuge tubes as "culture for isolation;" one for each of the two enrichments from the group of four students. Make sure to identify the sample from which each culture came.

b. Transfer 1.5 ml from one of the enrichment bags, collected from the smaller bag compartment, into the appropriately labeled 2-ml microcentrifuge tube. Repeat for the second enrichment. Use the 1-ml serological pipettes to perform the transfers.

c. Store the two tubes in the refrigerator for use during Session 4.

2. DNA Extraction

a. Label two sterile microcentrifuge tubes, one for each of the two enrichments from the group of four students.

b. Transfer 1 ml from the first enrichment bag, collected from the smaller bag compartment, into the appropriate microcentrifuge tube. Repeat with the second enrichment. Use 1-ml serological pipettes to perform the transfers.

c. Label two sterile microcentrifuge tubes, one for each control culture: surrogate control and negative control.

d. Transfer 1 ml of each control culture to the appropriate microcentrifuge tube. There will be four microcentrifuge tubes in total per group of four students: two control cultures and two food enrichments.

e. Centrifuge the tubes at 10,000–12,000 rpm for 2 min.

f. Pour off supernatant into a test tube for waste.

g. Add 180 µl of PBS solution to each microcentrifuge tube.

h. Vortex to resuspend the cell pellets thoroughly.

i. Repeat the washing steps, e–h.

j. Place all tubes (controls and food isolates) in the heating block (set at 100°C) and hold for 10 minutes to lyse cells. Alternatively, use a boiling water bath with the tubes placed in a floating rack.

k. Centrifuge the tubes at 10,000–12,000 rpm for 1 min. Collect the clear supernatant (cell lysate), which should contain the DNA that will serve as a template for PCR reaction, in a microcentrifuge tube.

l. Dilute a portion of the cell lysate (obtained in the previous step) using sterile water (1:10) in a new microcentrifuge tube.
Note: This step helps in diluting polymerase inhibitors originating from the food sample; however, appropriate concentration of template DNA in diluted lysate should be maintained. Concentration of template DNA can be checked spectrophotometrically (e.g., using Nanodrop spectrophotometer).

3. Reaction Mixtures Preparation

a. In a sterile microcentrifuge tube, mix 186 µl sterile water, 60 µl MyTaq mixture, 6 µl of primer stx_1F, 6 µl of primer stx_1R, 6 µl of primer stx_2F, 6 µl of primer stx_2R, 6 µl of primer eaeF, and 6 µl of primer eaeR.

b. Vortex the microcentrifuge tube briefly to mix its contents.

c. Label five sterile PCR tubes appropriately; two for the enrichments and three for the controls (surrogate control, negative control, and the positive control template provided by instructor).

d. Transfer 45 μl of the mixtures prepared in step 1 to each of the labeled PCR tubes.

e. Add 3 μl of diluted cell lysates (or the instructor-provided DNA for the positive control) to the appropriately labeled PCR tube; vortex the resulting PCR reaction mixture.

f. Place the prepared PCR tubes in thermocycler and record the location of each tube.

4. **Completing PCR in the Thermocycler**

a. Run the thermocycler at the following conditions: 95°C for 1 min followed by 30 cycles (95°C for 15 seconds, 58°C for 30 s, and 72°C for 90 s), then 72°C for 5 min, and finally 4°C until the reaction mixture is ready for storage (or for use in gel electrophoresis, if time permits).

b. Store the PCR products at –20°C until ready for analysis in the subsequent session.

SESSION 3: GEL ELECTROPHORESIS

Product of PCR, prepared in the previous session, will be separated by gel electrophoresis and identified on the gel using a UV source.

MATERIALS AND EQUIPMENT

Per Group of Four

- PCR products (from previous laboratory session)
- Pipettors and tips
- Plastic paraffin film (Parafilm) for mixing PCR product and loading dye
- Loading dye (EZ Vision dye)
- Ice bath

Class-Shared

- Electrophoresis equipment (chamber and power supply)
- Precast 1.2% agarose gel (prepared by instructor)
- Running Tris-acetate-EDTA (TAE) buffer
- 100 bp ladder – Will be loaded into the gel by instructor
- Gel documentation system (UV transilluminator, digital camera, data acquisition computer loaded with appropriate imaging software)
- Other common laboratory supplies and equipment

PROCEDURE

1. **Gel Electrophoresis**
   a. Label five spots on a piece of parafilm with sample names.
   b. Place 2 µl of loading dye near each label.
   c. Mix 10 µl of PCR product with the loading dye by pipetting in and out. Repeat for the remaining products.
   d. Load 10 µl of each mix into a separate well of the agarose gel.
   e. Record which sample was put into which well on the sheet provided.
   f. Run gel electrophoresis for approximately 1 h at 90V.
   g. Gel will be viewed using ultraviolet light.
2. **Bands Visualization**
   a. Carefully transfer the gel to the transilluminator of the gel documentation system.
   b. Adjust the camera settings appropriately and take a picture of the gel.
   c. Observe the picture of the gel and determine the sizes of separated bands in comparison with the DNA ladder.
   d. Compare the pictures of gel obtained with that of typical results displayed in Figure 13.4.
   e. Record your observations in Table 13.1.
   f. Discard the gel appropriately.

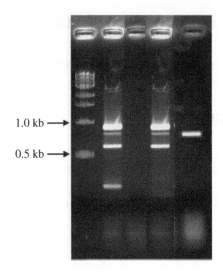

Figure 13.4 Gel showing multiplex PCR products resulting from *Escherichia coli* isolated from food. Lane 1 is the DNA molecular weight ladder and lanes 2–5 are for food isolates. The following are result interpretations. The smallest fragment (252 bp) represents a gene not analyzed in this exercise.

Band	Gene	Gene Product
890 bp	*eaeA*	Intimin
779 bp	*stx₂*	Shiga toxin 2
614 bp	*stx₁*	Shiga toxin 1
252 bp	*+93uidA*	β-glucuronidase

TABLE 13.1 {Add a descriptive title, including food and how results were obtained}

	Virulence genes			
Sample	*stx₁* present? (614 bp product)	*stx₂* present? (779 bp product)	*eae* present? (890 bp product)	Classification
Enrichment 1				
Enrichment 2				
E. coli O157:H7 (Instructor-provided)				
STEC surrogate				
Negtative control				

(*Add footnotes as required for this table*)

SESSION 4: ISOLATION

The previous sessions were dedicated to detecting STEC virulence genes in food enrichment without obtaining a pure culture of the source bacterium. For confirming PCR positive results, the targeted bacterium should be isolated and characterized. Isolating the pathogen facilitates investigating its characteristics, confirming its identity (to species and subspecies levels, if necessary), and tracking it through the food supply chain.

Isolation and identification steps are needed when the population in the enrichment tests positive for STEC virulence genes (Figures 13.1 and 13.2). Although most analyzed foods are not likely to contain STEC and thus should test negative by the PCR screening, instructors may ask students to complete the pathogen isolation and identification steps for training purposes. In the latter case, it will be appropriate to run this session simultaneously with Session 2, which will shorten the exercise considerably.

The goal for this portion of the exercise is to isolate and identify the strain responsible for the positive virulence gene PCR result (Figures 13.1 and 13.2). This can be accomplished by culturing the enrichment (saved in Session 2) on media selective for *E. coli* (e.g., EMB agar, Rainbow agar, CHROMagar STEC, or combinations of these media) and collecting typical isolates. These isolates are screened again, in a matrix format (Figure 13.5), for the presence of STEC virulence genes. The isolates that possess the virulence genes should be confirmed as *E. coli* using a battery of biochemical reactions. Isolates shown to carry STEC virulence genes and confirmed as *E. coli* are taken as proof that the original food sample contained STEC.

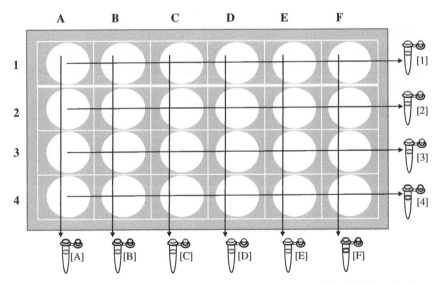

Figure 13.5   Matrix of 24 inculcated microfuge tubes, prepared for PCR-assisted isolation of Shiga toxin-producing *Escherichia coli* strains.

MATERIALS AND EQUIPMENT

Per Group of Two

- Refrigerated microcentrifuge tube containing the "culture for isolation" (see Session 2)
- *E. coli* OSU 315 (or other STEC surrogate control) in EC broth
- *E. coli* K12 (or other negative control strain) in EC broth
- Five EMB agar plates
- Four 9-ml tubes of peptone water
- Micropipettes and pipette tips

Class-Shared

- Common laboratory supplies and equipment (e.g., incubator set at 35°C)

PROCEDURE

1. **Food Sample**
   a. Label three EMB agar plates appropriately for the food sample.
   b. Prepare 10^{-1}–10^{-4} dilutions from the enrichment.
   c. Spread 100 µl from the enrichment, using a spreader, onto one EMB agar plate.
   d. On the two remaining plates, spread the 10^{-2} and 10^{-4} dilutions.
   e. Incubate all plates at 35°C for 24 hours.
   Note: If the period between this and subsequent session is more than one day, store the inoculated plates in a refrigerator and transfer them to the incubator 24 hours before the subsequent laboratory session.

2. **Control Strains**
   a. Label one EMB agar plate for each of the two control cultures.
   b. Using inoculation loop, prepare three-phase streaks from control culture broths.
   c. Incubate all plates at 35°C for 24 hours.

SESSION 5: ISOLATION (CONTINUED)

During the rest of the exercise, PCR-assisted isolation of STEC strain will be executed on one of the foods (analyzed by the group of four) that produced presumptive STEC results during PCR screening. Typical *E. coli* isolates from that food on EMB agar are transferred to EC broth and incubated in preparation for screening for STEC, and if needed, for confirming isolate identity as *E. coli*. Therefore, PCR-based protocol will be used in Sessions 5–7 to determine which isolates possess genes encoding Shiga toxins. Positive isolates will be confirmed in Session 8 as *E. coli* using biochemical tests.

MATERIALS AND EQUIPMENT

Per Group of Four

- Incubated EMB agar plates, obtained for a food that was presumptively STEC-positive by PCR screening (from the previous laboratory session)
- Incubated EMB agar plates that received control cultures
- 24 sterile microcentrifuge tubes, each containing 2 ml sterile EC broth.
- 24-well (4×6) microcentrifuge tube rack
- Sterile wooden toothpicks in a suitable container
- Micropipettor, 1000-μl capacity and sterile tips

Class-Shared

- Colony counter
- Other laboratory supplies and equipment (e.g., incubator, set at 35°C)

PROCEDURE

1. Examine the incubated EMB plates (food and control cultures) for typical *E. coli* colonies, i.e., colonies showing dark coloration and, ideally, distinctive metallic green sheen.
2. Using EMB plates prepared from food enrichments, mark and number up to 24 well-isolated, typical colonies. These colonies could be isolated from any of the plated dilutions of the enrichment.
3. Label each of the 24 microcentrifuge tubes with an assigned location in the microcentrifuge tube rack (Figure 13.5); for example, 4C, 3F, etc.
4. Assign each of the 24 colonies to a location in the microcentrifuge tube rack.
5. Using the sterile toothpicks, transfer part of colonies to each of the corresponding microcentrifuge tubes. Vortex each tube and return it to its location in the rack.
6. Incubate the rack of tubes at 35°C for 24 h.

SESSION 6: SCREENING ISOLATES FOR VIRULENCE GENES: PCR

MATERIALS AND EQUIPMENT

Per Group of Four

- Rack of up to 24 incubated microcentrifuge tubes (isolate cultures), prepared in Session 5
- Ten sterile microcentrifuge tubes
- Eleven PCR tubes
- Supplies for multiplex PCR, including the instructor-supplied DNA template for the positive control (see Session 2)

Class-Shared

- Supplies for multiplex PCR (see Session 2)
- Other common laboratory supplies and equipment

PROCEDURE

1. Combine the cultures in the 24 microcentrifuge tubes on the rack (a matrix of 4 × 6) as follows:
   a. Label 10 sterile microcentrifuge tubes as follows:
   b. Four tubes (1 to 4), one for each of the four rows of the rack (Figure 13.5).
   c. Six tubes (A to F), one for each of the six columns of the rack (Figure 13.5).
   d. Transfer 100 μL of each of the four cultures in column A; combine into the tube marked "A."
   e. Repeat step b for the tubes and columns marked B through F.
   f. Transfer 100 μl of each of the six cultures in row 1; combine into the tube marked "1."
   g. Repeat the previous step for the tubes and rows marked 2 through 4.
2. Prepare the 10 microcentrifuge tubes (1–4, and A–F) for multiplex PCR as follows:
   a. Centrifuge the tubes at 10,000–12,000 rpm for 2 min.
   b. Pour off supernatant into a test tube for waste.
   c. Add 180 μl of sterile PBS solution to each microcentrifuge tube.
   d. Vortex to resuspend the cell pellet thoroughly.
   e. Place all tubes in the heating block (set at 100°C) and hold for 10 minutes to lyse cells.
   f. Centrifuge the tubes at 10,000–12,000 rpm for 1 min. The clear supernatant (cell lysate) contains the DNA that serves as a template during PCR.
   g. Dilute a portion of the cell lysate (obtained in the previous step), at 1:10, using distilled water, in a new microcentrifuge tube.

3. Label 10 PCR tubes as follows: 1 to 4 and A to F. These match the numbers on the tubes containing the DNA templates, prepared in the previous steps.

4. For a positive control, use the DNA template of the pathogenic STEC culture that has been provided by instructor. Label the corresponding tube as "R."

5. Prepare the PCR reaction mixtures and run the thermocycler as described in Session 2.

6. Store the PCR tubes (containing the PCR products) at –20°C for analysis during the subsequent session.

7. Store the original rack of the incubated microcentrifuge tubes (isolate cultures) at 4°C. These cultures will serve as the source of isolates testing positive for STEC virulence factors, which will be subjected to biochemical testing during Session 8.

SESSION 7: SCREENING OF ISOLATES FOR VIRULENCE GENES: GEL ELECTROPHORESIS AND ISOLATE SELECTION

MATERIALS AND EQUIPMENT

Per Group of Four

- Eleven PCR tubes containing the PCR products (from the previous session)
- Rack of incubated microcentrifuge tubes (isolate cultures), prepared in Session 5
- Supplies for gel electrophoresis (see Session 3)
- Up to six TSA-YE plates

Class-Shared

- Supplies for gel electrophoresis (see Session 3)
- Other common laboratory supplies and equipment

PROCEDURE

1. Prepare gel and run electrophoresis as described in Session 3.
2. Visualize the bands as described in Session 3.
3. Determine which tube (or tubes) on the 24-tube rack is positive for STEC virulence genes. Here is an example for clarification: If PCR tubes 4 and C were positive for STEC virulence genes, then tube number 4C in the "isolate culture" rack is considered a potential STEC. Therefore, tube 4C serves as the source of culture to be confirmed as *E. coli* in the subsequent session.
4. For the cultures to be confirmed, streak for isolation from the appropriate microcentrifuge tubes onto labeled TSA-YE plates. If numerous isolates produced typical STEC banding pattern, choose three representing each original food sample for confirmation.
5. Incubate TSA-YE plates at 35°C for 24 h.

SESSION 8: ISOLATE IDENTITY CONFIRMATION: BIOCHEMICAL TESTING

In this session, isolates yielding presumptive positive results from PCR will be subjected to Gram staining and biochemical testing to confirm identity as *E. coli*. Biochemical tests include indole, catalase, and oxidase tests.

MATERIALS AND EQUIPMENT

Per Group of Four

- Incubated TSA-YE plates containing isolate cultures confirmed in the previous session to contain STEC virulence genes (maximum of three per sample)
- Sterile toothpicks
- Reagents and supplies for spot indole test: Indole reagent (1% p-dimethylaminocinnamaldehyde in 10% HCl), filter paper
- Reagents and supplies for catalase reaction: 6% hydrogen peroxide solution, microscope slide, disposable transfer pipette
- Reagents and supplies for oxidase reaction: oxidase test cards (or oxidase test strips)
- Reagents and supplies for Gram staining: Gram staining kit, microscope slides, microscope

PROCEDURE

Subject each presumptive positive culture to the following biochemical tests. Record results in Table 13.2.

1. **Spot Indole Test**
   a. Saturate a piece of filter paper with the indole reagent.
   b. Use sterile toothpick to remove a small portion of a bacterial colony from the TSA-YE plates.
   c. Rub the toothpick sample on the filter paper.
   d. Positive indole test is recorded when a blue-green color is observed within 30 seconds. Negative indole test result is recorded when no blue color is observed, or slightly pink color develops.
   e. Repeat with remaining cultures.

2. **Catalase Test**
   a. From the TSA-YE plate corresponding to a presumptive positive culture, using a sterile toothpick, transfer a small portion of one colony onto a microscope slide.
   b. Add a single drop of 6% hydrogen peroxide solution to the colony portion on the slide.
   c. Instant effervescence due to the production of oxygen bubbles indicates the presence of the catalase enzyme. No bubbles indicates a negative reaction.
   d. Repeat with remaining cultures.

TABLE 13.2 {*Add a descriptive title, including food and how results were obtained*}

| Isolate | Virulence genes | | | Cell morphology | Indole | Catalase | Oxidase | Identity |
	stx₁ present?	*stx₂* present?	*eae* present?					
(Example) 1E	Yes	Yes	Yes	G-negative rod	Positive	Positive	Negative	STEC

(*Add footnotes as required for this table*).

3. Oxidase Test

a. Using a sterile toothpick, transfer a portion of a colony from a TSA-YE plate onto a commercial oxidase card.

b. Color change within 20–30 seconds indicates a positive result, disregard color change occurring after this time.

c. Alternatively, transfer a portion of the colony to a piece of filter paper. Add one drop of oxidase reagent; wait 20–30 sec for a color change.

d. Repeat with remaining cultures.

Note: If oxidase test strip is used, appropriate testing procedure should be followed.

4. Gram Staining

Prepare a Gram stain from each culture according to the protocol conducted in earlier exercises.

DATA INTERPRETATION

E. coli strains are indole-positive, catalase-positive, oxidase-negative, Gram-negative, and rod-shaped bacteria. Isolates producing these reactions (in addition to typical morphology on EMB) can be confirmed as *E. coli*. *E. coli* isolates producing a positive result in multiplex PCR screening (presence of stx_1 or stx_2) with or without *eae*, can be confirmed as STEC. A food sample producing a minimum of one confirmed isolate can be considered positive for this pathogen.

Note that all STEC strains possess one or more of the Stx-coding genes. Some STEC strains also carry intimin-coding gene (*eae*). Intimin is a 90 kDa protein that is involved in the attaching and effacing mechanism of adhesion, thus strains producing this protein cause the attaching and effacing lesion. Therefore, the strains containing the combination of both virulence genes (*stx* and *eae*) are often associated with the most severe forms of STEC-induced disease.

QUESTIONS

1. Briefly describe a recent outbreak of foodborne infection caused by Shiga toxin-producing *Escherichia coli*. How would the analysis you learned in this exercise be useful in tracking the cause of such an outbreak?

2. The serotype of isolates analyzed in the exercise was not determined.

a. How could the serotype have been determined?

b. What is the relationship between the serotype and the virulence in *E. coli*?

c. Is this relationship between serotype and virulence coincidental or biologically meaningful?

3. After running an analysis similar to that described in this exercise, you obtained the following hypothetical results. Present these results in a gel schematic that meets the following description:

    a. Lane 1: DNA molecular weight ladder, representing a commercially available product

    b. Lane 2: A typical STEC

    c. Lane 3: STEC was present, but the sample produced a false-negative result

    d. Lane 4: *E. coli* K12

    e. Lane 5: *E. coli* K12 was used, but the sample produced a false-positive result

Discuss causes of false-positive and false-negative results and the ramifications for these incorrect results.

PART IV

CONTROL OF FOODBORNE MICROORGANISMS

Beneficial microorganisms have been applied for centuries in food processing. Lactic acid bacteria have been used in making fermented dairy and meat products, and acetic acid bacteria in making vinegar and in contributing to sauerkraut manufacture. Beneficial yeast and molds are used to make baked products, alcoholic beverages, and mold-ripened cheeses. Despite the benefits of these microorganisms, some control over their growth in fermented products is needed. Additionally, presence of these microorganisms in foods where they are not intended often causes these products to spoil. Many other microorganisms, such as *Pseudomonas* spp., cause serious food spoilage. For public health considerations, pathogenic microorganisms, generally, should not be present in ready-to-eat foods and ideally would not be present in raw ingredients. Therefore, control of foodborne microorganisms is necessary for ensuring safety and quality of foods, and for minimizing waste during food production, storage, and distribution.

Control of foodborne microorganisms involves preventing their growth by inhibitory factors, decreasing or eliminating their populations by lethal means, or applying treatments that have both microstatic (preventing growth) and microcidal (lethal) effects. Inhibiting growth may be sufficient in moderately processed foods (e.g., cured meats) where toxigenic microorganisms are the pathogens of concern. On the contrary, microbial elimination is needed for shelf-stable, low-acid foods such as most canned foods. Treatments that combine microstatic and microcidal factors include pasteurization; heat decreases the microbial population and refrigeration suppresses or prevents survivors from growing and compromising the safety or the quality of the product.

Analytical Food Microbiology: A Laboratory Manual, Second Edition. Ahmed E. Yousef,
Joy G. Waite-Cusic, and Jennifer J. Perry.
© 2022 John Wiley & Sons, Inc. Published 2022 by John Wiley & Sons, Inc.

Food processors use many approaches to control foodborne microorganisms; these include physical, chemical, and biological control methods (Table IV.1). This part of the book addresses the microbiological aspects of two methods, heat (Chapter 14) and biocontrol (Chapter 15).

TABLE IV.1 Antimicrobial factors and associated technologies used in control of food microbiota.

Type	Factor	Technology
Physical	Heating	Thermal processing
	Low temperature	Refrigeration
		Freezing
	Irradiation	Gamma radiation sterilization
		Ultraviolet decontamination
	Water activity reduction	Dehydration, drying, freeze drying
	High hydrostatic pressure	High hydrostatic pressure pasteurization
		Pressure-assisted thermal sterilization
	Electro- and electromagnetic technologies	Pulsed electric field processing
		Ohmic heating
		Microwave heating
		Radio-frequency heating
Chemical	Antimicrobial additives	Curing of meat products (nitrite)
		Preservation by lipophilic acids (e.g., benzoic and sorbic)
	Antimicrobial ingredients	Acidification (e.g., vinegar and phosphoric acid)
		Miscellaneous (e.g., phenolic ingredients)
	Sanitizers	Chlorine and chlorine dioxide technologies
		Peroxide technologies (hydrogen peroxide and peroxyacetic acid)
		Cationic sanitization (quaternary ammonium compounds)
		Ozone technology
		Others (iodine, alcohol, etc.)
Biological	Antimicrobial primary metabolites	Lactic and acetic fermentations
		Alcoholic fermentation
	Antimicrobial peptides	Biopreservation by bacteriocins
	Protective cultures	Antimicrobial culture biocontrol
	Bacteriophages	Bacteriophage control

REFERENCES

Lorenzo, J.M., Munekata, P.E., Dominguez, R., et al. (2018). Main groups of microorganisms of relevance for food safety and stability: General aspects and overall description. In: *Innovative Technologies for Food Preservation: Inactivation of Spoilage and Pathogenic Microorganisms* (ed. Barba, F.J., Sant'Ana, A.S., Orlien, V., and Koubaa, M.), 53–107. London: Elsevier.

Yousef, A.E. and Abdelhamid, A.G. (2019). Behavior of microorganisms in food: Growth, survival and death. In: *Food Microbiology: Fundamentals and Frontiers*, 5e (ed. Doyle, M., Diez-Gonzalez, F., and Hill, C.), 3–21. Washington, DC: American Society for Microbiology Press.

CHAPTER 14

THERMAL RESISTANCE OF MICROORGANISMS IN FOOD

MICROBIAL INACTIVATION KINETICS. INTRODUCING THERMAL PROCESSING CONCEPT.

INTRODUCTION

Heating is the most commonly used preservation method for food. When applied properly, heat kills pathogenic and spoilage organisms and inactivates enzymes that may degrade product quality. Despite these benefits, many thermally treated foods are less appealing to consumers than their raw counterparts. Therefore, emerging thermal processes are designed to improve product safety while minimizing adverse changes in a product's sensory and nutritional characteristics. For proper process design, the heat resistance of the microorganism(s) of concern in a particular product must be determined. This microorganism is considered the target of the thermal process being developed. The target organism is often a pathogen associated with the food product, but a problematic spoilage microorganism also may be targeted.

Determining Thermal Resistance of Targeted Microorganisms

Once the target organism(s) for a product has been identified, its thermal resistance must be determined in a systematic manner. Such a study would help in determining appropriate time–temperature combinations that meet the processing goal (e.g., pasteurization) for that particular food. Therefore, the chosen food is inoculated with the target organism, heat treated at the preferred temperature, and death of the organism is monitored as a function of the treatment time. This inactivation kinetics experiment is intended to determine appropriate processing parameters against the target microorganism in that food matrix. It should be kept in mind that resistance to heat varies not only with the target organism

Analytical Food Microbiology: A Laboratory Manual, Second Edition. Ahmed E. Yousef,
Joy G. Waite-Cusic, and Jennifer J. Perry.
© 2022 John Wiley & Sons, Inc. Published 2022 by John Wiley & Sons, Inc.

and the food into which it is inoculated, but also with the temperature tested. Therefore, inactivation kinetics are determined at temperatures that include or bracket those of interest to food processors. In the experimental section of this chapter, students will perform inactivation kinetic studies and determine effective time–temperature combinations to inactivate selected microorganisms in a liquid food product (milk).

Methods for Determining Thermal Resistance

Thermal treatment suitable for processing a food is reported as the temperature to which all particles of the food must be exposed and the time for which this temperature exposure must be maintained. The flaw in this requirement is that heat treatment is not instantaneous; there is an initial heating stage during which the product gradually shifts from its storage temperature to the holding (target) temperature. This portion of the "heating history" is referred to as the come-up time. Care must be taken in the design of thermal experiments to control and minimize the come-up time so that the inactivation achieved by the holding time–temperature combination can be quantified accurately. Considerations for thermal resistance experiments include maximizing the sample's contact with the heating medium, the thermal conduction characteristics of the sample's packaging material, and the consistency and accuracy of the heat source. Methods of testing thermal resistance vary in meeting these considerations. The following are setup options for thermal resistance studies:

- **Capillary tubes submerged in water or oil bath**
  Capillary tubes are the favored sample container for thermal inactivation studies. These are thin-walled glass tubes that hold small sample volumes (e.g., 0.2 ml). The capillary tubes are cleaned and sterilized, the inoculated liquid sample is introduced into tubes using a syringe with a fine needle, and the tubes are sealed carefully by heat-melting both ends. When this method is used, come-up time is considered negligible. The tubes are withdrawn from the heating bath at pre-determined time intervals and cooled instantaneously by submersion in an ice-water bath. After sanitizing their surfaces, the tubes are aseptically transferred to thick-walled test tubes that contain small volumes of sterile diluents. The capillary tubes are crushed carefully into the diluents using a suitable device (e.g., pestle-shaped glass rod). Counts of survivors in the tube contents are determined by appropriate dilution and plating on suitable recovery media.

  Use of capillary tubes maximizes the surface area-to-volume ratio of the sample to be treated, and these tubes can withstand a considerable range of temperatures. The major downfalls of capillary tubes are the difficulty in preparing and handling the tubes and their applicability only to foods in the liquid, but not in the solid or viscous, state. Capillary tubes are somewhat easy to load with liquid sample; however, care must be taken to minimize injury or lethality of the microorganism in the food sample while heat-sealing the ends of the tubes. Submersion and retrieval of the tubes from a water or oil bath may also be difficult. Releasing the treated sample from the capillary tube for analysis also requires some skill to consistently break the tubes without contaminating the sample while maximizing the recovery of the treated microorganism.

- **Small glass vials submerged in a water bath**

  Thin-walled glass vials with capacity of approximately 4 ml and Teflon-lined plastic screw caps are also suitable containers for use in thermal treatment studies. When half-filled, at least, and mounted in a suitable wire rack, these tubes can be submerged in the heating water bath. Full submersion of the sample-containing vials into the heating medium is crucial for uniform heating of treated sample. Unfortunately, the relatively large sample volume results in measurable temperature come-up time during heating; this should be accounted for during process design and data processing. Protracted come-up time makes it difficult to collect meaningful data when targeted temperatures cause rapid lethality. Additionally, the gaseous head space may impede the heat transfer to the top portion of the sample; therefore, it is advisable to fill the tube as much as practically possible. Nevertheless, these screw-top glass vials are easy to handle and fill with sample, and contents are easy to withdraw after heat treatment. The vials are most suitable for experiments designed to collect preliminary information about the thermal resistance of a microorganism in a liquid food. Glass vials will be used in this laboratory exercise.

- **PCR tubes in a thermocycler**

  PCR tubes are potentially useful containers for thermal treatment studies. These tubes have thin walls that will not significantly resist heat transfer. A thermocycler is the ideal heating apparatus for these tubes due to the perfect size and shape match between the tubes and the holes in the heating block of the equipment. Additionally, the thermocycler accurately heats these tubes to a predetermined temperature for a time specified by the operator, rapidly cools the treated sample, and holds the tubes cold until the analyst removes them for analysis. This system is likely to minimize errors in thermal resistance studies normally caused by protracted come-up and cooling-down times. However, the system is not ideal for this laboratory exercise considering that sample tubes need to be withdrawn at intervals during the course of the thermal treatment.

- **Microcentrifuge tubes held in a water bath**

  Microcentrifuge tubes are small (1–2 ml) and inexpensive containers for holding food during thermal treatment experiments. The tubes can be easily suspended in a flotation rack for heat treatments in a water bath; however, the inconsistency of heating applied to the sample due to the incomplete submersion of the tubes in the heating medium may cause irregularities in the results.

Calculating Thermal Resistance

After the data for kinetic experiments are collected, the $\log_{10}$ number of survivors is plotted (y-axis) as a function of treatment time (x-axis) to create a survivor plot. The "survivor plot" is used to determine thermal resistance at a specific temperature, measured as D-value.

D-value

Decimal reduction value (D-value) is a measure of the resistance of microorganisms to heat. A D-value refers to one specific organism at one specific temperature in one specific medium (or food product). Its value equals the time it takes (e.g., minutes), at a given temperature, to lethally reduce the population of a specific organism by one order of magnitude (i.e., one $\log_{10}$) in a particular food product. Experimental determination of D-value involves treating a population of the microorganism of concern with heat (at one desired temperature) over a predetermined time period and enumerating the population over the course of treatment. The $\log_{10}$ CFU/ml of survivors are plotted against treatment time to produce a survivor plot (which is often referred to as survivor curve); ideally, the experiment produces a linear plot (Figure 14.1).

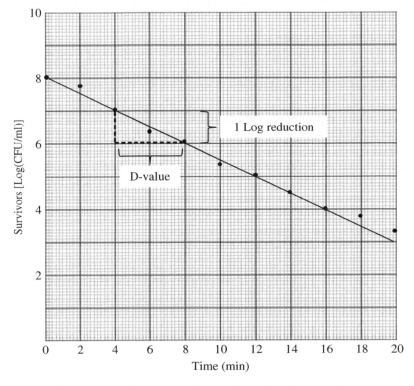

Figure 14.1 Illustration of a survivor plot of an organism.

The slope of the resulting survivor plot is measured, and D-value is calculated as $\frac{-1}{slope}$. This can be stated mathematically as:

$$D_T = \frac{-\Delta t}{\Delta \log N}$$

where, D_T is the decimal reduction time at the temperature T, t is the treatment time (e.g., minutes), and N is the population count (e.g., CFU/ml). The D_T is useful to calculate the total processing time needed to inactivate a certain population of the target microorganism in the food product at processing temperature T. If pasteurization of a food requires a minimum of 5-log reduction in the population of a specific targeted microorganism having a decimal reduction time of D_T, then pasteurization can be accomplished by heating this food at T for a total time of $5 \times D_T$. Here is a practical example: If the $D_{55°C}$ for *Campylobacter jejuni* in milk is 0.75 minutes, and the pasteurization process needs to accomplish an inactivation of 5 log CFU/ml of this pathogen, then the total thermal treatment time at 55°C should be a minimum of 3.75 minutes (0.75 minute × 5). Longer heating time may be applied commercially as an additional safety measure.

z-Value

When D-values for a given microorganism in a given food are measured at different temperatures, the relationship between these two parameters is of value to food processors. This relationship, in the form of z-value, can be represented by the following equation:

$$z = \frac{-\Delta T}{\Delta \log DT}$$

The log of D-values vs. treatment temperature constitutes the "thermal resistance plot," which is used to measure the temperature coefficient for microbial destruction (i.e., z-values). In this plot, the y-axis represents the log D-value, and the x-axis represents temperatures at which D-values were determined (Figure 14.2). The constructed thermal resistance plot shows points of equivalent lethality, i.e., each point along the line of the plot produces one log inactivation of the tested population for the time and temperature coordinates of this point. If this illustration produces a straight line, the slope of the line is used to calculate z-value. This parameter represents the change in the rate of microbial destruction with the change of temperature, that is, how much faster a microbe is killed when temperature is increased. Mathematically, z-value is the negative reciprocal of the slope of this line. Therefore, z-value is the temperature required to shift the D-value by 1-log. From this plot, the D-value at any given temperature can be determined and thus, the thermal treatment (temperature and time) necessary to achieve the desired log-reduction of the target organism can be calculated.

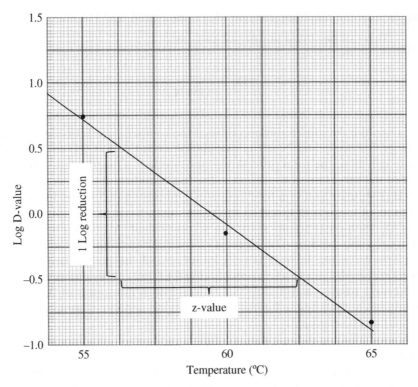

Figure 14.2 Illustration of a thermal resistance plot for an organism.

Challenges in Determining Microbial Thermal Resistance

Linearity of inactivation kinetics

Inactivation of microorganisms with heat has traditionally been described as a linear kinetic relationship. Provided this assumption is correct, it is easy to determine D-value from the slope of the survivor plot, and to use this value to predict survivor population after various heating times. However, linear inactivation is often observed only in a limited segment of the survivor plot. Linearity of the survivor plot seems to depend on the population density, suspending medium, the dose of the lethal factor (i.e., heating temperature), and many other factors. Similarly, the relationship between log D-values and temperature may not be linear, making it difficult to determine temperature–time combinations that produce similar lethality. For practical purposes, D- and z-values will be determined within the linear mid-section of the corresponding thermal destruction plots, if the plot is curved.

Food type affects thermal resistance

While microorganisms inherently vary in their thermal resistance, the surrounding medium (e.g., food) also has a large impact on the efficacy of heat treatments. The concentration and composition of ingredients in a medium influence inactivation of microorganisms by thermal treatments. The pH of the medium also impacts the efficacy of thermal treatments.

Physiological status affects thermal resistance of the test microorganism
Thermal resistance of a given microorganism depends on many factors, such as growth medium (broth vs. solid), stage of growth (exponential vs. stationary), and type of metabolites accumulated in the medium (e.g., acid vs. non-acid). These sources of variability should be avoided when thermal resistance of two different microorganisms is experimentally tested. For results consistency during this exercise, a specific procedure for preparing cell suspensions of test bacteria will be followed, as described later.

Difficulty in determining temperature come-up time
Time of heating inoculated samples (in the water bath) needed to reach the targeted temperature is called temperature come-up time. As soon as this time point in met, a sample is taken to represent the zero time at the targeted temperature. Come-up time should be determined, by instructors, before the experiment is performed. A thermocouple positioned in the center of the milk test vial can be used to determine the come-up time at each of the temperatures used in the exercise (i.e., 55, 60, and 65°C). Students should be provided with the come-up times before starting the exercise.

OBJECTIVES

1. Calculate D-values for selected foodborne microorganisms heated at the same temperature.
2. Compare the thermal resistance of the selected microorganisms in milk.
3. Determine z-values for a microorganism in milk.

MEDIA

Tryptic Soy Agar + Yeast Extract (TSA+YE)

This is a general-purpose rich medium. The addition of yeast extract to TSA allows the growth of fastidious organisms and aids in recovery of injured cells.

Tryptic Soy Broth (TSB)

Similar to TSA, TSB is a general-purpose rich medium for growing cultures.

PROCEDURE OVERVIEW

This laboratory will be completed in four laboratory sessions. During the first three sessions, the students will work in pairs, but they will work individually during the fourth session. During the first session, commercially sterile milk (UHT milk) will be inoculated with one of two foodborne bacteria (*Pseudomonas* sp. and *Enterococcus* sp.) and heat-treated, at a single temperature (55°C), for different heating times. Some groups will be analyzing *Pseudomonas* sp. while the other groups will analyze *Enterococcus* sp. A heat-sensitive *Pseudomonas fluorescens* strain and a heat-resistant strain of *Enterococcus faecium* are suitable for this laboratory exercise. Strains of *P. fluorescens* are generally sensitive to heat, whereas some strains of *E. faecium* (e.g., NRRL B-2354 or OSY31284) are heat-resistant. In the second session, milk will be inoculated with *E. faecium* only and heated at 60°C and 65°C. Some groups will be analyzing cultures heated at 60°C, while the other groups will analyze cultures heated at 65°C. Survivors of second-session heat treatments will be enumerated during the third laboratory session.

Data pooled from the three laboratory sessions will be analyzed during the fourth laboratory session. Survivor plots for both bacteria and thermal resistance plot for *E. faecium* will be constructed to determine D-values and z-value, respectively. D-value measurements will allow a comparison of the thermal resistance of the two organisms, whereas the z-value will show the temperature-dependence of D_T for *E. faecium*.

Culture Preparation for Consistent Thermal Resistance

Choosing strains that vary in thermal resistance is important for the success of this exercise. It is equally important that cell suspensions of these strains are prepared consistently to produce desirable results. The following procedure is recommended for preparing 20 ml cell suspension (containing 10^9 CFU/ml) of each of the strains described earlier:

- Frozen stock cultures of the heat-sensitive *P. fluorescens* (or *P. aeruginosa*) and the heat-resistant *E. faecium* strains are streaked on a suitable agar medium (e.g., TSA-YE) and the inoculated plates are incubated at 30°C for 24 hr. Additional transfer on the agar medium may be necessary if the strains are growing sluggishly.
- Incubated plates are checked for consistent colony morphology as an indicator of culture purity.
- For each strain, a small portion of a colony is transferred to 10 ml of a suitable liquid growth medium (e.g., TSB) in a 50-ml tube; inoculated tubes are incubated at 30°C for 24 hr with agitation at 200 rpm.
- Aliquot (0.5 ml) of each incubated culture is used to inoculate 200 ml TSB in a suitable flask and the mixture is incubated at 30°C for 24 hr with agitation at 200 rpm.
- After incubation, cells of each culture are harvested by centrifugation at 8000 rpm and 4°C for 10 minutes.
- The resulting cell pellets are suspended thoroughly in 20 ml buffered saline solution.
- The cell suspension should be suitable for use in the thermotolerance exercise.

Dilution schemes implemented in this exercise are based on the strains reported earlier and tested repeatedly at the food microbiology laboratories of Ohio State University. When other strains are chosen for this exercise or different procedures are used for preparing the cell suspensions, appropriate dilution schemes should be developed by instructors before use in teaching laboratories.

SELECTED REFERENCES

Claeys, W.L., Van Loey, A.M., and Hendrickx, M.E. (2002). Intrinsic time temperature integrators for heat treatment of milk. *Trends in Food Science and Technology* 13: 293–311.

Enright, J.B., Sadler, W.W., and Thomas, R.C. (1957). Pasteurization of milk containing the organism of Q fever. *American Journal of Public Health* 47: 695–700.

United States Food and Drug Administration. (2017). Grade "A" pasteurized milk ordinance. https://www.fda.gov/media/114169/download

Walker, G.C. and Harmon, L.G. (1966). Thermal resistance of *Staphylococcus aureus* in milk, whey, and phosphate buffer. *Applied Microbiology* 14: 584–590.

SESSION 1: INOCULATION AND THERMAL TREATMENTS OF TWO MICROORGANISMS

In this laboratory session, UHT milk will be inoculated with *Pseudomonas* sp. (e.g., *Pseudomonas fluorescens* or *P. aeruginosa*) or a heat-resistant strain of *Enterococcus faecium* and thermally treated at 55°C for up to 120 minutes. Students will work in groups of two; each group will work on one of the tested bacteria. Half of the class will test *Pseudomonas* and the other half will test *Enterococcus*. Tryptic soy agar, fortified with 0.6% yeast extract (TSA+YE), will be used to enumerate survivors of the heat treatment. The exercise is illustrated in Figure 14.3. Data collected will be used to compare thermal resistance of these organisms at this temperature.

MATERIALS AND EQUIPMENT

Per Student Pair

- Commercially sterile (UHT) milk: 25 ml in conical-bottom centrifuge tube (50-ml Falcon tube)

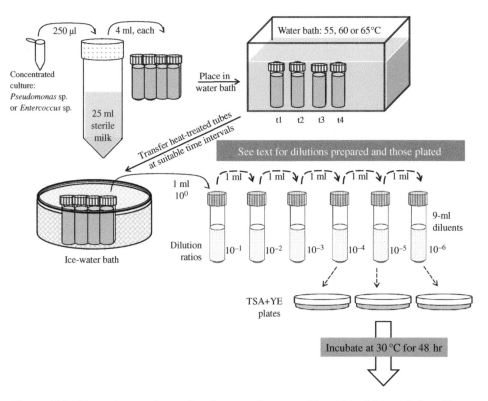

Figure 14.3 Procedure to determine thermotolerance of inoculated bacteria in milk showing an example of a dilution scheme; actual dilutions prepared depend on treated microorganism, initial population, and heating interval.

- Concentrated cell suspension in phosphate-buffered saline: 1 ml, 10^9 CFU/ml *Pseudomonas* sp. or *Enterococcus* sp., prepared by instructors as described earlier
- 9-ml peptone water tubes: 20 for *Pseudomonas* sp. or 26 for *Enterococcus* sp.
- Fifteen TSA+YE plates
- Five sterile glass vials (~ 4-ml capacity) with Teflon-lined screw caps
- One vial rack
- Ice-water bath
- Timer
- 5-ml disposable pipettes
- Pipette pump
- Laboratory tongs

Class-Shared

- Circulating water bath, set at $55 \pm 0.2°C$
- Other common laboratory supplies and equipment (e.g., disposable gloves, vortexer)

PROCEDURE

1. Inoculation

a. Vortex the provided cell suspension (*Pseudomonas* sp. or *Enterococcus* sp., depending on group assignment).

b. Transfer 250 µl of the cell suspension into the provided 25 ml sterile milk; this will achieve a concentration of approximately 10^7 CFU/ml milk.

c. Mix the contents of the inoculated tube.

2. Heat Treatment

a. Label (on the cap) five glass vials with the treatment times 0, 5, 10, 15, and 20 minutes for *Pseudomonas* sp. **or** five glass vials with the treatment times 0, 30, 60, 90, and 120 minutes for *Enterococcus* sp., **depending on group assignment**.

b. Aliquot 4 ml of inoculated milk into each of the vials; avoid pipetting milk foam, if present. Be sure to deposit milk directly in the bottom of the vial and avoid any shaking of vial contents.

c. Place the inoculated milk vials in the vial rack.

d. Place the rack in the water bath ($55 + 0.2°C$). **Make sure the vials are totally immersed in the hot water.**

e. Once the come-up time is achieved (provided by instructor), start the timer; remove the "Time 0" vial and place it (totally immersed) in the ice-water bath.

f. Remove remaining vials at times 5, 10, 15, and 20 minutes for *Pseudomonas* sp. and 30, 60, 90, and 120 minutes for *Enterococcus* sp.; place removed tubes, immediately, in ice-water bath (totally immersed).

3. **Serial Dilutions and Plating**

a. Label peptone water tubes for the following dilutions at the stated sampling times:

Pseudomonas sp.		*Enterococcus* sp.	
Time (min)	**Dilution[a]**	**Time (min)**	**Dilution[a]**
0	$10^{-1} - 10^{-6}$	0	$10^{-1} - 10^{-6}$
5	$10^{-1} - 10^{-5}$	30	$10^{-1} - 10^{-6}$
10	$10^{-1} - 10^{-4}$	60	$10^{-1} - 10^{-5}$
15	$10^{-1} - 10^{-3}$	90	$10^{-1} - 10^{-5}$
20	$10^{-1} - 10^{-2}$	120	$10^{-1} - 10^{-4}$

[a]*These dilution schemes are based on cell populations in the suspensions prepared by course instructors, and the predicted thermal resistance of tested bacteria.*

b. Perform appropriate serial dilutions of each sample using the labeled peptone water tubes.

c. Label TSA+YE plates (one only per dilution) for the three highest dilutions for each treatment time. For example, for the Time-0 vial, label plates receiving each of these dilutions: 10^{-4}, 10^{-5}, and 10^{-6}; one plate per dilution.

d. Spread-plate 100 µl of appropriate dilution onto each plate.

e. Invert plates and incubate at 30°C for 48 hours.

SESSION 2: HEAT TREATMENTS OF *Enterococcus* sp. AT 60 AND 65°C

Students will complete the experiments performed in the first laboratory session and run additional heat treatments. Therefore, "Part 1" includes colony counting of plates incubated since the previous laboratory session, and "Part 2" involves heat treatments of *Enterococcus* sp. only at two additional temperatures (60 and 65°C). Half of the class tests 60°C and the other half tests 65°C.

MATERIALS AND EQUIPMENT

Per Student Pair

- Previously incubated TSA+YE plates
- Commercially sterile (UHT) milk: 25 ml in conical-bottom centrifuge tube (50-ml Falcon tube)
- Concentrated cell suspension in phosphate-buffered saline: 1 ml, 10^9 CFU/ml *Enterococcus* sp. prepared by instructors as described earlier.
- 9-ml peptone water tubes: 26 for the 60°C treatment and 20 for the 65°C treatment
- Fifteen TSA+YE plates
- Five sterile glass vials (~ 4-ml capacity) with Teflon-lined screw caps
- One vial rack
- Ice-water bath
- Timer
- 5-ml disposable pipettes
- Pipette pump
- Tongs

Class-Shared

- Water bath, set at $60 \pm 0.2°C$
- Water bath, set at $65 \pm 0.2°C$
- Other common laboratory supplies and equipment (e.g., disposable gloves, vortexers, and colony counters)
- Class computer (optional)

PROCEDURE

Part 1: Colony Counting

1. Do a preliminary estimate of the number of colonies on all the plates, regardless of whether the counts would be used in calculations.
2. Use the decision tree described in Chapter 3 to determine the rules applicable to the colony counts on the plates prepared by the group.

TABLE 14.1 Surviving populations of *Pseudomonas* sp. and *Enterococcus* sp. when inoculated milk was heated at 55°C for up to 120 minutes. Data collected by multiple groups within the class.

Bacterium	Group initials	Treatment Time (min)	Dilution	Colony count	Survivors CFU/ml	Survivors Log CFU/ml
Pseudomonas sp.	AB/CD	0				
		5				
		10				
		15				
		20				
Pseudomonas sp.	etc.	0				
		5				
		10				
		15				
		20				
Enterococcus sp.	MN/OP	0				
		30				
		60				
		90				
		120				
Enterococcus sp.	etc.	0				
		30				
		60				
		90				
		120				

Add appropriate footnotes.

3. Record the colony counts in Table 14.1, using data only from plates that fit the criteria of the counting rules. Alternatively, enter these data in the class computer loaded with appropriate software.

4. Determine the population count (CFU/ml) in milk at each heating time point.

Part 2: Heat Treatments

1. **Inoculation**
   a. Vortex the provided suspension of *Enterococcus* sp.
   b. Transfer 250 μl of the cell suspension into the provided 25 ml sterile milk; this will achieve a concentration of approximately 10^7 CFU/ml milk.
   c. Mix the contents of the inoculated tube.

2. **Heat treatment**
   a. Label (on the cap) five glass vials with the treatment times 0, 30, 60, 90, and 120 minutes for tubes heated at 60°C, **or** five glass vials with the treatment times 0, 5, 10, 15, and 20 minutes for tubes heated at 65°C, **depending on group assignment**.
   b. Aliquot 4 ml of inoculated milk into each of the vials; avoid pipetting milk foam, if present. Be sure to deposit milk directly in the bottom of the vial and avoid any shaking of vial contents.
   c. Place the vials, containing inoculated milk, in the vial rack.
   d. Place the rack in the appropriate water bath; **make sure vials are totally immersed in the hot water**.
   e. Once the come-up time is achieved (provided by instructor), start the timer; remove the Time 0 vial and place (totally immersed) in ice-water bath.
   f. Remove remaining vials at times 30, 60, 90, and 120 minutes for the tubes in the 60°C-water bath, or 5, 10, 15, and 20 minutes for the tubes in the 65°C-water bath; place the removed vials, immediately, in ice-water bath (totally immersed).

3. **Serial dilutions and plating**
   a. Label peptone water tubes for the dilutions and time intervals shown in the following table:

Heating at 60°C		Heating at 65°C	
Time (min)	**Dilution[a]**	**Time (min)**	**Dilution[a]**
0	$10^{-1} - 10^{-6}$	0	$10^{-1} - 10^{-6}$
30	$10^{-1} - 10^{-6}$	5	$10^{-1} - 10^{-5}$
60	$10^{-1} - 10^{-5}$	10	$10^{-1} - 10^{-4}$
90	$10^{-1} - 10^{-5}$	15	$10^{-1} - 10^{-3}$
120	$10^{-1} - 10^{-4}$	20	$10^{0} - 10^{-2}$

[a]*These dilution schemes are based on cell population in the suspension prepared by course instructors, and the predicted thermal resistance of the tested bacterium.*

b. Perform appropriate serial dilutions of each sample using the labeled peptone water tubes.

c. Label TSA+YE plates (one only per dilution) for the three highest dilutions for each treatment time. For example, for the Time-0 vial, label plates to receive the following dilutions: 10^{-4}, 10^{-5}, and 10^{-6}; one plate per dilution.

d. Spread plate 100 μl of appropriate dilution onto each plate.

e. Invert plates and incubate at 30°C for 48 hours.

SESSION 3: COUNTING *ENTEROCOCCUS* SP. SURVIVORS

In this laboratory session, *Enterococcus* sp. survivors in the inoculated milk (that was heat-treated in session two) will be determined.

MATERIALS AND EQUIPMENT

Per Student Pair

- Incubated plates
- Colony counter
- Graph paper

Class-Shared

- Other common laboratory supplies and equipment
- Class computer (optional)

PROCEDURE

1. Do a preliminary estimate of the number of colonies on all plates, regardless of whether the counts would be used in calculations.
2. Use the decision tree discussed in Chapter 3 to determine the rules applicable to colony counts on the plates prepared by the group.
3. Record the colony counts in Table 14.2, using data only from plates that fit the criteria of the counting rules. Alternatively, enter these data in the class computer loaded with appropriate software.
4. Determine the population count (CFU/ml) in milk at each heating time point.

TABLE 14.2 {Add a descriptive title, describing the results of heating *Enterococcus* sp. at 60°C and 65°C, including how results were obtained}

Temperature	Group initials	Treatment Time (min)	Dilution	Colony count	Survivors CFU/ml	Log CFU/ml
60°C	AB/CD	0				
		30				
		60				
		90				
		120				
60°C	etc.	0				
		30				
		60				
		90				
		120				
65°C	MN/OP	0				
		5				
		10				
		15				
		20				
65°C	etc.	0				
		5				
		10				
		15				
		20				

Table should include relevant footnotes.

SESSION 4: DETERMINATION OF D-VALUES AND Z-VALUE

During this session, D-values for *Pseudomonas* sp. and *Enterococcus* sp. in milk heated at 55°C will be determined. Additionally, D-values for *Enterococcus* sp. in milk heated at 60°C and 65°C will be determined. The D-values for *Enterococcus* sp. at the three temperatures tested will be used to determine the z-value for this bacterium.

Determining D-valuves at 55°C

1. Use the data collected in Table 14.1 to construct a survivor plot for *Pseudomonas* sp. or *Enterococcus* sp. (heated at 55°C) as follows:
   a. Use graphing paper, similar to that shown in Figure 14.4, to construct a survivor plot: Draw and label the y-axis to represent the survivor populations (log CFU/mL) and x-axis to represent heating times (minutes), based on the data reported in Table 14.1.

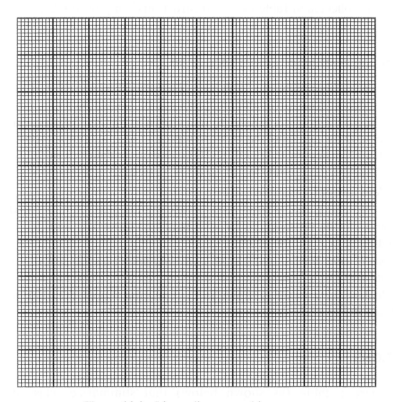

Figure 14.4   Linear-linear graphing paper.

b. Add a figure title; make sure you indicate the product tested and the treatment temperature applied.

c. Plot each data point (log population and heating time) reported in Table 14.1.

2. Visually check the plotted data for linearity. If most of the data points can be represented by a straight line, draw the line that approximates the best-fit for the data.

Note: If using the computer to plot the data points, use the software to determine the best-fit line.

3. Calculate the D-values using the equation reported earlier and the illustration shown in Figure 14.1.

4. Record the D-value in Table 14.3. Alternatively, enter these data in the spreadsheet of the class computer.

*Note: If survivor plot is S-shaped curve, utilize **only** the linear portion of the survivor plot to calculate the D-value.*

Determining D-values at 60°C and 65°C

Use the data collected in Table 14.2 to construct survivor plots for *Enterococcus* sp. heated at 60°C and 65°C. Follow the procedure described in the previous section.

Determining z-value for *Enterococcus* sp.

1. Use the data collected in Table 14.3 to construct a thermal resistance plot for *Enterococcus* sp. (heated at 55, 60, and 65°C) as follows:

a. Use graphing paper, similar to that shown in Figure 14.4, to construct the plot: Draw and label the y-axis to represent D-value (log D_T) and x-axis to represent heating temperature (°C), based on the data reported in Table 14.3.

b. Plot each data point (log D_T and T°C) reported in Table 14.3.

c. Add an appropriate figure title.

2. Visually check the plotted data points for linearity. If most of the data points can be represented by a straight line, draw the line that approximates the best-fit for the data.

Note: If using the computer to plot the data points, use the software to determine the best-fit line.

3. Calculate the z-value as shown in the equation reported earlier and illustrated in Figure 14.2.

QUESTIONS

1. Prepare a table (as shown below) that summarizes class data for *Enterococcus* sp. survivors at different heating temperatures. Use these

TABLE 14.3 {Add a descriptive title including food and how results were obtained}

D_T			
55°C	60°C	65°C	z-value

Table should include relevant footnotes.

data to calculate D-values for this strain at the three tested temperatures.

Temperature	Time	Log CFU/ml				Average Log CFU/ml
		Group AB&YZ[a]	Group CD&WX	Group EF&UV	etc.	
55°C						
60°C						

[a]*Initials of group members.*

2. Use the D-values at different temperatures (as calculated in Question 1) to construct the thermal resistance plot for the tested *Enterococcus* strain. Based on the plotted data, calculate z-value for this strain.

3. Supposed a pasteurization process was designed to target the *Enterococcus* strain tested in this exercise. Based on the data collected, what treatment time is needed to carry out that pasteurization at 62.5°C? Show how you obtained the answer.

4. What were the microorganisms of concern used originally to determine pasteurization parameters (i.e., temperature and time) for the following foods: liquid milk, liquid egg, and apple juice?

CHAPTER 15

PRODUCTION OF BACTERIOCINS IN MILK

BACTERIOCIN BIOASSAY. MICRO-DILUTION SERIES FOR POPULATION ENUMERATION.

INTRODUCTION

What are Bacteriocins?

Some bacterial strains produce antimicrobial peptides that are capable of inactivating other bacteria. Among these peptides are bacteriocins, which act against bacteria closely related to the producer. Bacteriocins represent a category of antimicrobial peptides that are produced inside the producer cell through ribosomal biosynthesis, and they are often released in the medium that has been used to grow the producer bacteria. Before their release from the producer cells, some bacteriocins undergo post-translational modifications, whereas others are excreted in an unmodified form. A bacteriocin typically has a narrow activity spectrum, meaning that it is only inhibitory to a limited number of species. There are bacteriocins active against some Gram-positive bacterial pathogens. Pediocin and nisin, for example, are active against *Listeria monocytogenes* and *Clostridium botulinum*, respectively. Other bacteriocins (e.g., colicin) act against some Gram-negative bacterial pathogens, such as Shiga toxin-producing *Escherichia coli*.

Bacteriocins and Food Preservation

Bacteriocins produced by lactic acid bacteria (LAB) have been used in food preservation. Nisin, a commonly used bacteriocin, is produced by some strains of *Lactococcus lactis*. Nisin has been approved in the US and many other countries

Analytical Food Microbiology: A Laboratory Manual, Second Edition. Ahmed E. Yousef,
Joy G. Waite-Cusic, and Jennifer J. Perry.
© 2022 John Wiley & Sons, Inc. Published 2022 by John Wiley & Sons, Inc.

as a food additive. Pediocin, produced by strains of *Pediococcus* spp., is also used in preservation but at much smaller scale compared to nisin. There are other promising bacteriocins under investigation and some of these are likely to be used as preservatives in the future.

In addition to their use in production of bacteriocins as food additives, some LAB can be used to produce food ingredients with antimicrobial characteristics. Developing food ingredients with antimicrobial activity may be a beneficial trend in enhancing food safety. These ingredients not only improve a food's safety, but they may also enhance its sensory attributes. This exercise consists of fermenting milk with a bacteriocin-producing lactic acid bacterium and quantifying the amount of bacteriocin released into the milk. The resulting fermented milk containing the antimicrobial may be suitable for use as an ingredient in formulating cheese spreads or ranch-style salad dressings; however, commercial usage of such ingredients may require regulatory approval.

LAB share common metabolic and physiological characteristics. They are Gram-positive, acid-tolerant, non-respiring, non-sporulating bacteria. As the name implies, LAB are capable of fermenting sugars into lactic acid. These bacteria lack catalase, hence the antimicrobial metabolic byproduct H_2O_2 may accumulate in LAB-fermented media. Growth of LAB may be inhibited by aerobic incubation, which promotes the accumulation of H_2O_2. Therefore, incubation anaerobically or under modified atmosphere conditions often enhances colony formation by LAB.

Bacteriocin Bioassay

The amount of a bacteriocin released during fermentation is often measured semi-quantitatively by testing the antimicrobial activity of the fermenting medium. Most methods of measuring bacteriocin levels rely on using a sensitive bacterium (indicator or target) and determining the degree of inactivation or inhibition of this indicator by the medium containing the bacteriocin. These methods are described as "bioassays" because biological indicators are used in measuring the activity. When a reference standard of the bacteriocin of interest is available, the bioassay may be used to measure bacteriocin concentration with reasonable accuracy. Considering that preparations with known nisin concentration are commercially available, determining the concentration of this bacteriocin in fermented media is feasible. Unfortunately, most bacteriocins are not available commercially. In these cases, bioassays determine bacteriocin activity in relative units, i.e., in activity arbitrary units. Examples of bioassays with or without reference standard are presented.

Bioassay without bacteriocin standard

When no standard bacteriocin preparation is available, dilution-based bioassay is a popular approach for measuring relative antimicrobial activity in a medium (e.g., food) fermented by a bacteriocin producer. The fermented medium can be referred to as fermentate. This bioassay involves diluting the antimicrobial-containing medium (normally a two-fold dilution series), mixing the dilutions with an antimicrobial-sensitive bacterium (i.e., indicator) in a suitable culture

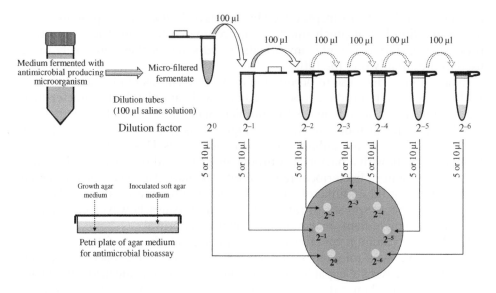

Figure 15.1 Bioassay for quantifying antimicrobial activity in a fermentate without bacteriocin standard.

medium, incubating the mixture, and determining the highest dilution that inhibits the growth of the indicator (Figure 15.1). The higher the dilution inhibiting indicator's growth, the more potent the antimicrobial (e.g., bacteriocin) is in the fermentate. This dilution-based bioassay is often referred to as the "critical dilution bioassay method," which conceptually is similar to the "minimum inhibitory concentration" technique for measuring antibiotic potency.

A version of the dilution-based bioassays, called "spot-on-lawn method," is often used in bacteriocin activity measurement. This method involves spotting small volumes (e.g., 5 or 10 μl) of the bacteriocin-containing dilutions onto a layer of soft agar medium seeded with a suitable indicator bacterium (Figure 15.1). The spotted plates are incubated and resulting areas of inhibition, observed on indicator lawn where the dilutions were spotted, are determined. The higher the dilution resulting in an inhibition area, the more potent the bacteriocin is in the original culture or fermentate. For a successful bacteriocin bioassay, it is important to choose a suitably sensitive indicator. The more sensitive the indicator bacterium is to the tested antimicrobial agent, the more accurate the measurement of activity of the agent will be.

As just described, the highest dilution factor generating an area of inhibition indicates the strength (and the quantity) of the antimicrobial agent. The bacteriocin activity is, therefore, proportional to the reciprocal of the highest dilution factor producing a detectable area of inhibition (i.e., the critical dilution). This critical dilution factor will be symbolized as DF_c. Using this information, a relative bacteriocin activity value, known as activity arbitrary units (AU), is calculated as follows:

$$AU \text{ per ml} = \frac{1000}{DF_c \times \mu l \text{ spotted}}$$

For example, if a fermentate was diluted $1/2^1$ to $1/2^6$; these correspond to dilution factors, DF, of 2^{-1} to 2^{-6} (Figure 15.1). Suppose 5 μl of each dilution was spotted on an indicator-seeded soft agar layer, and this resulted in areas of inhibition (i.e., clear circles) at all dilutions except $1/2^6$. Based on this information, DF_c should be $1/2^5$ (i.e., 2^{-5}), and the bacteriocin activity in the fermentate is calculated as follows:

$$\text{Bacteriocin activity} = \frac{1000}{2^{-5} \times 5} = 12,800 \text{ AU per ml}$$

Bioassay with bacteriocin reference standard

The concentration of a bacteriocin in a fermentate (or any medium) can be determined with reasonable accuracy if a preparation having known concentration of this bacteriocin (i.e., a bacteriocin reference standard) is available. To complete this bioassay, the analyst first should determine analytically the relationship between the concentration of the bacteriocin in the standard and the degree of indicator inhibition caused by different concentrations. This relationship between bacteriocin concentration and indicator inhibition is referred to as "dose-response relationship," or, in layman's terms, the "standard curve." The analyst should also determine simultaneously the degree of inhibition associated with the bacteriocin-containing medium or fermentate and compare this inhibition with that depicted in the standard curve. This comparison allows for determining the concentration of the bacteriocin in the fermentate.

To determine the dose-response relationship, dilutions of the bacteriocin are prepared and a small volume (e.g., 5 μl) of each dilution is spotted onto a layer of soft-agar medium seeded with the indicator bacterium (Figure 15.2). A sample of the fermented medium likely to contain that bacteriocin is spotted on the same seeded agar plate or on a similarly prepared duplicate plate. After incubating the plates, circular areas of no growth (i.e., clear agar) appear where the solutions of the antimicrobial agent are spotted (Figure 15.3). A dose-response plot is constructed to reveal the mathematical relationship between bacteriocin concentration (i.e., the dose) and area or diameter of the observed inhibition (i.e., the response). It has been noted by many analysts that the diameter of the inhibition area is almost linearly proportional to the $\log_{10}$ bacteriocin concentration. Considering that the bacteriocin is often diluted in two-fold (not decimal) ratios, $\log_2$ bacteriocin concentration is used instead of $\log_{10}$ concentration (Figure 15.4). Other mathematical functions may adequately describe the relationship between bacteriocin concentration and inhibition area or diameter. To determine the concentration of the bacteriocin in the fermented medium, the diameter of the inhibition area resulting from the sample of that medium is matched with a similar point on the standard dose-response plot; thus, the corresponding bacteriocin concentration can be determined.

This exercise demonstrates how a strain of *Lactococcus lactis* can be used to ferment milk and produce a bacteriocin (nisin). The concentration of the nisin produced will be determined using a bioassay with nisin standard.

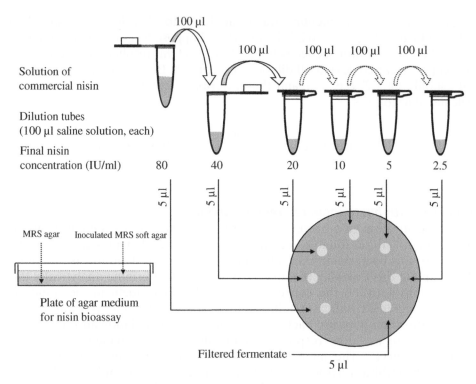

Figure 15.2 Bioassay for quantifying antimicrobial activity in milk fermentate against a standard of pure nisin.

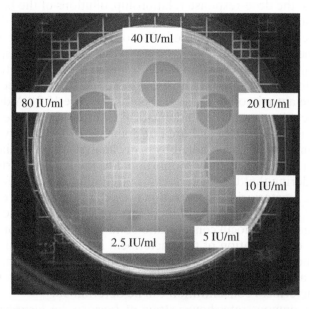

Figure 15.3 Inhibition areas resulting from spotting dilutions of nisin stock solution onto a soft MRS agar seeded with sensitive *Lactobacillus* sp. (courtesy of Patsy Leung).

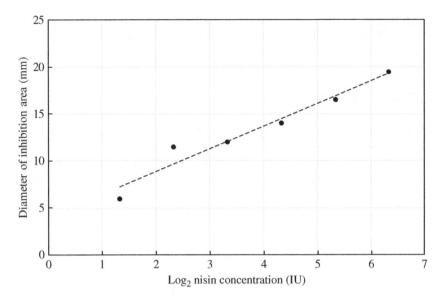

Figure 15.4 Dose response plot depicting the linear relationship between $\log_2$ nisin concentration in international units (IU) and diameter of inhibition area in millimeters (mm). These data were gathered from the results of bioassay illustrated in Figures 15.1 and 15.2.

OBJECTIVES

1. Produce nisin (a bacteriocin) during milk fermentation with a nisin-producing culture.
2. Practice measuring bacteriocin concentration using a dilution-based bioassay.
3. Practice population enumeration using micro-dilutions coupled with spread-plating.

MEDIA

Lactobacilli de Man, Rogosa, & Sharpe (MRS) Media

MRS broth
This medium supports growth of most LAB, particularly the slow-growing strains of lactobacilli. The medium is nutritionally rich, containing peptone, beef extract, yeast extract, and glucose. Fastidious microorganisms such as LAB grow well on this medium. The medium is somewhat selective for lactobacilli since it contains Tween 80 and sodium acetate.

MRS agar
MRS agar is prepared from MRS broth by addition of agar to a final concentration of 1.5%.

MRS soft agar

The medium is prepared from MRS broth by addition of agar to a final concentration of 0.75%.

PROCEDURE OVERVIEW

Commercial nonfat dry milk (milk powder) reconstituted at a 10% rate (wt/vol) in water will be used as the growth medium. Alternatively, commercial ultra-high temperature sterilized milk (UHT milk) could be used. Sterile milk will be inoculated with the bacteriocin-producing cultures at 1% (vol/vol). A nisin-producing culture (e.g., *Lactococcus lactis* subsp. *lactis* ATCC 11454) will be used to ferment the sterile milk. Alternatively, other bacteriocin-producing LAB may be used. For example, some *Enterococcus* spp. can ferment milk and produce potent enterocins.

Inoculated tubes will be incubated at 30°C for 24 hr. Most LAB cultures will ferment the lactose in the milk and produce organic acids that will cause denaturation of some of the milk proteins and the milk will coagulate; therefore, the test tube contents will be mixed vigorously to disrupt the curd. The fermented milk will be assayed for LAB population count, pH, and bacteriocin activity. In preparation for the bacteriocin assay, a portion of the liquefied curd will be transferred to a sterile microfuge tube and centrifuged. The clear supernatant (i.e., whey) will be collected, micro-filtered to remove remaining bacteria, and the filtrate will be bioassayed for bacteriocin activity. Successful determination of antimicrobial activity will depend on the selection of the target bacterium (bacteriocin-sensitive indicator). *Pediococcus pentosaceus* will be used in this exercise as an indicator for nisin antimicrobial activity, but other suitable indicators (often a LAB) may be used. If enterocin-producing bacteria are used in the exercise, a suitable indicator should be selected for the bioassay.

To facilitate quantification of the bacteriocin produced in milk, commercial nisin will be used to prepare a dose-response curve (i.e., a standard curve). When the bacteriocin activity in milk is compared with the standard curve results, it will be possible to determine the antimicrobial activity of the fermentate in nisin-equivalent units. An overview of this procedure is depicted graphically in Figure 15.5.

Organization

Students will work in groups of two. Group members will share responsibility in preparing the fermented milk, completing the nisin bioassay in the fermentate, and determining the antimicrobial concentration using a dose-response curve. In the first laboratory session, milk tubes will be prepared and inoculated with the LAB culture and inoculated tubes will be incubated. During the second laboratory session, coagulated milk will be analyzed for LAB population count and prepared for bacteriocin bioassay. The remainder of the fermented milk will be used to measure pH. During the third session, incubated plates will be used to determine the population of LAB, and the bulk of the bacteriocin bioassay will be completed. Results of the bacteriocin bioassay will be examined during the fourth exercise session.

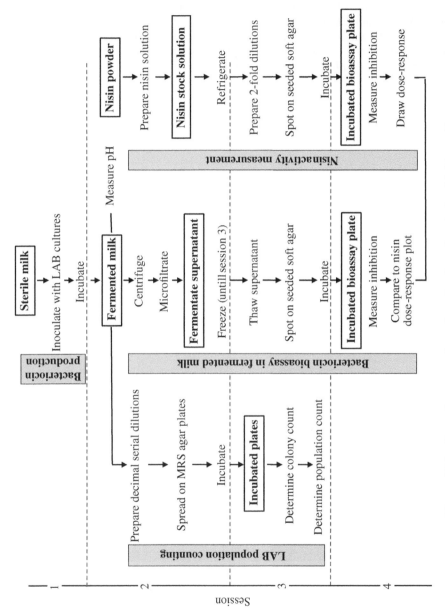

Figure 15.5 Procedure overview of production of antimicrobial ingredients.

SELECTED REFERENCES

Cotter, P.D., Ross, R.P., and Hill, C. (2013). Bacteriocins – a viable alternative to antibiotics? *Nature Reviews Microbiology* 11: 95–105.

Huang, E., Hussein, W.E., Campbell, E.P., and Yousef, A.E. (2021). Applications in food technology: antimicrobial peptides. In: *Biologically Active Peptides* (ed. F. Toldrá F and J. Wu). Amsterdam: Elsevier (in press).

Papagianni, M., Avramidis, N., Filioussis, G. (2006). Determination of bacteriocin activity with bioassays carried out on solid and liquid substrates: Assessing the factor "indicator microorganism." *Microbial Cell Factories* 5: 30.

SESSION 1: FERMENTATION

Sterile milk will be prepared and fermented with the LAB strain.

MATERIALS AND EQUIPMENT

Per Group of Two

- 1 ml LAB culture: This is a 16-hr culture of *Lactococcus lactis* subsp. *lactis* ATCC 11454 (a nisin producer) in MRS broth. Alternatively, other nisin-producing strains can be used.
- One 50-ml conical centrifuge tube (Falcon tube) containing 25 ml sterile reconstituted skim milk (or UHT milk)
- Micropipetters (1000 μl) and sterile pipette tips

Class-Shared

- Incubator set at 30°C
- Other common laboratory supplies and equipment

PROCEDURE

1. Label the milk tube with appropriate information.
2. Vortex the LAB culture.
3. Dispense 250 μl of LAB culture into the appropriately labeled milk tube, vortex the inoculated milk.
4. Incubate the inoculated tube at 30°C for 24 hr.

SESSION 2: ANALYSIS OF FERMENTED MILK

Fermented milk will be analyzed for LAB population count and pH. Additionally, an aliquot of fermented milk will be prepared for the bacteriocin assay to be completed during the third and fourth sessions.

MATERIALS AND EQUIPMENT

Per Group of Two

- Incubated milk tube
- 10 ml uninoculated sterile milk (same batch of milk used in Session 1)
- A test tube containing 10 ml sterile diluent (phosphate-buffered saline solution)
- Six MRS agar plates
- One sterile 0.2-μm syringe microfilter, 2-ml (or 3-ml) syringe
- Beaker containing sterile microcentrifuge tubes (~1.5 ml capacity) and covered with aluminum foil to protect tubes from contamination
- Sterile Pasteur pipette
- Micropipetters (1000 μl and 100 μl) and sterile pipette tips

Class-Shared

- Microcentrifuge
- Other common laboratory supplies and equipment

PROCEDURE

1. **LAB Population Count**
   a. Vortex the fermented milk test tube to break the curd and to homogenize the sample.
   b. Prepare 10^{-1} to 10^{-7} micro-dilutions:
      i. Label seven sterile microcentrifuge tubes with appropriate dilution factors.
      ii. Dispense 900 μl diluent into each of the microcentrifuge tubes.
      iii. Vortex the fermented milk, transfer 100 μl into the microcentrifuge tube receiving the 10^{-1} dilution, and vortex the diluted sample.
      iv. Prepare the remaining dilutions, until the 10^{-7} dilution is reached; remember to change tips between dilutions.
   c. Dispense 100 μl of the 10^{-5}, 10^{-6}, and 10^{-7} dilutions onto appropriately labeled MRS agar plates, two plates per dilution.
   d. Spread inocula on MRS agar plates with sterile cell spreaders.
   e. Incubate the plates at 30°C for 48 hr.
2. **Preparing Fermentate Supernatant for Bacteriocin Bioassay**
   a. Dispense 1 ml of the fermented homogenized milk into a sterile microcentrifuge tube.

b. Centrifuge the milk at 3000 rpm for 20 min. A relatively clear supernatant fluid should separate from the milk solids.

c. Open the syringe microfilter package, exposing only the side of the filter to be fitted to the syringe barrel.

d. Remove the syringe plunger and any tip-guard or needle. Screw the filter onto the syringe barrel (Figure 15.6).

e. Remove the remainder of the microfilter package and place the sterile end of the microfilter onto the sterile microcentrifuge tube held on a microcentrifuge tube rack.

f. Using a sterile Pasteur pipette, carefully transfer the supernatant of centrifuged milk into the syringe barrel of the microfiltration unit.

g. Attach the syringe plunger and push it gently to drive the supernatant through the filter. Filtration may be stopped when 150–250 µl of sterile filtrate is collected in the microcentrifuge tube.

h. Freeze the filter-sterilized supernatant at –20°C for use in a subsequent laboratory session for bacteriocin bioassay.

3. **pH Measurement**

a. Insert the pH electrode into the provided unfermented milk and record sample pH.

b. Insert the pH electrode into the remainder of the homogenized fermented milk and measure the pH. Record the pH value observed.

   Note: pH meter should be calibrated by a student or instructor before use. The electrode used should have a diameter suitable for the tube size used. The electrode should be rinsed clean and dry-blotted before and after use.

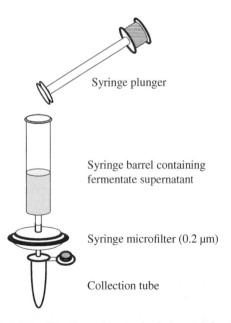

Syringe plunger

Syringe barrel containing fermentate supernatant

Syringe microfilter (0.2 µm)

Collection tube

Figure 15.6   Microfiltration of bacteriocin-containing fermentate.

SESSION 3: LACTIC ACID BACTERIA COUNT (CONTINUED) AND BACTERIOCIN BIOASSAY

During this session, incubated MRS agar plates will be used to determine LAB population in the fermented milk. Additionally, frozen supernatant of fermented milk, prepared during Session 2, will be used for bacteriocin bioassay, with nisin used as the reference standard.

MATERIALS AND EQUIPMENT

Per Group of Two

- Incubated MRS agar plates (prepared in Session 2)
- Frozen filtered supernatant of fermented milk (prepared in Session 2)
- Two test tubes, each containing 9 ml sterile diluent (phosphate-buffered saline solution)
- 5 ml overnight culture (~16-hr culture) of *Pediococcus pentosaceus* or another suitable nisin-sensitive indicator bacterium (e.g., an *Enterococcus* sp.)
- Two test tubes each containing 9 ml sterile molten MRS soft agar
  Note: Remove from the water bath only when ready to use.
- Two MRS agar plates, prepared 2–3 days in advance and held protected from contamination
  Note: Freshly prepared agar plates tend to impact the solidification of the soft agar which may cause the spotted bacteriocin solution to diffuse uncontrollably.
- Microcentrifuge tube containing 120 µl nisin solution (80 international units [IU]/ml), prepared by instructor during a previous session
  Note: 1 µg of pure nisin = 40 IU
- Beaker containing sterile microcentrifuge tubes (~1.5 ml capacity) and covered with aluminum foil
- Micropipetters (1000 µl, 100 µl, 20 µl) and sterile pipette tips
- Other common laboratory supplies

Class-Shared

- Microcentrifuge
- Water bath, set at ~50°C
- Other common laboratory supplies and equipment (e.g., colony counters)

PROCEDURE

1. **Population Count**
   a. Inspect the incubated MRS agar plates and determine those containing colonies in the countable range (see counting rules, Chapter 3).
   b. Count colonies and determine LAB population count as CFU/ml milk.

TABLE 15.1 Population count and pH of milk fermented (or unfermented) with a lactic acid bacterium.

Group initials	pH of milk		Population count (CFU/ml)
	Fermented	Unfermented	

[Add appropriate footnotes]

    c. Record these results in Table 15.1 and in the class computer worksheet, if available.

2. Preparing Soft Agar-Culture Overlay

    a. Prepare a 10^{-1} dilution of the sensitive bacterial strain (*P. pentosaceus*) culture as follows:

        i. Transfer 1 ml of the culture into one of the test tubes containing 9 mL sterile diluent.

        ii. Vortex the contents of the tube.

    b. Prepare the mixture of the nisin-sensitive indicator and the soft agar as follows:

        i. Transfer 1 ml of 10^{-1} dilution prepared in the previous step into one of the 9 ml molten MRS soft agar-tubes that has been pre-cooled in a water bath to about 50 °C.

        ii. Mix gently by rolling between hands.

        Note: Vigorous shaking incorporates air bubbles in the agar which, when poured, produces an undesirable rough surface for bacteriocin spotting.

    c. Pour the contents of the inoculated soft-agar tube into the previously prepared plate of regular MRS agar. Do not agitate or swirl the plate.

    d. Repeat the two previous steps (steps b and c) with the second soft-agar tube and the MRS agar plate.

    e. Let the soft-agar overlay solidify by leaving the covered plates on the bench undisturbed. Although solidification of the agar overlay takes only a few minutes at room temperature (~22°C), it is advisable to leave these plates for ~30 min before spotting the filtrates; this extra time minimizes diffusion of spots and results in regular and round inhibition areas that are easy to measure.

3. Preparing Nisin Standard Solution

    a. Prepare a series of two-fold dilutions of the standard nisin solution (80 IU/ml). Five dilution tubes are required (Figure 15.2). Notice that in a two-fold dilution, the volume of diluent in the tube equals the volume being transferred.

Note: If pure nisin is not available, commercial nisin preparations may be used. In the latter case, concentration of nisin in the product should be considered when preparing the standard solutions.

b. Place 100 μl of the diluent into five separate sterile microcentrifuge tubes. Label the tubes with these calculated nisin concentrations: 40, 20, 10, 5, and 2.5 IU/ml.

c. To the 40-IU tube, add 100 μl of the original standard solution (80 IU/ml).

d. Vortex the 40-IU tube, then transfer 100 μl of the tube contents to the 20-IU tube.

e. Repeat this two-fold dilution procedure until completing the 2.5-IU tube.

4. **Spotting**

a. Appropriately label the bottom of plates containing the MRS agar and culture overlay. Divide the plate into 7 sections and label the locations to be spotted with the nisin solution or the fermented milk supernatant. One spot of the fermentate will be applied to a designated location on the plate (Figure 15.2).

b. Apply 5 μl (using a 20-μl micropipetter) of the filtered fermented milk supernatant (after thawing) onto the surface of the soft agar-culture overlay. Keep the pipetter vertical to produce as round and even a spot as possible. Avoid agitating the plate to prevent spot spreading.

c. Repeat step b using the nisin standard solutions (2.5–80 IU/ml). Evenly space the spots on the plate.

d. Repeat steps a–c using the second MRS soft agar-culture overlay plate (duplicate testing).

e. Keep the plates undisturbed until all spots are absorbed into the agar; this process may take up to 60 minutes.

f. Incubate the plates at 35°C for 48 hr.

SESSION 4: BACTERIOCIN BIOASSAY (CONTINUED)

MATERIALS

Per Group of Two

- Incubated soft MRS agar plates (Session 3)
- 10-cm ruler (transparent plastic rulers with 0.5-mm divisions are ideal)

Class-Shared

- Common laboratory supplies and equipment (e.g., colony counters)

PROCEDURE

1. Inspect the incubated soft MRS agar plates using a colony counter and observe areas of inhibition. Spotted nisin solutions should produce distinct, clear areas of inhibition.
2. Determine if the spots from the filtered fermented milk supernatant produced inhibition.
3. Measure diameter of all inhibition areas using the ruler. If the inhibition areas are not round, take two diameter readings, one perpendicular to the other, and average the two readings.
4. Record diameter readings in Table 15.2.
5. Using graph paper with linear axes, plot $\log_2$ nisin concentration (x-axis) against diameter of inhibition area (y-axis) for the nisin standard solutions. Label this plot as "dose-response plot." For guidance, see Figure 15.4.
6. Calculate the average diameter of the inhibition areas for the fermented milk supernatant (from the two incubated plates).
7. Using the dose-response plot, determine the $\log_2$ concentration of nisin that corresponds to the diameter observed for fermented milk. The anti-log of this value is the estimated nisin concentration (or nisin-equivalent concentration) in the fermented milk supernatant.

TABLE 15.2 Diameter of inhibition areas resulting from spotting 5 μl of nisin standard solution and filtered fermented milk supernatant.

Nisin standard		Filtered fermented milk supernatant	
Concentration (IU/ml)[a]	Diameter (mm)	Spot number	Diameter (mm)
80		1	
40		2	
20		Average inhibition area diameter	
10			
5			
2.5			

[a] International units per milliliter of nisin standard solutions

TABLE 15.3 Bacteriocin concentrations (IU/ml)^a in samples of fermented milk.

Group initials [nisin producer]	Concentration (IU nisin/ml)^a

^{a} International units of nisin per milliliter of milk

8. Record results in Table 15.3.

QUESTIONS

1. A stock solution of a known bacteriocin was prepared to contain 100 μg/ml. A series of five two-fold dilutions were prepared of the stock solution. These dilutions, along with the original stock solution, were spotted (5 μl per spot) on a soft agar medium seeded with *Listeria innocua*, which was used as a suitably sensitive indicator; the agar plate was prepared as described in this exercise. After incubation of the plate and measuring the diameters of the inhibition areas, the following results were obtained:

Bacteriocin concentration (μg/ml)	Diameter of inhibition areas (mm)
100	20
50	15
25	12
12.5	10
6.25	8
3.125	5

   a. Use the provided data to prepare a dose-response plot, where the y-axis represents diameters of inhibition areas and x-axis represents $\log_2$ bacteriocin concentrations.

   b. Describe the dose-response relationship as observed in the plot.

2. Suppose a culture of a microorganism was spotted (5 μl) on the same soft agar plate described in Question 1. After incubation, this spot produced an

inhibition area with 17 mm diameter. Assuming this culture produced the same bacteriocin described in Question 1, what is the concentration of the bacteriocin in the culture?

3. In addition to the bacteriocins listed in this chapter, there are many other known bacteriocins. List one of these bacteriocins and its most sensitive pathogenic target. How would you change the procedure described in this exercise to test the activity of that bacteriocin?

APPENDIX I

LABORATORY EXERCISE REPORT

Laboratory reports may be prepared for some or all the laboratory exercises of this book. Taking time outside of the laboratory to prepare and organize reports assists in confirming that students understand the rationale and concepts of their laboratory training. A laboratory report should be concise and contain information relevant to the exercise. In some exercises, a "laboratory worksheet," instead of a report, may be prepared. Laboratory worksheets mainly consist of tables filled in during laboratory activities, with brief comments on the gathered data. Worksheet structure may be expanded to train students how to prepare a complete laboratory report. Therefore, questions pertinent to different sections of a typical laboratory report may be included. Using this expanded definition of a worksheet, students may be required to answer these questions and submit them along with the annotated and completed worksheet.

OUTLINE

The laboratory report should include these sections:

- Cover page
- Introduction
- Objectives
- Methods
- Results
- Discussion
- References

Analytical Food Microbiology: A Laboratory Manual, Second Edition. Ahmed E. Yousef, Joy G. Waite-Cusic, and Jennifer J. Perry.
© 2022 John Wiley & Sons, Inc. Published 2022 by John Wiley & Sons, Inc.

Students also may be asked to address specific questions related to the exercise and include their answers in relevant sections of the report.

Cover Page

The cover page should include the exercise title, student name (and class partner's name in case of working in groups), and date of submission. The laboratory exercise title should be descriptive and informative, yet concise. It should indicate the method used and the specific food analyzed by the student preparing the report. The title should help the reader understand the purpose of the exercise. It will not be acceptable to use the same title provided in the book or class handouts, or to refer generically to the report as "class data."

Introduction

The introduction should include brief comments on the importance of the organism(s) or method(s) studied in the exercise. Additionally, the following questions may be addressed in the introduction: Does the food support growth or survival of the microorganism subjected to analysis? If growth is not supported, does the food tested inhibit or inactivate such a microorganism? Under which conditions would the microorganism proliferate in the food analyzed? If there are standards stating an unacceptable level of this microorganism in food, what are these standards?

Objectives

The objectives should be brief, explaining in one sentence why the laboratory exercise was performed. Although a set of objectives has been included with each laboratory exercise, students should summarize these in their own words.

Methods

Since detailed methodologies are included already in each exercise, the "Methods" section of the laboratory report should be very short. Cite the book as the source of the procedure and indicate which method was followed. Additionally, this section covers the following:

- Information about the sample used in the analysis. These may include (a) name, (b) method of packaging (e.g., canned, vacuum-packaged), (c) storage conditions (e.g., refrigerated, frozen, shelf-stable), (d) package size, and (e) sell-by date.
- Discrepancies between the method presented in the book and actual procedures performed in the laboratory.

Results

This section should include data collected by the student, if working individually, or by the student group. Students may occasionally be asked to present and comment on class data, i.e., data collected by all individuals or all groups in the class. Data should be presented in tables and/or graphs, when possible (see the Data Presentation section on how tables and figures are prepared). Assign numbers and titles to all tables and graphs. Include footnotes with tables to explain units and abbreviations (e.g., CFU/ml) or to explain irregularities. Label all axes on graphs and assign units to all numbers. When results require calculations, show one example of each type of calculation and include units.

Discussion

In this section, results are explained; therefore, students should refer to the contents of tables and figures presented in the "Results" section, without repeating these results. If the food is judged to be of poor microbiological quality, explain what could have been the cause of the contamination (e.g., cross-contamination, under-processed). Make plausible assumptions about the microbiological quality and safety of food tested, if asked by instructors.

If instructors inoculated the food with a targeted microorganism before analysis, the discussion should be prepared differently than when unaltered food is used in the analysis. In this case, instructors may provide specific questions to assist students in preparing a meaningful discussion.

References

Students should cite published data that explain, support, or oppose the results of their exercise. Before citation, it is important that the students read the published original research and summarize its findings in their own words. "Original research" means data that were gathered during executing experiments by the authors of the published work. Therefore, review articles, books, chapters in edited books, magazine articles, social media, etc., should not be cited in the laboratory report, unless explicitly permitted by the instructors. For citations and listing of references, it is recommended that guidelines of the American Society for Microbiology are followed.

DATA PRESENTATION

The most common ways to present microbiological data are tables and figures. Well-structured tables and figures improve the readability of the report. Each table or figure should be informative and self-explanatory. The reader should be able to understand the contents of a table or a figure without the need to search for the details in the rest of the report.

Tables

Tables are an accurate representation of the data collected. A table consists of a title, column headings, and contents (Table 1). The title is convention-ally placed on top of the table and refers to the following:

(a) Data being presented in table's contents (e.g., aerobic mesophilic plate count)

(b) Variables tested by these data (e.g., source of salad, considering that different students analyzed samples from different stores)

The heading of a column should always describe the contents of this column. Footnotes define terms or explain conditions.

Table A3.1. Aerobic mesophilic plate count of bagged fresh-cut salad, collected from different grocery stores in Columbus, Ohio, and analyzed by students during spring semester, 20xx.

Student initials	Colony count[a]	Dilution counted[b]	Population			
			CFU/g[c]	Log_{10} CFU/g	Average log_{10} CFU/g	Standard deviation
AB	45,31	10^{-3}	3.8×10^5	5.6		
CD	73,95	10^{-3}	8.4×10^5	5.9		
EF	145,193	10^{-2}	1.7×10^5	5.2	5.7	0.41
GH	21,23	10^{-4}	2.2×10^6	6.3		
IJ	57,49	10^{-3}	5.3×10^5	5.7		

[a]Counts on duplicate plates

[b]Volume spread-plated was 0.1 mL

[c]Colony forming units per gram

Note that tables prepared for a teaching laboratory report are often more detailed than those that appear in published research papers. However, even in a research setting, preparing a detailed table will serve as an accurate record of exercise results, and it should help in preparing the needed concise table if the results of the analysis are to be published.

Figures

Microbiological data can be presented graphically in many kinds of figures. The data in Table A3.1, for example, can be presented as a bar graph, with x-axis labeled by student initials (or store number) and y-axis representing the population count (log_{10} CFU/g). Conventionally, the independent variable is presented on the x-axis, whereas the y-axis is assigned to the dependent variable. As is the case with tables, the figure should have a title which clearly states what dependent and independent variables are being presented. Unlike tables, the figure's title is placed at the bottom of the figure. In professionally prepared reports or research papers, a set of data can be presented as a table or a figure, but not both. Although this redundancy is avoided in scientific writing, it may be allowed in laboratory exercise reports. Figure 15.4 (Chapter 15) is an example of a lines graph that could be prepared from the exercises included in this book.

Scientific Notation and Significant Digits

It is important that population counts are presented in scientific notations. For example, if the calculated population in a given food is six thousand, five hundred CFU/g, this number should be presented as 6.5×10^3 CFU/g. Note that the number contains only two significant digits, the six and the five. This should be the case with microbial populations counted by plating techniques. As you can see in Table A3.1, many of the agar plates produced colony counts with only two significant digits; therefore, the final population should not contain more than two significant digits. It is important to note that the two significant digit rule applies only to the final population count, but not to colony counts on plates or to intermediate calculations leading to population counts. Large numbers resulting from these calculations should be rounded to two significant digits.

MISCELLANOUS WRITING ISSUES

Microorganisms' Nomenclature

- The scientific name of a microorganism includes the genus name and the species epithet (e.g., *Staphylococcus aureus*). If only the genus of an isolate is known, the species epithet is replaced with the word "species" or its abbreviation, "sp."
- Microorganisms' names should be italicized (e.g., *Listeria monocytogenes*). However, if the whole sentence, which includes the organism's name, is italicized, use non-italicized font for the organism (e.g., *Ready-to-eat foods in the United states should not contain* Listeria monocytogenes *at any detectable level.*).
- Note that the first letter of the genus should be capitalized, but not that of the species epithet.

Grammar and Style

Reports written with correct grammar, using simple sentences and sound style, are easier for instructors to evaluate than those written otherwise. The following is some advice based on the observations of the book authors:

- Numbers under ten should be spelled out (e.g., There were five illnesses), and sentences should not begin with numerals (e.g., Eighty-six people became ill. . .).
- Proper nouns should be capitalized. Examples include names of people, cities, companies, and institutions.
- Avoid writing in first person (I, me, we, etc.) or in second person (he/she, you, they, etc.). Third person is most appropriate for professional writing (e.g., When examining these data, *one* may conclude that. . .).

Plagiarism

Although classmates may share the same data set, prepared reports should represent an individual's effort. Copying other's work is plagiarism. The "other's work" could be that of a classmate, published work (books, papers, etc.), or contents posted on the internet. Plagiarism is a violation of the Student Conduct Code at most institutions and can result in serious consequences, including expulsion. Plagiarism-checking programs are widely available, and some instructors use them regularly before grading reports.

BACTERIAL AND FUNGAL STRAINS RECOMMENDED FOR USE AS CONTROL ORGANISMS

This list includes bacterial and fungal species and strains that are useful as controls in the exercises presented in this book. Most strains are available for purchase from commercial culture collections, which have been indicated in the descriptions below. Other strains may be used as controls; however, their behavior should be verified before use in classroom exercises.

General Cultural Handling Methods

Bacterial and fungal cultures should be cryogenically preserved for long-term storage. Typically, cultures are prepared for long-term storage by culturing in a non-selective broth (e.g., tryptic soy broth) under favorable incubation conditions and then mixing with glycerol (25–35% v/v) and storing at –80°C in a cryogenic vial. To revive cultures, an aliquot (10 μl) is transferred from the cryogenic vial to a suitable non-selective growth media (broth or agar) and incubated under favorable conditions. Best practice is to confirm identity using a selective-differential media or molecular method.

Specific Strains

Aspergillus sp.

Chapter 9: Examination and Enumeration of Foodborne Fungi
Control culture for microscopy

Analytical Food Microbiology: A Laboratory Manual, Second Edition. Ahmed E. Yousef,
Joy G. Waite-Cusic, and Jennifer J. Perry.
© 2022 John Wiley & Sons, Inc. Published 2022 by John Wiley & Sons, Inc.

Enterococcus faecium NRRL B-2354

Chapter 4: Practicing Basic Techniques
Part of mixed culture for plate count exercise
Chapter 14: Thermal Resistance of Microorganisms in Food
Heat-resistant strain

Enterococcus faecalis

Chapter 11: *Listeria monocytogenes*
Negative control
Control for CAMP test

Escherichia coli K12

Chapter 4: Practicing Basic Techniques
Part of mixed culture for plate count exercise
Chapter 8: Detection and Enumeration of *Enterobacteriaceae* in Food
Positive control
Chapter 9: Examination and Enumeration of Foodborne Fungi
Control culture for microscopy
Chapter 13: Shiga toxin-producing *Escherichia coli*
Negative control

Escherichia coli O157:H7 ATCC 43888 or OSU 315

Chapter 13: Shiga toxin-producing *Escherichia coli*
Surrogate control – does not carry *stx* genes

Escherichia coli O157:H7 EDL933

Chapter 13: Shiga toxin-producing *Escherichia coli*
Pathogenic positive control

Lactococcus lactis subsp. *lactis* ATCC 11454

Chapter 15: Production of Bacteriocins in Milk
Bacteriocin (nisin)-producing strain

Listeria innocua

Chapter 11: *Listeria monocytogenes*
Positive/Negative control

Listeria monocytogenes

Chapter 11: *Listeria monocytogenes*
Positive control
Control for CAMP test

Pediococcus pentosaceus OSU

Chapter 15: Production of Bacteriocins in Milk
Bacteriocin (nisin)-sensitive strain (indicator)

Penicillium sp.

Chapter 9: Examination and Enumeration of Foodborne Fungi
Control culture for microscopy

Pseudomonas fluorescens

Chapter 7: *Pseudomonas* spp. and Other Spoilage Psychrotrophs
Positive control
Chapter 14: Thermal Resistance of Microorganisms in Food
Heat sensitive strain

Rhizopus sp.

Chapter 9: Examination and Enumeration of Foodborne Fungi
Control culture for microscopy

Saccharomyces cerevisiae

Chapter 9: Examination and Enumeration of Foodborne Fungi
Control culture for microscopy

Salmonella Typhimurium ATCC BAA-2828 (non-pathogenic)

Chapter 12: *Salmonella*
Positive control

Staphylococcus aureus (a strain that is coagulase-positive and produces staphylococcal enterotoxin A, at least, should be selected)

Chapter 10: *Staphylococcus aureus*
Positive control
Chapter 11: *Listeria monocytogenes*
Control for CAMP test

Staphylococcus epidermidis

Chapter 10: *Staphylococcus aureus*
Negative control

Streptococcus agalactiae

Chapter 11: *Listeria monocytogenes*
Control for CAMP test

APPENDIX III

MICROBIOLOGICAL MEDIA

This list includes microbiological media relevant to the exercises presented in this book. Compositions of these media are based on formula published in various sources, including the Food and Drug Administration's Bacteriological Analytical Manual (https://www.fda.gov/food/laboratory-methods-food/media-index-bam), the Food Safety Inspection Service's Microbiology Laboratory Guidebook (https://www.fsis.usda.gov/news-events/publications/microbiology-laboratory-guidebook), International Organization for Standardization, ISO 07.100.30 Food Microbiology (https://www.iso.org/ics/07.100.30/x/), and the Difco™ and BBL™ Manual: Manual of Microbiological Culture Media, Second Edition, 2009 (Becton, Dickinson and Company, Sparks, MD). Information on proprietary media were sourced from the manufacturer's websites (e.g., Oxoid.com).

MEDIA PREPARATION GUIDELINES

Liquid media are typically referred to as broths, whereas solid media are referred to as agars. In solid media, agar is commonly used as the gelling agent at a concentration of approximately 1.5–2.0% (15–20 g/l). Soft or semi-solid agar are formulated with lower concentrations of agar. Most broth and agar media with the same name (e.g., MRS broth and MRS agar) have identical formulations except for the addition of agar. However, there are rare exceptions (e.g., Tryptic Soy Broth and Tryptic Soy Agar) where the formulations differ slightly. Typically, media recipes are written with the ingredient list organized with general ingredients (non-selective carbon and nitrogen sources) in descending order by mass first, followed by unique or selective ingredients in descending order. If the

Analytical Food Microbiology: A Laboratory Manual, Second Edition. Ahmed E. Yousef, Joy G. Waite-Cusic, and Jennifer J. Perry.
© 2022 John Wiley & Sons, Inc. Published 2022 by John Wiley & Sons, Inc.

medium is an agar, then agar is usually the last ingredient listed. If the medium contains additives, supplements, or special solutions, that ingredient or formulation is listed separately. Typically, recipes are written to produce a total volume of 1 liter. However, some chromogenic and/or expensive media may be formulated for smaller volumes (e.g., 100 ml).

There are three primary steps in media preparation: (i) preparation of the solution; (ii) sterilizing the media; and (iii) aliquoting the media into a suitable format. These steps do not always occur in the same order and some steps may be interrupted. Most media are prepared from a pre-formulated powder by dissolving in a suitable volume of distilled water following the instructions on the bottle. The solution is heated to boiling to facilitate the solubility of the agar and other ingredients and to pre-heat the media prior to sterilization. Most media are sterilized by using an autoclave (121°C for at least 15 min); however, some ingredients used in media are sensitive to autoclaving, and these media may be sterilized by boiling. Other ingredients are very sensitive to heat (e.g., many antibiotics), so these ingredients are sterilized by filtration (0.22 μm) prior to adding to media that has been previously autoclaved and tempered to 48–50°C. Broths are typically dissolved in a bulk solution and then aliquoted into tubes or small bottles prior to sterilization, whereas agars are sterilized as a bulk solution and tempered to 50°C prior to pouring into Petri plates and allowed to cool and solidify. A standard Petri plate (100 mm diameter) is typically filled with about 15–20 ml of agar solution. This allows for sufficient media to cover the surface of the Petri plate, prevent the media from drying out during incubation, and provide consistent thickness to observe differential reactions. For certain assays, the desirable format for agar may be in a tube (slant or stab), in which case the hot agar solution would be aliquoted into the tubes prior to sterilization.

Many microbiological media recipes include information about the pH of the media. For most media, following the simple preparation instructions results in media with the proper pH. However, some media require a pH adjustment, which is often clearly indicated in the instructions. If media are not performing as expected, checking the pH of the prepared media would be a good point to start troubleshooting. The sterilization procedure can cause a change in pH, so measurement before and after sterilization may be necessary.

Media may be prepared in advance and stored in the refrigerator (4°C) for extended periods of time prior to use. If planning to store prepared, filled Petri plates, they should be properly dried before bagging to reduce condensation, which encourages mold growth. Some selective media lose activity over time (e.g., Bismuth Sulfite Agar); in this case, follow special instructions for storage and use these media within the appropriate timeframe.

Media Reconstitution and Autoclaving

- Always read the label on the container of the medium before use. Verify and record the medium identity, the manufacturer, and the expiration date.
- Accurately weigh or measure all ingredients and add them in the manner and sequence recommended by medium manufacturer, or the protocol being used.
- Use distilled or deionized water to reconstitute powdered or concentrated media.

- If the medium contains agar, stir continuously and heat to boiling before dispensing into bottles. Agar melts (as evident by medium clearing) in boiling aqueous solution, but it solidifies at much lower temperatures. Agar is generally used at 15 g/l (1.5%), although this may vary depending on the medium and use. If agar is not fully melted, media may not solidify upon cooling.

- Any medium requiring sterilization should be placed in a heat resistant container. Never fill a container more than 3/4 full – this minimizes potential overflow if bubbling occurs. Screw the caps on loosely to prevent pressure buildup. In general, the cap should be approximately 1/2 turn from tight.

- Mark racks of tubes or media bottles with indicating tape (autoclave tape) to verify sterilization.

- Agar sets (turns into a gel) at about 45°C. Therefore, if autoclaved or steamed media are to be used immediately (i.e., within few hours of making), they should be held in a water bath set at 50°C until used.

- Sterilization is often achieved by holding liquid or agar media at 121°C for 15 minutes in the autoclave. Large loads of material or larger volume containers may require more autoclaving time to ensure all items are completely sterilized.

- Dry materials (e.g., glassware, stoppers, etc.) require longer autoclaving time (30 min) to become sterile. Many autoclaves have a "dry" or "gravity" setting that is useful for sterilizing dry materials because it includes a processing stage that helps to minimize condensation.

- Proper pressure exhaust rate should be observed, which depends on the materials being autoclaved. Liquids should be exhausted more slowly than empty glassware so that the change in pressure does not cause liquid to boil over.

MEDIA RECIPES

Antibiotic Plate Count Agar (APCA)

(Plate Count Agar is also called Standard Methods Agar or Tryptone Glucose Yeast Agar)

Medium used in: Chapter 9: Fungi
Medium (per liter):

Pancreatic digest of casein	5 g
Yeast extract	2.5 g
Dextrose	1 g
Agar	15 g

Supplements (per liter):

Chlortetracycline	100 mg
Chloramphenicol	100 mg

The pH of the finished media will be 7.0 ± 0.2.

Preparation instructions: Prepare medium base in 1 l of distilled water and autoclave. Temper to 50°C before adding antibiotic solutions (2 ml of each antibiotic solution, containing 500 mg/10 ml). Mix and pour complete medium (15–20 ml) into Petri plates.

Baird-Parker Agar

Medium used in: Chapter 10: *Staphylococcus aureus*
Medium base (per liter):

Tryptone	10 g
Beef extract	5 g
Yeast extract	1 g
Glycine	12 g
Sodium pyruvate	10 g
Lithium chloride • 6 H_2O	5 g
Agar	20 g

The pH of the final medium will be 6.9 ± 0.2.
Addition (per liter):

Egg yolk	15 g
Potassium tellurite	0.15 g

Note: Addition is suspended in a total volume 50 ml.
Preparation instructions: Prepare medium base in 950 ml of distilled water and autoclave. Temper to 50°C before mixing in the addition (50 ml). Pour complete medium (15–18 ml) into Petri plates.

Bismuth Sulfite (BS) Agar (Wilson and Blair)

Medium used in: Chapter 12: *Salmonella*
Medium (per liter):

Peptone	10 g
Beef extract	5 g
Dextrose	5 g
Na_2HPO_4	4 g
$FeSO_4$	0.3 g
Bismuth sulfite indicator	8 g
Brilliant green	25 mg
Agar	20 g

The pH of the finished media will be 7.7 ± 0.2.
Preparation instructions: Prepare medium base in 1 l of distilled water. Mix thoroughly and heat with agitation. Boil for 1 minute to sterilize and create a uniform suspension (there will be a precipitate that doesn't dissolve). DO NOT AUTOCLAVE. Temper to 50°C. Mix gently by swirling to distribute precipitate before pouring (20 ml) into Petri plates.
Note: This medium loses selectivity over time (selectivity decreases in 48 hours). Prepare the media the day before use and store in the dark.

Brain-Heart Infusion Agar with Yeast Extract (BHI-YE)

Medium used in: Chapter 10: *Staphylococcus aureus*
Medium base* (per liter):

Pancreatic digest of casein	16 g
Brain-heart infusion (solids)	8 g
Peptic digest of animal tissue	5 g
NaCl	5 g
Dextrose	2 g
Na_2HPO_4	2.5 g
Agar	13.5 g

*formulations may differ by manufacturer.
Addition (added as a powder during base preparation):

Yeast extract	5 g

The pH of the finished media will be 7.4 ± 0.2.

Preparation instructions: Prepare medium base and yeast extract in 1 l of distilled water and autoclave. Temper to 50°C. Pour complete medium (15–20 ml) into Petri plates.

Brilliance *Listeria* Agar (BLA) (Oxoid)

Medium used in: Chapter 11: *Listeria monocytogenes*
Medium base (per liter):

Peptone	18.5 g
Yeast extract	4 g
NaCl	9.5 g
Sodium pyruvate	2 g
$LiCl_2$	15 g
Maltose	4 g
X-glucoside chromogenic mix	0.2 g
Agar	14 g

Additive (per liter):

Lecithin solution	40 ml

Supplements (per liter):

Nalidixic acid	26 mg
Polymyxin B	10 mg
Ceftazidime	6 mg
Amphotericin	10 mg

The pH of the finished media will be 7.2 ± 0.2.

Preparation instructions: Prepare medium base in 1 l of distilled water and autoclave. Temper to 50°C. Add additive and supplements and mix well. Pour complete medium (15–20 ml) into Petri plates.

Shelf life: Prepared medium can be stored in the refrigerator for up to two weeks before use.

Buffered Peptone Water (BPW)

Medium used in: Chapter 12: *Salmonella*
Medium (per liter):

Peptone	10 g
NaCl	5 g
Na_2HPO_4	3.5 g
KH_2PO_4	1.5 g

The pH of the finished media will be 7.2 ± 0.2.
Preparation instructions: Prepare medium in 1 l of distilled water and autoclave. Cool to ambient temperature before use.

Dichloran Rose Bengal Chloramphenicol (DRBC) Agar

Medium used in: Chapter 9: Fungi
Medium (per liter):

Dextrose	10 g
Bacteriological peptone	5 g
KH_2PO_4	1 g
$MgSO_4 \bullet 7H_2O$	0.5 g
Chloramphenicol	0.1 g
Rose Bengal	0.025 g
Dichloran	0.002 g
Agar	15 g

The pH of the finished media will be 5.6 ± 0.2.
Preparation instructions: Prepare medium in 1 l of distilled water and autoclave. Temper to 50°C and pour (15–20 ml) into Petri plates.
Note: Rose Bengal dye is light sensitive and produces inhibitory compounds when exposed to light. After the DRBC agar has solidified it should be stored in the dark prior to use.

Dichloran 18% Glycerol (DG18) Agar

Medium used in: Chapter 9: Fungi
Medium base (per liter):

Dextrose	10 g
Peptone	5 g
KH_2PO_4	1 g
$MgSO_4 \bullet 7H_2O$	0.5 g
Dichloran	1 g
Chloramphenicol	0.1 g
Agar	15 g

Addition (per liter):

Glycerol	220 g or 174 ml

The pH of the finished media will be 5.6 ± 0.2 with a water activity of 0.955.

Preparation instructions: Prepare medium base in **800 ml** of distilled water. Add glycerol and bring final volume to 1 liter. Sterilize by autoclaving. Temper to 50°C and pour (15–20 ml) into Petri plates.

E. coli Broth with Antibiotics (EC-broth+VCC)

Medium used in: Chapter 13: STEC

Medium base (per liter):

Trypticase or tryptose	20 g
Lactose	5 g
NaCl	5 g
K_2HPO_4	4 g
KH_2PO_4	1.5 g
Bile salts No. 3	1.5 g

Supplements (per liter):

Vancomycin	8 mg
Cefixime	0.05 mg
Cefsulodin	10 mg

The pH of the finished media will be 6.9 ± 0.2.

Preparation instructions: Prepare medium base in 1 l of distilled water and autoclave. Cool to room temperature and add filter-sterilized supplements.

Enterobacteriaceae Enrichment Broth, Mossel (Neogen)

Medium used in: Chapter 8: *Enterobacteriaceae*

Medium (per liter):

Pancreatic digest of gelatin	10 g
Dextrose	5 g
Na_2HPO_4	8 g
KH_2PO_4	2 g
Oxgall	20 g
Brilliant green	15 mg

The pH of the finished media will be 7.2 ± 0.2.

Preparation instructions: Prepare medium base in 1 l of distilled water. Mix thoroughly and heat with agitation. Bring to a boil and hold at 100°C in a water bath or flowing steam for 30 minutes. DO NOT AUTOCLAVE. Cool rapidly in cold water.

Eosin Methylene Blue (EMB) Agar, Levine

Medium used in: Chapter 13: STEC
Medium (per liter):

Pancreatic digest of gelatin	10 g
Lactose	10 g
K_2HPO_4	2 g
Eosin Y	0.4 g
Methylene blue	0.065 g
Agar	15 g

The pH of the finished media will be 6.8 ± 0.2.

Preparation instructions: Prepare medium base in 1 l of distilled water and autoclave. Temper to 50°C. Mix well to ensure even distribution of precipitate before pouring (12–20 ml) into Petri plates.

Note: Eosin Y dye is light sensitive and produces inhibitory compounds when exposed to light. After the EMB agar has solidified it should be stored in the dark prior to use.

Glucose Agar (also called Glucose OF Medium) (Neogen)

Medium used in: Chapter 8: *Enterobacteriaceae*
Medium (per liter):

Enzymatic digest of casein	2 g
K_2HPO_4	0.3 g
Glucose (dextrose)	10 g
NaCl	5 g
Bromothymol blue	0.08 g
Agar	4 g

Preparation instructions: Prepare medium base in 1 l of distilled water and aliquot into tubes (10 ml). Autoclave and cool to ambient temperature before use.

Half-Fraser Broth (also called Demi-Fraser Broth)

Medium used in: Chapter 11: *Listeria monocytogenes*
Fraser Medium base (per liter):

Pancreatic digest of casein	5 g
Proteose peptone no. 3	5 g
Beef extract	5 g
Yeast extract	5 g
NaCl	20 g
Na_2HPO_4	12 g
KH_2PO_4	1.35 g
Esculin	1 g
$LiCl_2$	3 g

<u>Half-Fraser Additives (per liter):</u>

Ferric ammonium citrate	500 mg
Nalidixic acid	10 mg
Acriflavine hydrochloride	12.5 mg

Note: Some manufacturers sell the medium base with some or all of the additives included in the pre-mixed powder.

The pH of the finished media will be 7.2 ± 0.2.

<u>Preparation instructions:</u> Dissolve media in 1 l of distilled water and autoclave. Temper to 50°C before adding reconstituted supplement.

Lactobacilli deMan, Rogosa, & Sharpe (MRS) Agar (Neogen, Oxoid)

Medium used in: Chapter 15: Production of Bacteriocins in Milk
<u>Medium (per liter):</u>

Peptone	10 g
Beef extract	10 g
Yeast extract	5 g
Dextrose (glucose)	20 g
Sodium acetate • $3H_2O$	5 g
K_2HPO_4	2 g
Triammonium citrate	2 g
$MgSO_4 • 7H_2O$	0.2 g
$MnSO_4 • 4H_2O$	0.05 g
Tween 80	1.08 g
Agar	15 g

The pH of the finished media will be 6.2 ± 0.2.

<u>Preparation instructions:</u> Prepare medium base in 1 l of distilled water and autoclave. Temper to 50°C and pour (12–20 ml) into Petri plates.

Lactobacilli deMan, Rogosa, & Sharpe (MRS) Soft Agar

Medium used in: Chapter 15: Production of Bacteriocins in Milk
<u>Medium (per liter):</u>

Peptone	10 g
Beef extract	10 g
Yeast extract	5 g
Dextrose (glucose)	20 g
Sodium acetate • $3H_2O$	5 g
K_2HPO_4	2 g
Triammonium citrate	2 g
$MgSO_4 • 7H_2O$	0.2 g
$MnSO_4 • 4H_2O$	0.05 g
Tween 80	1.08 g
Agar	7.5 g

The pH of the finished media will be 6.2 ± 0.2.

Preparation instructions: Prepare medium base in 1 l of distilled water and aliquot into tubes (10 ml). Autoclave and temper to 50°C. Pour as an overlay onto previously prepared MRS agar.

Lactobacilli deMan, Rogosa, & Sharpe (MRS) Broth

Medium used in: Chapter 15: Production of Bacteriocins in Milk
Medium (per liter):

Peptone	10 g
Beef extract	10 g
Yeast extract	5 g
Dextrose (glucose)	20 g
Sodium acetate • $3H_2O$	5 g
K_2HPO_4	2 g
Triammonium citrate	2 g
$MgSO_4$ • $7H_2O$	0.2 g
$MnSO_4$ • $4H_2O$	0.05 g
Tween 80	1.08 g

The pH of the finished media will be 6.2 ± 0.2.
Preparation instructions: Prepare medium base in 1 l of distilled water and autoclave. Cool to ambient temperature before use.

Modified Oxford (MOX) Agar

Medium used in: Chapter 11: *Listeria monocytogenes*
Oxford Medium base (per liter):

Pancreatic digest of casein	8.9 g
Proteose Peptone No. 3	4.4 g
Yeast extract	4.4 g
Tryptic digest of beef heart	2.7 g
Starch	0.9 g
NaCl	4.4 g
Esculin	1 g
Ferric ammonium citrate	0.5 g
$LiCl_2$	15 g
Agar	15.3 g

Modified Oxford Supplement (per liter):

Colistin sulfate	10 mg
Moxalactam	20 mg

Note: There are several variations of the combination of antibiotics used in Modified Oxford Agar. Some formulations may include the antibiotics as part of the powdered base medium and other formulations may have them as separate supplements that are added as solutions after autoclaving.
The pH of the finished media will be 7.0 ± 0.2.

Preparation instructions: Prepare medium base in 1 l of distilled water and autoclave. Temper to 50°C before adding reconstituted Modified Oxford supplement. Mix gently and pour (12–20 ml) into Petri plates.
*Note: This medium is light-sensitive. Prepared plates stored in the dark and
refrigerated prior to use.*

Motility Medium

Medium used in: Chapter 11: *Listeria monocytogenes*
Medium (per liter):

Pancreatic digest of casein	10 g
Beef extract	3 g
NaCl	5 g
Agar	4 g

The pH of the finished media will be 7.4 ± 0.2.
Preparation instructions: Prepare medium base in 1 l of distilled water and
bring to a boil. Dispense 8–10 ml into test tubes with caps. Sterilize by autoclaving and cool to ambient temperature before use.

Peptone Water (1%)

Medium used in: Chapter 5: Aerobic Mesophilic Count
Medium (per liter):

Peptone	10 g
NaCl	5.0 g

The pH of the finished media will be 7.2 ± 0.2.
Preparation instructions: Dissolve peptone in 1 l of distilled water. Sterilize by
autoclaving and cool to ambient temperature before use.

Phosphate Buffered Saline

Medium used in: Chapter 14: Thermal Resistance of Microorganisms in Food
Solution (per liter):

NaCl	8.0 g
KCl	0.2 g
Na_2HPO_4	1.42 g
KH_2PO_4	0.24 g

Preparation instructions: Dissolve the salts in 800 ml of distilled water in 1-l
volumetric flask, adjust the pH to 7.4, and add distilled water until the volume
is 1 l. Autoclave and cool to ambient temperature before use.

Pseudomonas Isolation Agar (PIA)

Medium used in: Chapter 7: *Pseudomonas*
Medium (per liter):

Peptone	20 g
$MgCl_2$	1.4 g
K_2SO_4	10 g
Irgasan	0.025 g
Agar	13.6 g

Additive:

Glycerol	20 ml

The pH of the finished media will be 7.0 ± 0.2.
Preparation instructions: Prepare medium base in **980 ml** of distilled water. Add glycerol and bring final volume to 1 liter. Sterilize by autoclaving. Temper to 50°C and pour (15–20 ml) into Petri plates.

Rappaport-Vassiliadis (RV) Broth (Soya Peptone Formula)

Medium used in: Chapter 12: *Salmonella*
There are multiple versions of RV broth with slightly different formulations that are used as part of the secondary selective enrichment for the detection of *Salmonella* in foods. This version is very similar to the versions recommended by FSIS's MLG and ISO; however, soya peptone (or enzymatic digest of soya) is substituted for tryptone (pancreatic digest of casein). FDA recommends a similar formula to that of FSIS; however, the preparation instructions indicate separate preparation of concentrated magnesium chloride and malachite green solutions prior to mixing with the prepared base.

Medium (per liter):

Soya peptone	5 g
NaCl	8 g
KH_2PO_4	1.6 g
$MgCl_2 \bullet 6H_2O$	40 g
Malachite green	0.04 g

Note: This powdered medium is extremely hygroscopic ($MgCl_2 \bullet 6H_2O$) and must be protected from moisture!
The pH of the finished media will be 5.2 ± 0.2.
Preparation instructions: Prepare medium base in 1 l of distilled water. Dispense 10 ml aliquots into tubes. Sterilize by autoclaving at **115°C for 15 minutes**.
Note: This media loses selectivity over time. Store in the refrigerator and use within 1 month.

Tetrathionate (TT) Broth

Medium used in: Chapter 12: *Salmonella*

There are multiple versions of TT broth with significantly different formulations that are used as part of the secondary selective enrichment for the detection of *Salmonella* in foods. The two most common formulations are included here. The main selective agents are common to both formulations; however, their quantities differ. The type and quantity of non-selective ingredients also differ between the formulations.

FDA formulation

TT Medium base (per liter):

Polypeptone	5 g
Bile salts	1 g
Calcium carbonate	10 g
Sodium thiosulfate • 5H$_2$O	30 g

Addition: Iodine-Potassium Iodide (I$_2$-KI) Solution

Potassium iodide	5 g
Iodine, resublimed	6 g
Distilled water (sterile)	20 ml

Preparation instructions: Dissolve potassium iodide in 5 ml of sterile distilled water. Add iodine – it takes extensive agitation and time to dissolve the iodine. Dilute to 20 ml with sterile distilled water.

Addition: Brilliant Green Solution

Brilliant green	0.1 g
Distilled water (sterile)	100 ml

The pH of the finished complete medium will be 8.4 ± 0.2.

Preparation instructions: Mix TT medium base in 1 l distilled water. The calcium carbonate doesn't completely dissolve so you will notice a significant white precipitate. To sterilize, heat base to boiling. DO NOT AUTOCLAVE. Prepared TT base can be stored in the refrigerator.

On the day of use, add 20 ml of I$_2$-KI solution and 10 ml of brilliant green solution to the prepared 1L of TT base. Mix thoroughly before dispensing 10 ml into sterile tubes.

Note: This medium loses selectivity over time.

FSIS formulation (Hajna – Hajna & Damon, 1956)

TT Medium base (per liter):

Tryptone	18 g
Yeast extract	2 g
Dextrose	0.5 g
d-Mannitol	2.5 g
Sodium desoxycholate	0.5 g

NaCl	5 g
Sodium thiosulfate	38 g
Calcium carbonate	25 g
Brilliant green	0.01 g

Addition: Iodine-Potassium Iodide (I_2-KI) Solution

Potassium iodide	8 g
Iodine, resublimed	5 g
Distilled water (sterile)	40 ml

<u>Preparation instructions:</u> Dissolve potassium iodide in 20 ml of sterile distilled water. Add iodine – it takes extensive agitation and time to dissolve the iodine. Dilute to 40 ml with sterile distilled water.
The pH of the finished complete medium will be 7.6 ± 0.2.

<u>Preparation instructions:</u> Mix TT medium base in 1 l distilled water. The calcium carbonate doesn't completely dissolve so you will notice a significant white precipitate. To sterilize, heat base to boiling. DO NOT AUTOCLAVE. Prepared TT base can be stored in the refrigerator up to six months prior to use.
On the day of use, add 4% of the I_2-KI solution to the needed quantity of TT base. Mix thoroughly before dispensing 10 ml into sterile tubes.
Note: This medium loses selectivity over time.

Thioglycolate Agar

Medium used in: Chapter 6: Spores

<u>Medium (per liter):</u>

Pancreatic digest of casein	15 g
Yeast extract	5 g
Dextrose	5.5 g
NaCl	2.5 g
L-Cystine	0.5 g
Sodium thioglycolate	0.5 g
Agar	20 g

The pH of the finished medium will be 7.1 ± 0.2.
<u>Preparation instructions:</u> Prepare medium base in 1 l of distilled water and aliquot into tubes (10 ml). Autoclave and temper to 50°C. Prepare it shortly before use. Pour as an overlay onto previously prepared agar.

Tributyrin Agar (Sigma Aldrich)

Medium used in: Chapter 7: *Pseudomonas*
<u>Medium (per liter):</u>

Special peptone	5 g
Yeast extract	3 g
Agar	12 g

Addition:

Neutral tributyrin 10 g

The pH of the finished media will be 7.5 ± 0.2.

Preparation instructions: Dissolve media in 1 l of distilled water and autoclave. Temper to 80°C before adding neutral trybutyrin. Pour complete medium (20 ml) into Petri plates.

Tryptic Soy Agar (TSA)

Medium used in: Chapter 5: Aerobic Mesophilic Count
Medium (per liter):

Trypticase peptone	15 g
Phytone peptone	5 g
NaCl	5 g
Agar	15 g

The pH of the finished media will be 7.3 ± 0.2.

Preparation instructions: Dissolve media in 1 l of distilled water and autoclave. Temper to 50°C and pour (12–20 ml) into Petri plates.

Tryptic Soy Agar + Blood (TSA-Blood) (Also Called Blood Agar)

Medium used in: Chapter 10: *Staphylococcus aureus*, Chapter 11: *Listeria monocytogenes*
TSA Medium base (per liter):

Trypticase peptone	15 g
Phytone peptone	5 g
NaCl	5 g
Agar	15 g
Additions:	
Defibrinated sheep blood	50 ml

Note: Other types of blood (i.e., defibrinated horse blood) can be used in place of sheep blood. Different blood types may produce more distinct hemolysis.

The pH of the finished media will be 7.3 ± 0.2.

Preparation instructions: Prepare TSA medium base in 1 l of distilled water and autoclave. Temper to 50°C before adding blood supplement. Pour complete medium (20 ml) into Petri plates.

Tryptic Soy Agar with Yeast Extract (TSA-YE)

Medium used in: Chapter 11: *Listeria monocytogenes*; Chapter 13: STEC
TSA Medium base (per liter):

Trypticase peptone	15 g
Phytone peptone	5 g
NaCl	5 g
Agar	15 g

Addition:

Yeast extract 3–6 g*

Formulations vary between 0.3 and 0.6% of yeast extract in final medium.

The pH of the finished media will be 7.3 ± 0.2.

Preparation instructions: Dissolve TSA media base and yeast extract in 1 l of distilled water and autoclave. Temper to 50°C and pour (12–20 ml) into Petri plates.

Tryptic Soy Broth (TSB)

Medium used in: Chapter 5: Aerobic Mesophilic Plate Count

Medium (per liter):

Trypticase or tryptone (pancreatic digest of casein)	17 g
Phytone (papaic digest of soya meal)	3 g
NaCl	5 g
K_2PO_4	2.5 g
Dextrose	2.5 g

The pH of the finished media will be 7.3 ± 0.2.

Preparation instructions: Dissolve TSB media in 1 l of distilled water and autoclave.

Tryptic Soy Broth with Salt and Pyruvate (TSB-SP)

Medium used in: Chapter 10: *Staphylococcus aureus*

TSB Medium (per liter):

Trypticase or tryptone (pancreatic digest of casein)	17 g
Phytone (papaic digest of soya meal)	3 g
NaCl	5 g
K_2HPO_4	2.5 g
Dextrose	2.5 g

Additions:

| NaCl | 95 g |
| Sodium pyruvate | 10 g |

The pH of the finished media will be 7.3 ± 0.2.

Preparation instructions: Dissolve TSB media and additions in 1 l of distilled water and autoclave.

Tryptone Glucose Extract (TGE) Agar

Medium used in9: Chapter 6: Spores
<u>Medium (per liter):</u>

Beef extract	3 g
Enzymatic digest of casein (tryptone)	5 g
Dextrose (glucose)	1 g
Agar	15 g

The pH of the finished media will be 7.0 ± 0.2.
<u>Preparation instructions:</u> Dissolve media in 1 l of distilled water and autoclave. Temper to 50°C and pour (12–20 ml) into Petri plates.

Violet Red Bile Glucose (VRBG) Agar

Medium used in: Chapter 8: *Enterobacteriaceae*
<u>Medium (per liter):</u>

Peptone	7 g
Yeast extract	3 g
NaCl	5 g
Dextrose (glucose)	10 g
Bile salts no. 3	1.5 g
Neutral Red	0.03 g
Crystal Violet	2 mg
Agar	15 g

The pH of the finished media will be 7.4 ± 0.2.
<u>Preparation instructions:</u> Dissolve media in 1 l of distilled water and let stand for a few minutes. Mix thoroughly and adjust pH to 7.4 ± 0.2. Heat while stirring and boil for 2 min. DO NOT AUTOCLAVE. Temper to 50°C and pour (12–20 ml) into Petri plates.

Xylose Lysine Desoxycholate (XLD) Agar

Medium used in: Chapter 12: *Salmonella*
<u>Medium (per liter):</u>

Yeast extract	3 g
NaCl	5 g
L-lysine	5 g
Lactose	7.5 g
Sucrose	7.5 g
Xylose	3.75 g
Sodium desoxycholate	2.5 g
Ferric ammonium citrate	0.8 g
Sodium thiosulfate	6.8 g
Phenol red	0.08 g
Agar	15 g

The pH of the finished media will be 7.4 ± 0.2.

<u>Preparation instructions:</u> Prepare medium base in 1 l of distilled water. Mix thoroughly and heat with agitation. Bring just to a boil. DO NOT AUTOCLAVE. Temper to 50°C and pour (20 ml) into Petri plates.

Note: May be stored up to 30 days at 4°C prior to use.

APPENDIX IV

SUPPLIES AND EQUIPMENT AVAILABILITY

The design and furnishing of a microbiology laboratory vary considerably according to the specialization and function of the laboratory (e.g., research specialty, regulatory, quality assurance). The subject of this book deals with teaching and training in food microbiology laboratories. Throughout the book, specific equipment and supplies needed for the completion of each exercise have been presented by laboratory session; however, there are common pieces of equipment and/or consumable supplies that are expected to be available to analysts during laboratory sessions, regardless of their presence or absence in the exercise supply list.

In addition to the equipment in the teaching laboratory, which is used directly by students, there are many others needed for microbiological media preparation, dispensing, sterilization, and disposal. These include autoclaves, hotplates, dishwashers, glassware storage cabinets and drying racks, media and reagent storage cabinets, top-loading and analytical balances, freezers for culture stocks, media dispensers, repeater pipettes, and others. Supplies that are needed but not used directly by students include dry media, reagents, large flasks, graduated cylinders, glassware detergents, autoclave gloves, autoclave tapes, and others. These items are preferably housed in a service laboratory or a media preparation area. Both the service and teaching laboratories should contain first-aid kits, fire extinguishers, and other safety equipment and supplies.

For a food microbiology teaching laboratory, the following is a list of equipment that should be consistently accessible and supplies that should be stocked in excess to have on hand as needed. This is not meant to be a comprehensive equipment and supply list for the book and should be considered a supplement to the lists presented in each chapter.

Analytical Food Microbiology: A Laboratory Manual, Second Edition. Ahmed E. Yousef,
Joy G. Waite-Cusic, and Jennifer J. Perry.
© 2022 John Wiley & Sons, Inc. Published 2022 by John Wiley & Sons, Inc.

EQUIPMENT

Balance, top loading

Food sample preparation requires accurate measurement of analytical sample mass. Top-loading balances with 500-g capacity and accuracy of 0.1 g are sufficient for this purpose. Balance platforms should be large enough to accommodate a container into which food can be weighed. Access to multiple balances in each laboratory is advisable to prevent delays in executing the exercises.

Bunsen burner

One Bunsen burner per pair of students should be available at all times. This equipment is imperative for proper aseptic technique.

Calculator

Hand-held, simple, sanitizable calculators are suitable for on-the-bench quick calculations. These should be provided to students for use in the laboratory and should be sanitized after use.

Colony counter

The dark-field colony counter is a useful piece of equipment in any microbiology laboratory, and it is essential in food microbiology teaching laboratories. Students use it to count colonies on agar media and to view colony morphology. In a food microbiology teaching laboratory setting, it is advisable to provide one colony counter per six students.

Computer or computing tablet

In most exercises, there is a need to collect data from all students or groups before they leave the teaching laboratory. A desktop computer loaded with a customized worksheet is ideal for this task. To avoid long waiting times for data entry, it is recommended that students are provided with tablets, connected wirelessly to the class computer. This facilitates data collection while minimizing traffic in the laboratory. Additionally, cloud software allows students to enter their data simultaneously from these tablets without the need for a desktop computer. When computers or tablets are used within the microbiology teaching laboratory, they should be equipped with a waterproof sanitizable keyboard and pointing devices (mouse or touchscreen). These parts should be sanitized at the end of each laboratory session.

Homogenizer or blender

Analytical food samples must be mixed thoroughly with suitable diluents (i.e., homogenized) before preparation of serial dilutions. The most commonly used homogenization equipment is a pedal homogenizer (referred to throughout this book as a stomacher). Along with the homogenizer, sterile bags to hold food being homogenized (stomacher bags) are needed. Alternatively, a blender (base and jar) may be used to homogenize food samples.

Incubator

For all exercises, it is expected that an incubator with sufficient space has been set and equilibrated to the appropriate temperature before the start of the laboratory session. Most exercises in this book require incubation of inoculated media at 35–37 °C. There are instances in which alternative or additional incubation temperatures are needed; therefore, it is suggested that analysts have access to a minimum of two functioning incubators. Most incubators hold incubated items statically, but few others provide shaking of liquid media during incubation.

Micropipette

The use of micropipettes (as opposed to serological pipettes) for liquid handling supports accurate volume transfers and reduces the likelihood of environmental contamination. Several exercises include microbiological techniques that cannot be accomplished without these tools. Ideally, each pair of student analysts has access to a dedicated set of micropipettes, having 10, 100, and 1000 µl capacities. These micropipettes should be maintained properly thorough periodic cleaning, decontamination, and calibration.

Microscope

A compound light microscope is needed to view microorganisms on stained slides. The microscope should be equipped with objective lenses having various magnifications, including a 100× oil immersion lens. In order to prevent delays during laboratory sessions, it is preferable that each student pair be allocated a separate microscope. Occasionally, phase-contrast microscopes and dissection microscopes (stereoscopes) are needed. Presence of a digital imaging microscope (camera-equipped microscope) is beneficial for taking pictures of cellular morphology.

Microtiter plate reader

The enzyme-linked immunoassay (ELISA), included in the *Staphylococcus aureus* exercise (Chapter 10), requires the use of a microtiter plate reader to measure light absorbance of individual wells in a 96-well plate format. The equipment is also useful if exercises include measuring the minimum inhibitory concentration of antimicrobials.

Molecular analysis equipment

A thermocycler and electrophoresis unit should be sufficient to serve 10–12 students in exercises requiring PCR runs. Running PCR analysis also requires a gel documentation station, which is made of UV trans-illuminator, digital camera, computer, and appropriate software.

pH meter

Generally, it is beneficial to have a pH meter in the food microbiology teaching laboratory, but specifically it is needed when growth of acid-producing bacteria requires monitoring.

Refrigerator

Depending on the schedule of laboratory sessions, it is likely that refrigeration of inoculated microbiological media, before incubation, is needed. Additionally, refrigeration may be needed for short-term storage of food samples before analysis, and storage of intermediate products of analysis between laboratory sessions. A designated refrigerator for the storage of biohazardous materials should be used for this purpose.

Spectrophotometer

Spectrophotometers are useful for estimating cell densities of liquid cultures, typically at a wavelength of 600 nm. Most spectrophotometers require specific cuvettes or tubes for holding samples during measurements.

Vortex

Several exercises involve the preparation of serial dilutions. Dilution tubes should be thoroughly mixed by vortexing in order to ensure accurate bacterial enumeration. In addition to preparing serial dilutions, mixing is needed in other tasks. It is recommended that a vortex be shared between a group of two, to no more than four students.

Water bath or heating block (dry bath)

Water baths are needed for many of the exercises in this book. These are used to temper molten agar (50°C), provide stable incubation temperature for critical enzyme reactions, support consistent thermal treatment of cell suspensions, etc. Thermostatically controlled, circulating water baths are ideal for these tasks. For accurate temperature control, a water bath with both heating and refrigeration capabilities is needed. In certain applications, heating blocks (dry baths) are suitable replacements for water baths.

SUPPLIES

Bacteriological needles and loops

Both loops and needles will be necessary for inoculation of microbiological media. If these are the traditional, reusable variety, students should be instructed regarding proper flaming. Alternatively, disposable versions of these tools are readily available. If disposable versions are used, it is important to instruct students on how to handle these (bags typically contain several disposable loops or needles) in order to minimize inadvertent contamination of the remaining loops/needles.

Biohazard boxes and containers

Small biohazard containers should be placed on benches to dispose of small, contaminated items such as pipette tips, used glass slides and cover slips, and serological pipettes. Large biohazard cardboard boxes, properly lined with plastic liners, are needed for the disposal of cultures, incubated Petri plates, contaminated food, and other items.

Disposable gloves, all sizes

It is suggested that disposable gloves are used in every exercise, regardless of whether pathogenic cultures are being handled. Latex gloves are typically used, but nitrile gloves may be used if users' skin is sensitive to latex.

Markers

Permanent parkers are needed for writing on labels or disposable supplies, but not directly on glassware. Wax pencils are needed for marking or labeling glass slides.

Micropipette tips (sterile)

A dedicated box of tips for each pipettor size should be provided to each pair of students. These should be routinely re-sterilized to mitigate potential contamination from one laboratory session to the next. Filter-barrier tips are recommended for use during molecular assays to minimize contamination.

Paper towels

A ready supply of disposable towels should be maintained at all times for hand drying and cleaning of spills.

Rulers

Plastic, transparent, sanitizable, small rulers are suitable for use in teaching laboratories. These are used to measure diameters of colonies or inhibition areas.

Safety goggles

Sanitizable safety goggles should be supplied for use in the laboratory.

Sample preparation supplies

Food samples utilized for each exercise may vary significantly in their properties, requiring various steps for proper analytical sampling. Therefore, analysts should have access to multiple cutting boards, knives, forceps, scissors, spoons, and plastic food storage containers for partitioning and mixing food samples. Ideally, knives, forceps, etc., can be sterilized by dipping in ethanol and passing through a flame. Spray bottles of alcohol/ethanol should also be available for use to decontaminate cutting boards and plastic containers.

Sanitizer
Sanitizer bottles are needed for use on bench tops before and after each session as well as for use in the event of inadvertent spills of biohazardous materials. Ethanol (70%), quaternary ammonium compounds, and hypochlorous acid solutions are commonly used sanitizers.

Serological pipettes

Disposable serological pipettes are needed in some cases, such as the transfer of 1 ml of food homogenate, from the stomacher bag to the first dilution tube. When the food sample to be analyzed is a liquid, larger serological pipettes (i.e., 10 or 25 ml) are the best tools for measuring the analytical sample.

Spread-plating supplies

Spread plating is performed in nearly all exercises. Students should have access to cell spreaders, either disposable or reusable. If glass or metal spreaders are used, a closed container of ethanol of a size suitable for dipping the spreader should also be provided. Students should be instructed to keep these jars closed to prevent evaporation and inadvertent fires. Ethanol jars should be replaced regularly as these can become contaminated with bacterial spores.

Staining supplies

Several staining procedures are utilized in the exercises contained in this book. It is expected that each pair of analysts has access to glass slides, clothespins, wax pencils, immersion oil, lens cleaner, bibulous paper, and lens paper. Ideally, a staining rack that can be set over a sink is available to be shared by two to four students. It is advisable that Gram staining reagents be kept available to each student pair.

Stomacher bags and holders

These are sterile sturdy plastic bags for holding food during homogenization in the stomacher. They are also used to hold samples during enrichments. Suitable holders should be available to keep these bags in upright positions during weighing food or incubating enrichments. The holder can be as simple as a light wire basket or a specialized multi-bag holder. Specialized filter bags are useful for analyzing samples that can clog micropipette tips or serological pipets.

Striker or lighter
Strikers with replaceable flint are recommended.

Tape

Both label tape (for labeling dilution tubes) and less costly masking tape (for taping Petri plate stacks together during incubation) should be available. Transparent tapes also are used in microbiology teaching laboratories, specifically in the fungi exercise.

Tube racks

Racks of various sizes, particularly test tubes racks, should be available. Additionally, racks for microcentrifuge tubes and PCR tubes are needed in some exercises. Floating racks are beneficial for holding small tubes during heating in water baths.

INDEX

Page numbers referring to figures shown in *italics*
Page numbers referring to tables shown in **bold**

Analytical Food Microbiology: A Laboratory Manual, Second Edition. Ahmed E. Yousef,
Joy G. Waite-Cusic, and Jennifer J. Perry.
© 2022 John Wiley & Sons, Inc. Published 2022 by John Wiley & Sons, Inc.

Printed and bound by CPI Group (UK) Ltd, Croydon, CR0 4YY

27/10/2024

14580263-0005